A Classification of Igneous Rocks
and Glossary of Terms

A Classification of Igneous Rocks and Glossary of Terms

Recommendations of the
International Union of Geological Sciences
Subcommission on the Systematics
of Igneous Rocks

Prepared by
R. W. Le Maitre (Editor),
P. Bateman, A. Dudek, J. Keller,
J. Lameyre, M. J. Le Bas, P. A. Sabine,
R. Schmid, H. Sørensen, A. Streckeisen,
A. R. Woolley and B. Zanettin

BLACKWELL SCIENTIFIC PUBLICATIONS

OXFORD LONDON EDINBURGH BOSTON MELBOURNE

© 1989 by International Union of
Geological Sciences and published for them
by Blackwell Scientific Publications
Editorial Offices:
Osney Mead, Oxford OX2 0EL
  (*Orders:* Tel: 0865 240201)
8 John Street, London WC1N 2ES
23 Ainslie Place, Edinburgh EH3 6AJ
3 Cambridge Center, Suite 208
  Cambridge, Massachusetts 02142, USA
107 Barry Street, Carlton
  Victoria 3053, Australia

First published 1989

Printed and bound in Great Britain by
Redwood Burn Ltd,
Trowbridge, Wilts.

DISTRIBUTORS

USA
  Publishers' Business Services
  PO Box 447
  Brookline Village
  Massachusetts 02147
  (*Orders*: Tel: (617) 524 7678)

Canada
  Oxford University Press
  70 Wynford Drive
  Don Mills
  Ontario M3C 1J9
  (*Orders*: Tel: (416) 441-2941)

Australia
  Blackwell Scientific Publications
  (Australia) Pty Ltd
  107 Barry Street
  Carlton, Victoria 3053
  (*Orders*: Tel: (03) 347 0300)

British Library
Cataloguing in Publication Data

International Union of Geological Sciences,
Subcommission on the Systematics of Igneous Rocks
A classification of igneous rocks and glossary of terms.
1. Igneous rocks. Classification
I. Title II. Le Maitre, R. W.
532'.1'012

ISBN 0-632-02593-X

Library of Congress
Cataloging-in-Publication Data

International Union of Geological Sciences.
  Subcommission on the Systematics of Igneous Rocks.
    A classification of igneous rocks and glossary
    of terms:
  recommendations of the International Union of
  Geological Sciences Subcommission on the Systematics
  of Igneous Rocks / compiled by
  Roger W. Le Maitre.
          p.   cm.
    Bibliography: p.
    Includes index.
    ISBN 0-632-02593-X
    1. Rocks, Igneous - Nomenclature.
    2. Rocks, Igneous - Classification.
    I. Le Maitre, R. W. (Roger Walter)
    II. Title.
  QE461.I547 1989
  552' . 1'012-dc20

Contents

Foreword

In the early summer of 1958 Ernst Niggli asked Theo Hügi and me if we would be willing to collaborate in revising Paul Niggli's well-known book "Tabellen zur Petrographie und zum Gesteinsbestimmen" which had been used as a text for decades at the Federal Polytechnical Institute of Zürich. We agreed and I was placed in charge of the classification and nomenclature of igneous rocks. Quite soon I felt that the scheme used in the Niggli Tables needed careful revision but, as maybe twelve other classification schemes had already been published, Eduard Wenk warned that we should not propose an ominous thirteenth one; instead he proposed that it would be better to outline the inherent problems of igneous rock classification in an international review article and should present a provisional proposal, asking for comments and replies. This was dangerous advice!

However, the article was written (Streckeisen, 1964), and the consequence was an avalanche of replies, mostly consenting, and many of them with useful suggestions. It thus became clear that the topic was of international interest and that we had to continue. A short report (Streckeisen, 1965) summarised the results of the inquiry. Subsequent discussions with colleagues from various countries led to a detailed proposal (Streckeisen, 1967), which was widely distributed. This was accompanied by a letter from Professor T.F.W. Barth, President of the International Union of Geological Sciences

(IUGS), who emphasized the interest in the undertaking, and asked for comments.

The IUGS Commission on Petrology then established a Working Group on Rock Nomenclature, which made arrangements to discuss the nomenclature of magmatic rocks at the International Congress in Prague; the discussion was fixed for August 21, 1968. For this meeting a large amount of documentation was provided; it contained an Account of the previous work, a Report of the Petrographic Committee of the USSR, a Report of the Geological Survey of Canada, and comments from colleagues throughout the world. But political events prevented the intended discussion.

At this stage, Professor T.F.W. Barth, as President of the IUGS, suggested the formation of an International Commission. The Subcommission on the Systematics of Igneous Rocks was formed, under the IUGS Commission on Petrology, to deliberate the various problems of igneous rock nomenclature and to present definite recommendations to the IUGS.

The Subcommission began its work in March, 1970. This was done by way of correspondence with subsequent meetings for discussions and to make decisions. Tom Barth suggested beginning with plutonic rocks, as this was easier; his advice was followed.

It was agreed that plutonic rocks should be classified and named according to their modal mineral contents and that the QAPF double triangle should serve for their presentation. A difficulty arose in discussing the nomenclature of granites; the most frequent granites were named adamellite in England and quartz-monzonite in America. With energetic intervention, A.K. and M.K. Wells (see Contribution No. 12, Appendix A) advocated that a logical classification would demand that quartz-monzonite in relation to monzonite must have the same status as quartz-syenite to syenite and quartz-diorite to diorite. On this critical point, Paul Bateman made an inquiry concerning this topic among leading American geologists (see Contribution No. 21, Appendix A) with the result that 75% of the respondents declared themselves willing to accept quartz-monzonite as a term for field 8*, i.e. for a straight relation to monzonite.

In this relatively short period of time, the Subcommission had, therefore, discussed the problems of plutonic rocks, so that in spring 1972 at the Preliminary Meeting in Berne it made recommendations, which were discussed and, with some modifications, accepted by the Ordinary Meeting in Montreal, August 1972. A special working group had been set up to discuss the charnockitic rocks and presented its recommendations in 1974.

Work then started on the problems of the classification of volcanic and pyroclastic rocks. The latter was dealt with by a special working group set up under the chairmanship of Rolf Schmid. After much discussion and some lengthy questionnaires this group published a descriptive nomenclature and classification of pyroclastic rocks (Schmid, 1981).

The first problem to be addressed for volcanic rocks was whether their classification and nomenclature should be based on mineralogy or chemistry. Strong arguments were put forward for both solutions. However, crucial points were the fine-grained nature and presence of glass that characterise many volcanic rocks, which means that modal contents can be extremely difficult to obtain. Similarly, the calculation of modes from chemical analyses was considered to be too troublesome or not sufficiently reliable. After long debate the Subcommission decided on the following important principles:-

(a) if modes are available, volcanic rocks should be classified and named according to their position in the QAPF diagram;

(b) if modes are not available, chemical parameters should be used as a basis for a chemical classification which, however, should be made comparable with the mineralogical QAPF classification (Grenoble, 1975).

Various methods of chemical classification were considered and tested by a set of combined modal and chemical analyses. Finally, the Subcommission agreed to use the total alkali – silica (TAS) diagram that Roger Le Maitre had elaborated and correlated with the mineralogical QAPF diagram (Le Maitre, 1984) by using the CLAIR and PETROS databases. After making minor modifications, the TAS diagram was accepted by the Subcommission (Le Bas et al., 1986).

Later on, Mike Le Bas started work on distinguishing the various types of volcanic nephelinitic rocks using normative parameters; similar work is also underway to distinguish the various volcanic leucitic rocks. An effort to classify high-Mg volcanic rocks (picrite, meimechite, komatiite) united USSR and western colleagues at the closing meeting in Copenhagen in 1988.

The intention to compile a glossary of igneous rock names, which should contain recommendations for terms to be abandoned, and definitions of terms to be retained, had already been expressed at the beginning of the undertaking (Streckeisen, 1973, p. 27). A first approach was made in October 1977 by a questionnaire which contained a large number of igneous rock terms. Colleagues were asked whether, in their opinion, the terms were of common usage, rarely used today, or almost never been used. More than 200 detailed replies were received, and almost all heartily advocated the publication of a Glossary, hoping it would be as comprehensive as possible. At the final stage of our undertaking, Roger Le Maitre has taken over the heavy burden of compiling the Glossary, for which he will be thanked by the entire community of geologists.

The work of the Subcommission began with the Congress of Prague and will end with that of Washington, a space of 20 years. During this time, 49 circulars, containing 145 contributions and comments amounting to some 2000 pages, were sent to members and interested colleagues — a huge amount of knowledge, stimulation, ideas and suggestions. Unfortunately only part of this mass of knowledge was able to be incorporated in the final documents. However, all the documents will be deposited in the British Museum (Natural History) in London, so that they will be available for use in the future.

Within this period of time, a large number of geologists have been collaborating as colleagues, whether as members of the Subcommission or of working groups, as contributors of reports and comments, as guests of meetings, and in other ways: in short, a family of colleagues from many different countries and continents, united in a common aim.

On behalf of the Subcommission, I thank all those colleagues who have helped by giving advice, suggestions, criticisms and objections, and I am grateful for the continual collaboration we have enjoyed.

Albert Streckeisen,
Berne, Switzerland.
November 1988.

Chairman's Preface

Excellent as the contents of this book are, it would be foolish to imagine that the subject of the systematics of igneous rocks, as dealt with by the IUGS Subcommission, is now closed. As new rock types are discovered or new features appreciated, so will modifications or possibly new classifications have to be made to accommodate them.

The future work of the Subcommission lies not only in keeping the systematics up-to-date and internationally relevant, but also in looking further into some specialised areas in which the present day nomenclature is proving inadequate. One field is that of the potassic rocks, particularly the lamproites, together with the lamprophyric and kimberlitic rocks. Another is that of the basaltic rocks which will also need attention in the near future. At the same time, the Subcommission will maintain a broad overview on the whole subject and will be pleased to receive comments and contributions (including corrections to this book) from all interested persons. For convenience, these should be sent to the Secretary of the Subcommission, Dr. A. R. Woolley, Department of Mineralogy, British Museum (Natural History), Cromwell Rd., London, SW7 5BD, U.K.

Mike Le Bas,
Chairman, IUGS Subcommission on
    the Systematics of Igneous Rocks,
University of Leicester, U.K.
March, 1989.

Editor's Preface

As the member of the Subcommission given the responsibility for compiling and producing this publication, I would like to thank personally all those people who made my task not only easier but possible. In particular my sincere thanks to a small group of dedicated volunteers who spent many hours of their own time checking nearly 800 source references for validity, in libraries throughout the world, and in patiently answering my innumerable and persistent queries. Without their extensive help and dedication the compilation and checking of the bibliography would not have been possible. Those involved were:- Aldo Cundari (Australia), Arnost Dudek (Czechoslovakia), Marjorie Heggen (Australia), Jörg Keller (FDR), Cornelia Kluth (FDR), Marie Alice Lançon (France), Jean Lameyre (France), Mike Le Bas (UK), Nicolai Mikhailov (USSR), Henning Sørensen (Denmark), Albert Streckeisen (Switzerland), and Alan Woolley (UK).

These people, together with the following, also contributed in various other important ways, such as helping to compile nearly 1600 glossary entries, proof reading various editions of the text, and by late stage discussions on classification problems with specific rock types:- Dan Barker (USA), Giuliano Bellieni (Italy), Jose Brändle (Spain), Felix Chayes (USA), Tony Crawford (Australia), Slava Efremova (USSR), Ulrich Knittel (FDR), Eric Middlemost (Australia), Ian Muir (UK), Nick Rock (Australia), Peter Sabine (UK), Rolf Schmid (Switzerland), David Slancy

(Australia), V. Széky-Fux (Hungary), Maurice Wells (UK), John Wilkinson (Australia).

I would also like to thank Clive Bishop, Keeper of Minerals at the British Museum (Natural History) for allowing me access to their library during a visit to London when I was able to check many of the more obscure references; and Ralph Cameron (Australia) for helping with the classical derivations of certain of the rock names.

Leo Flynn and Rodney Van Cooten of the University of Melbourne Computing Services also deserve a special mention for their friendship and patience in helping me to get the maximum out of a Macintosh computer and for helping to create some of the special characters used in the Glossary and Bibliography. The software used to produce this book was Borland Reflex® Plus for maintaining the linked database of rock descriptions, references, contributors and journals; Microsoft® Word for creating and editing the text; Claris MacDraw II® for producing the diagrams; Microsoft® Excel for producing the histograms; and Aldus PageMaker® for producing camera-ready copy for the publisher.

Finally, I would like to sincerely thank my wife, Valerie, who spent many lonely evenings and weekends watching her husband give all his attention to a Macintosh computer.

Roger Le Maitre,
University of Melbourne, Australia.
February, 1989.

A. Introduction

This book is essentially in two parts. The first is chapter **B. Classification and Nomenclature of Igneous Rocks** which contains a summary of all the previously published recommendations, together with a few new ones, made by the International Union of Geological Sciences (IUGS) Subcommission on the Systematics of Igneous Rocks. This is the first time that all these recommendations have been presented as a single logical classification scheme.

The Subcommission was set up after the International Geological Congress meeting in Prague in 1968 as the result of an earlier investigation into the problems of igneous rock classification that had been undertaken by Professor Albert Streckeisen from 1958–1967 (Streckeisen, 1967). He was appointed the first Chairman of the Subcommission, a position he held from 1969 to 1980 and was followed by Bruno Zanettin (1980-1984, Italy), and Mike Le Bas (1984-present, UK). Secretaries of the Subcommission were V. Trommsdorff (1970-75, Switzerland), R. Schmid (1975-80, Switzerland), G. Bellieni (1980-84, Italy), and A.R. Woolley (1984-present, UK).

During this time 49 circulars were distributed containing a total of 145 contributions from petrologists throughout the world. All of these documents have now been deposited in the Department of Mineralogy at the British Museum (Natural History) in London, so that they will be available for future use. A list of their contents is given in **Appendix A**. These circulars formed the basis for discussions at official meetings held in Bern (1972), Montreal (1972), Grenoble (1975), Sydney (1976), Prague (1977), Padova (1979), Paris (1980), Cambridge (1981), Granada (1983), Moscow (1984), London (1985), Freiburg im Breisgau (1986), and Copenhagen (1988) during which many of the official recommendations were finalised.

Records of the Subcommission also indicate that 419 people from 49 countries participated in the formulation of the recommendations in various ways. Of these, 51 were official members of the Subcommission representing their countries at various times; 129 were members of various working groups that were periodically set up to deal with specific problems; 40 were members of one special working group only, set up with Rolf Schmid as Chairman, to look at the nomenclature of pyroclastic rocks; 176 wrote to the Subcommission with comments, recommendations etc.; and 23 attended various meetings as guests. The names of all these people, together with the name of the country which they were either representing or were living in when correspondence took place, and the ways in which they participated, are given in **Appendix B**. For convenience, two lists are given; one is a simple alphabetical list of people together with their country; the other is a list of people grouped alphabetically by country together with their type of participation.

The Subcommission would like to thank

1

sincerely all these people for their involvement and dedication in formulating an acceptable international classification scheme which bridges linguistic, cultural, and political boundaries.

The second part of the book consists of chapters **C. Glossary of Terms** and **D. Bibliography**. The glossary contains a brief description of 1586 igneous rock names and some terms, of which 297 (19%) have been approved and defined by the Subcommission. These terms are given in bold capitals and also listed separately in **Appendix** C. Accompanying each entry is the reference where each term was originally used, its derivation or type locality, and details of where further information can be found in certain standard petrological texts. The bibliography gives the full details of the 787 references together with the names of the rocks that each contains.

Although the concept of a glossary was mentioned by the Subcommission over 10 years ago (Streckeisen, 1976) it was not until late 1986 that the work was started in earnest. The original idea for the glossary was that it should only include those names that were recommended for use by the Subcommission. However, like Topsy "it just grew", until it became obvious that for it to be really useful it should be as complete as possible. We hope that we have achieved this aim and that it will become the initial reference for information on all igneous rock names. As far as the Subcommission is aware, this is the most complete glossary of igneous rock names that has ever been published.

B. Classification and Nomenclature

This chapter is a summary of all the published recommendations agreed to by the IUGS Subcommission on the Systematics of Igneous Rocks together with some other decisions taken since the meeting in Copenhagen.

B.1. PRINCIPLES

Throughout its deliberations on the problems of classification the Subcommission has been guided by the following principles, most of which have been detailed by Streckeisen (1973 and 1976):-

(a) for the purposes of classification and nomenclature the term "igneous rock" is taken to mean "Massige Gesteine" in the sense of Rosenbusch, which in English can be translated as "**igneous or igneous-looking**". Igneous rocks may have crystallised from magmas or may have been formed by cumulate, deuteric, metasomatic or metamorphic processes. Such arguments as to whether charnockites are igneous or metamorphic rocks are, therefore, irrelevant in this context.

(b) the **primary classification** of igneous rocks should be based on their **mineral content** or **mode**. If a mineral mode is impossible to determine, because of the presence of glass, or because of the fine-grained nature of the rock, then other criteria may be used, e.g. chemical composition, as in the TAS diagram.

(c) the term **plutonic rock** is taken to mean an igneous rock with a phaneritic texture (i.e. a relatively coarse-grained rock in which the individual crystals can be distinguished with the naked eye) which is presumed to have formed at considerable depth. Many rocks that occur in orogenic belts have suffered some metamorphic overprinting, so that it is left to the discretion of the user as to whether to use an igneous or metamorphic term to describe the rock (e.g. gneissose granite or granitic gneiss).

(d) the term **volcanic rock** is taken to mean an igneous rock with an aphanitic texture (i.e. a relatively fine-grained rock in which most of the individual crystals cannot be distinguished with the naked eye), often containing glass, and which is presumed to have been associated with volcanic activity. Such rocks may have been erupted onto the surface of the earth, as lava flows, or may have been intruded at a high level as dykes, sills, plugs, etc.

(e) rocks should be named according to what they are, and not according to what they might have been. Any manipulation of the raw data used for classification should be justified by the user.

(f) any useful classification should correspond with natural relationships.

(g) the classification should follow as closely as possible the historical tradition so that well established terms, for

3

example granite, basalt, andesite, etc., are not redefined in a drastically new sense.

(h) the classification should be simple and easy to use.

(i) all official recommendations should be published in English, and that any translation or transliteration problems should be solved by members in their individual countries. However, publications by individual Subcommission members, in languages other than English, were encouraged in order to spread the recommendations to as wide an audience as possible.

B.2. MODAL PARAMETERS

The primary modal classifications of plutonic rocks and volcanic rocks are based on the relative proportions of the following mineral groups for which volume modal data must be determined:-

Q = quartz, tridymite, cristobalite

A = alkali feldspar, including orthoclase, microcline, perthite, anorthoclase, sanidine, and albite (An_0 to An_5)

P = plagioclase (An_5 to An_{100}) and scapolite

F = feldspathoids or foids including nepheline, leucite, kalsilite, pseudoleucite, sodalite, nosean, haüyne, cancrinite, analcime, etc.

M = mafic and related minerals, e.g. mica, amphibole, pyroxene, olivine, opaque minerals, accessory minerals (e.g. zircon, apatite, sphene, etc.), epidote, allanite, garnet, melilite, monticellite, primary carbonate, etc.

Groups Q, A, P, and F comprise the **felsic** minerals, while the minerals of group M

are considered to be **mafic** minerals, from the point of view of the modal classifications.

The sum of Q+A+P+F+M must, of course, be 100%. Notice, however, that there can never be more than four non-zero values, as the minerals in groups Q and F are mutually exclusive, i.e. if Q is present, F must be absent, and vice versa.

B.3. NOMENCLATURE

During the work of the Subcommission it was quickly realised that the classification schemes would rarely go beyond the stage of assigning a general **root name** to a rock. As such root names are often not specific enough, especially for specialist use, the Subcommission encourages the use of additional qualifiers which may be added to any root name.

These additional qualifiers may be mineral names (e.g. biotite granite), textural terms (e.g. porphyritic granite), chemical terms (e.g. Sr-rich granite), genetic terms (e.g. anatectic granite), tectonic terms (e.g. post-orogenic granite) or any other terms that the user thinks are useful or appropriate. For general guidance on the use of qualifiers the Subcommission makes the following points:-

(a) the addition of qualifiers to a root name must not conflict with the definition of the root name. That means that a biotite granite, porphyritic granite, Sr-rich granite, and post-orogenic granite must still be granites in the sense of the classification. Quartz-free granite, however, would not be permissible because the rock could not be classified as a granite, if it contained no quartz.

(b) the user should define what is meant by the qualifiers used if they are not self explanatory. This applies particularly to geochemical terms, such as Sr-rich or Mg-poor, when often no indications are given of the threshold values above or below which the term is applicable.

(c) if more than one mineral qualifier is used the mineral names should be given in order of increasing abundance (Streckeisen, 1973, p.30; 1976, p.22), e.g. a hornblende biotite granodiorite should contain more biotite than hornblende. Notice that this is the opposite of the convention often adopted by metamorphic petrologists.

(d) the use of the suffix **-bearing**, as applied to mineral names, has not been able to be consistently defined, as it is sometimes used with different threshold values. For example, in the QAPF classification, 5% Q in Q+A+P is used as the upper limit of the term quartz-bearing, while 10% F in A+P+F is used as the upper limit of term foid-bearing. The value of 10% is also used for plagioclase-bearing ultramafic rocks (Fig. B.6). For glassy rocks, however, 20% is taken as the limit as can be seen below.

(e) for volcanic rocks containing glass, the amount of glass should be indicated by using the following prefixes (Streckeisen, 1978, 1979):-

% glass	Prefix
0-20	glass-bearing
20-50	glass-rich
50-80	glassy

For rocks with more than 80% glass special names such as **obsidian, pitchstone**, etc. are used. Furthermore, for volcanic rocks, which have been named according to their chemistry using the TAS diagram, the presence of glass can be indicated by using the prefix **hyalo-** with the root name e.g. hyalo-rhyolite, hyalo-andesite etc. For some rocks special names have been given, e.g. **limburgite** = hyalo-nepheline basanite

(f) the prefix **micro-** should be used to indicate that a plutonic rock is finer-grained than usual, rather than giving the rock a special name. The only exceptions to this are the long established terms **dolerite** and **diabase** (= microgabbro) which may still be used. These two terms are regarded as being synonymous. The use of diabase for Palaeozoic or Precambrian basalts or for altered basalts of any geological age should be avoided.

(g) the prefix **meta-** should be used to indicate that an igneous rock has been metamorphosed, e.g. meta-andesite, meta-basalt etc., but only when the igneous texture is still preserved and the original rock type can be deduced.

(h) volcanic rocks which cannot have a complete mineral mode determined and have not yet been analysed, may be named provisionally, following the terminology of Niggli (1931, p.357), by using their visible minerals (usually phenocrysts) to assign a name which is preceded by the prefix **pheno-** (Streckeisen, 1978, p.7; 1979, p.333). Thus a rock containing phenocrysts of sodic plagioclase in a cryptocrystalline matrix may be provisionally called pheno-andesite. Alternatively the provisional "field" classifications could be used (Fig. B.17)

(i) the colour index **M'** is defined (Streckeisen, 1973, p.30; 1976, p.23) as **M** – muscovite, apatite, primary carbonates, etc., as muscovite, apatite, and primary carbonates are considered to be colourless minerals for the purpose of the colour index. This enables the terms leucocratic, mesocratic, melanocratic, and ultramafic to be defined in terms of the following ranges of colour index:-

	range of M'
leucocratic	0–35
mesocratic	35–65
melanocratic	65–90
ultramafic	90–100

Note that these terms are applicable only to rocks and should not be used to describe minerals.

B.4. USING THE CLASSIFICATION

One of the problems of classifying igneous rocks is that they cannot all be classified sensibly by using only one system. For example, the modal parameters required to adequately define a felsic rock, composed of quartz and feldspars, are very different from those required to define an ultramafic rock, consisting of olivine and pyroxenes. Similarly, lamprophyres have usually been classified as a separate group of rocks. Also modal classifications cannot be applied to rocks which contain glass or are too fine-grained to have their modes determined, so that other criteria, such as chemistry, have to be used in these cases.

As a result several classifications have to be presented, each of which is applicable to a certain group of rocks, e.g. pyroclastic rocks, lamprophyres, plutonic rocks, etc. This, however, means that one has to decide which of the classifications are appropriate for the rock in question. To do this in a consistent manner, so that different petrologists will arrive at the same answer, a hierarchy of classification had to be agreed upon. The basic principle involved in this was that the "special" rock types (e.g. lamprophyres, pyroclastic rocks, etc.) must be dealt with first so that anything that was not regarded as a "special" rock type would be classified in either the plutonic or volcanic classifications, which after all contain the vast majority of igneous rocks. The sequence that should be followed is shown diagrammatically on the chart supplied with the book and is as follows:-

(a) if the rock is considered to be of pyroclastic origin go to section B.5 Pyroclastic rocks and Tephra.

(b) if the rock contains more than 50% of modal carbonate go to section B.6 Carbonatites.

(c) if the rock is considered to be a lamprophyre, lamproite or kimberlite go to section B.7 Lamprophyric rocks. The criteria for making this decision are given in section B.7.

(d) if the rock contains more than 10% of modal melilite go to section B.8 Melilitic rocks.

(e) if the rock is considered to be charnockitic go to section B.9 Charnockitic rocks. The criteria for making this decision are given in section B.9.

(f) if the rock is considered to be plutonic, as defined in section B.1, go to section B.10 Plutonic rocks.

(g) if the rock is considered to be volcanic,

as defined in section B.1, go to section B.11 Volcanic rocks.

(h) if you get to this point, either the rock is not igneous or you have made a mistake.

B.5. PYROCLASTIC ROCKS AND TEPHRA

This classification was prepared by a small working group of the IUGS Subcommission, chaired by Rolf Schmid, after circulating and analysing the results of six questionnaires sent to more than 150 geologists throughout the world (Schmid, 1981).

It should be used only if the rock is considered to have had a pyroclastic origin i.e. was formed by disruption as a direct result of volcanic action. It specifically excludes rocks formed by the autobrecciation of lava flows, because the lava flow itself is the direct result of volcanic action, not its brecciation.

The nomenclature and classification is purely descriptive and thus can easily be applied by non-specialists. By defining the term "pyroclast" in a broad sense (see section B.5.1), the classification can be applied to air fall, flow and surge deposits as well as to lahars, subsurface and vent deposits (e.g. hyaloclastites, intrusion and extrusion breccias, tuff dykes, diatremes, etc.). The terms used in the classification solely describe the granulometric state of the rocks or deposits. Combined with other terms, however, compositional or genetic information may be included.

The grain size limits used for subdividing pyroclasts and pyroclastic deposits should be regarded as provisional until there is international agreement on the granulometric divisions of sedimentary rocks. When indicating the grain size of a single pyroclast or the middle grain size of an assemblage of pyroclasts the general terms "mean diameter" and "average pyroclast size" are used, without defining them explicitly, as grain size can be expressed in several ways. It is up to the user of this nomenclature to specifiy the method by which grain size was measured in those cases where it seems necessary to do so.

B.5.1. Pyroclasts

Pyroclasts are defined as fragments generated by disruption as a direct result of volcanic action. Note that this excludes particles formed by autobrecciataion of lava flows, because the flow itself is the direct result of volcanic action, not its brecciation.

The fragments may be individual crystals, crystal fragments, glass fragments or rock fragments. Their shapes acquired during disruption or during subsequent transport to the primary deposit must not have been altered by later redepositional processes. If they have the fragments are called "reworked pyroclasts", or "epiclasts" if their pyroclastic origin is uncertain. The various types of pyroclasts are mainly distinguished by their size (see Table B.1):-

Bombs — pyroclasts whose mean diameter exceeds 64mm. and a shape or surface (e.g. bread-crust surface), which indicates that they were in a wholly or partly molten condition during their formation and subsequent transport.

Blocks — pyroclasts whose mean diameter exceeds 64mm. and whose angular to subangular shape indicates that they were solid during their formation.

Lapilli — pyroclasts of any shape with an mean diameter of 64mm to 2mm.

Ash grains — pyroclasts with an mean diameter of less than 2mm. They may be further divided into **coarse ash grains** (2mm to 1/16mm) and **fine ash grains** (less than 1/16mm). The fine ash grains may also be called **dust grains.**

B.5.2. Pyroclastic Deposits

Pyroclastic deposits are defined as an assemblage of pyroclasts which may be unconsolidated or consolidated. They must contain more than 75% by volume of pyroclasts, the remaining materials generally being of epiclastic, organic, chemical sedimentary, or authigenic origin. When they are predominantly consolidated they may be called **pyroclastic rocks** and when predominantly unconsolidated they may be called **tephra.** The following terms are applicable to unimodal and well-sorted pyroclastic rocks (Table B.1):-

Agglomerate — a pyroclastic deposit whose average pyroclast size exceeds 64mm. and in which rounded pyroclasts predominate.

Pyroclastic breccia — a pyroclastic rock whose average pyroclast size exceeds 64mm. and in which angular pyroclasts predominate.

Lapilli tuff — a pyroclastic rock whose average pyroclast size is 64mm to 2mm.

Tuff or **ash tuff** — a pyroclastic rock whose average pyroclast size is less than 2mm. It may be further divided into **coarse (ash) tuff** (2mm to 1/16mm) and **fine (ash) tuff** (less than 1/16mm). The fine ash tuff may also be called **dust tuff.** Tuffs and ashes may be further qualified by their fragmental composition as shown in Fig. B.1, i.e a **lithic tuff** would contain a predominance of rock fragments, a **vitric tuff** would contain a predominance of pumice and glass fragments, and a **crystal tuff** would contain a predominance of crystal fragments.

Table B.1. Classification and nomenclature of pyroclasts and well-sorted pyroclastic deposits based on clast size (after Schmid, 1981, Table 1).

Clast size in mm.	Pyroclast	Pyroclastic deposit	
		Mainly unconsolidated: tephra	Mainly consolidated: pyroclastic rock
64	bomb, block	agglomerate bed of blocks or bomb, block tephra	agglomerate pyroclastic breccia
2	lapillus	layer, bed of lapilli or lapilli tephra	lapilli tuff
1/16	coarse ash grain	coarse ash	coarse (ash) tuff
	fine ash grain (dust grain)	fine ash (dust)	fine (ash) tuff (dust tuff)

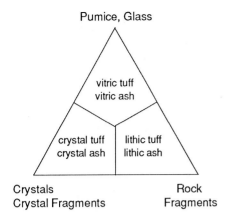

Fig. B.1. Classification and nomenclature of tuffs and ashes based on their fragmental composition (after Schmid,1981, Fig. 1).

Any of these terms for pyroclastic deposits may also be further qualified by the use of any other suitable prefix, e.g. air-fall tuff, flow tuff, basaltic lapilli tuff, lacustrine tuff, rhyolitic ash, vent agglomerate, etc. The terms may also be replaced by purely genetic terms, such as hyaloclastite or base-surge deposit, whenever it seems appropriate to do so.

Polymodal or **poorly-sorted pyroclastic deposits** in which there is more than one dominant size fraction of clasts, should be named using the appropriate combination of terms given in Table B.1. Some examples could be:-

ash lapilli tuff – where lapilli > ash
lapilli ash tuff – where ash > lapilli
bed of lapilli and ash, lapilli ash tephra
　– where ash > lapilli

B.5.3. Mixed Pyroclastic-Epiclastic Deposits

For rocks which contain both pyroclastic and normal clastic (epiclastic) material the Subcommission suggested that the general term **tuffites** can be used within the limits given in Table B.2. Tuffites may be further divided according to their average grain size by the addition of the term "tuffaceous" to the normal sedimentary term e.g. tuffaceous sandstone.

Table B.2. Terms to be used for mixed pyroclastic-epiclastic rocks (after Schmid,1981, Table 2).

Average clast size in mm.	Pyroclastic		Tuffites (mixed pyroclastic –epiclatic)	Epiclastic (volcanic and/or nonvolcanic)
64	Agglomerate, pyroclastic breccia		Tuffaceous conglomerate, tuffaceous breccia	Conglomerate, breccia
	Lapilli tuff			
2	(Ash) tuff	coarse	Tuffaceous sandstone	Sandstone
1/16		fine	Tuffaceous siltstone	Siltstone
1/256			Tuffaceous mudstone, shale	Mudstone, shale
Amount pyroclastic material	100% to 75%		75% to 25%	25% to 0%

B.6. CARBONATITES

This classification should be used only if the rock contains more than 50% modal carbonates (Streckeisen, 1978 and 1979). Carbonatites may be either plutonic or volcanic in origin. Mineralogically the following classes of carbonatites may be distinguished:-

Calcite-carbonatite — where the main carbonate is calcite. If the rock is coarse-grained it may be called sovite; if medium- to fine-grained, **alvikite**

Dolomite-carbonatite — where the main carbonate is dolomite. This may also be called **beforsite**.

Ferrocarbonatite — where the main carbonate is iron-rich.

Natrocarbonatite — essentially composed of sodium, potassium, and calcium carbonates. At present this unusual rock type is only found at Oldoinyo Lengai volcano in Tanzania.

Qualifications, such as dolomite-bearing, may be used to emphasise the presence of a minor constituent (<10%). Similarly, igneous rocks containing less than 10% of carbonate may be called calcite-bearing ijolite, dolomite-bearing peridotite, etc.

Igneous rocks with between 10% and 50% carbonate minerals may be called calcitic ijolite or carbonatitic ijolite, etc.

Note that the terms leucocratic and melanocratic should not be applied when describing carbonatites, as all primary carbonate minerals fall into group M.

If the carbonatite is too fine-grained for an accurate mode to be determined or if the carbonates are complex Ca–Mg–Fe solid solutions then the chemical classification shown in Fig. B.2 can be used.

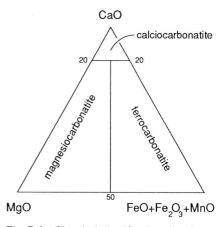

Fig. B.2. Chemical classification of carbonatites using wt% oxides (Woolley & Kempe, 1989).

B.7. LAMPROPHYRIC ROCKS

This classification should be used only if the rock is considered to be a lamprophyric rock, which the Subcommission considers to consist of lamprophyre, lamproite and kimberlite. As this is a somewhat subjective decision, upon which not all persons will agree, the Subcommission has discussed the features that it considers to be diagnostic of these rocks. Although the general criteria given below were originally designed for lamprophyres (Streckeisen, 1978 and 1979), they have been slightly modified so that they are also applicable to lamproites and kimberlites.

Some of the general characteristics common to lamprophyres, lamproites and kimberlites are that:-

(a) they normally occur as dykes (lamprophyres and lamproites), pipes (kimberlites), or small extrusions (lamproites), and are not simply textural varieties of common plutonic or volcanic rocks.

(b) feldspars and/or feldspathoids, when present, are restricted to the groundmass. They are absent in kimberlites.

(c) hydrothermal alteration of olivine, pyroxene, biotite, and plagioclase, when present, is common.

(d) calcite, zeolites, and other hydrothermal minerals may appear as primary phases.

At a more detailed level, the three types can be characterised mineralogically as follows:-

(a) **lamprophyres** are porphyritic, mesocratic to melanocratic (M'=35–90) rocks but rarely ultramafic (M'=90–100). They usually contain essential biotite (or Fe-phlogopite) and/or amphibole, as well as clinopyroxene and olivine, and sometimes melilite. They may be classified according to their felsic minerals as shown in Table B.3.

(b) **lamproites** are similar to lamprophyres

but also contain unusual phases such as K-Ti-richterite, priderite, wadeite, jeppite and Fe-orthoclase. Leucite may be present.

(c) **kimberlites** are ultramafic and consist of major amounts of serpentinised olivine with variable amounts of phlogopite, orthopyroxene, clinopyroxene, carbonate and chromite. Characteristic accessory minerals include pyrope garnet, monticellite, rutile and perovskite.

Similarly, the three types can be characterised chemically as follows:-

(a) **lamprophyres** tend to have contents of K_2O and/or Na_2O, H_2O, CO_2, S, P_2O_5, Ba, etc., that are relatively high compared with other rocks of similar composition.

(b) **lamproites** are similar to lamprophyres but tend also to have molar $(Na_2O+K_2O)/Al_2O_3$ greater than 1, i.e. they are peralkaline.

Table B.3. Classification and nomenclature of lamprophyres based on their light-coloured minerals, alkali feldspar (or), plagioclase (pl), feldspar (feld) and feldspathoid (foid) (after Streckeisen, 1978, p.11). The characteristic mafic minerals are also listed.

Light-coloured constituents		Predominant mafic minerals			
feldspar	foid	biotite, diopsidic augite, (±olivine)	hornblende, diopsidic augite, (±olivine)	amphibole, (barkevikite, kaersutite), Ti-augite, olivine, biotite	melilite, biotite, ±Ti-augite, ±olivine, ±calcite
or > pl	—	minette	vogesite	—	—
pl > or	—	kersantite	spessartite	—	—
or > pl	feld > foid	—	—	sannaite	—
pl > or	feld > foid	—	—	camptonite	—
—	glass or foid	—	—	monchiquite	polzenite
—	—	—	—	—	alnöite

(c) **kimberlites** are generally undersaturated, ultrabasic rocks (SiO_2 25%-35%) with low Al_2O_3 (<5%). The ratio of Na_2O/K_2O is very low (<0.5) and molar $(Na_2O+K_2O)/Al_2O_3$ is generally less than 1.

Some of these characteristics are taken from the data of Bergman (1987), Mitchell (1985, 1986) and Rock (1987).

B.8. MELILITIC ROCKS

This classification should be used only if the rock contains more than 10% modal melilite (Streckeisen, 1978 and 1979) which is regarded as a mafic mineral belonging to group M. These rocks are subdivided into ultramafic and non-ultramafic melilitic rocks.

B.8.1. Ultramafic Melilitic Rocks (M>90%)

These are classified according to their mineral content as shown in Fig. B.3. Notice that the general name for plutonic melilitic rocks is **melilitolite**, and for volcanic melilitic rocks is **melilitite**. If olivine is greater than 10% the term olivine is attached to the name.

The following special varietal names may also be used:-
uncompahgrite = pyroxene melilitolite
kugdite = olivine melilitolite
olivine uncompahgrite = olivine pyroxene melilitolite.

B.8.2. Melilitic Rocks with M<90%

For rocks in this group (non-ultramafic melilitic rocks) the appropriate name in either the plutonic or volcanic QAPF classification must be used and qualified with the term melilite. For example, if the rock contains more than 10% melilite and falls into the nephelinite field of the volcanic QAPF diagram, it should be called melilite nephelinite, etc.

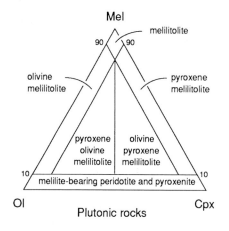

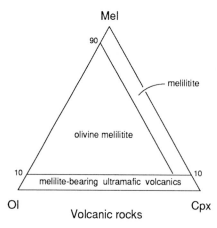

Fig. B.3. Classification and nomenclature of melilitic igneous rocks based on melilite (Mel), olivine (Ol), and clinopyroxene (Cpx) (after Streckeisen, 1978, Fig. 5).

B.9. CHARNOCKITIC ROCKS

This classification should be used only if the rock is considered to belong to the charnockitic suite of rocks which is characterised by the presence of hypersthene (or fayalite plus quartz) and, in many of the rocks, perthite, mesoperthite or antiperthite (Streckeisen, 1974, 1976). They are often associated with norites and anorthosites and are closely linked with Precambrian terranes.

Although many show signs of metamorphic overprinting, such as deformation and recrystallisation, they conform to the group of "igneous and igneous-looking rocks" and have, therefore, been included in the classification scheme.

The classification is based on the QAP triangle, i.e. the upper half of the QAPF double triangle (Fig. B.4), and the general and special names that may be applied to the fields are given in Table B.4.

However, as one of the characteristic of charnockites is the presence of various types of perthite, this raises the problem of how to distribute the perthites between A and P. The Subcommission has, therefore, recommended that the perthitic feldspars should be distributed between A and P in the following way:-

Perthite — assign to A as the major component is alkali feldspar.

Mesoperthite — assign equally between A and P as the amounts of the alkali feldspar and plagioclase (usually oligoclase or andesine) components are roughly the same.

Antiperthite — assign to P as the major component is andesine with minor albite as the alkali feldspar phase.

To distinguish those charnockitic rocks that contain mesoperthite it is further suggested that the prefix m-, being short for mesoperthite, could be used e.g. **m-charnockite**.

Table B.4. Special terms used for charnockitic rocks (from Streckeisen, 1974, p355).

QAPF field	General term	Special term
2	hypersthene alkali feldspar granite	alkali feldspar charnockite
3	hypersthene granite	charnockite (3b farsundite)
4	hypersthene granodiorite	opdalite or charno-enderbite
5	hypersthene tonalite	enderbite
6	hypersthene alkali feldspar syenite	—
7	hypersthene syenite	—
8	hypersthene monzonite	mangerite
9	monzonorite (hypersthene monzodiorite)	jotunite
10	norite (hypersthene diorite), anorthosite (M<10)	—

B.10. PLUTONIC ROCKS

This classification should be used only if the rock is considered to be plutonic i.e. it is assumed to have formed at considerable depth and has a relatively coarse-grained texture in which the individual crystals can easily be seen with the naked eye. There is, of course, a gradation between plutonic rocks and volcanic rocks and the

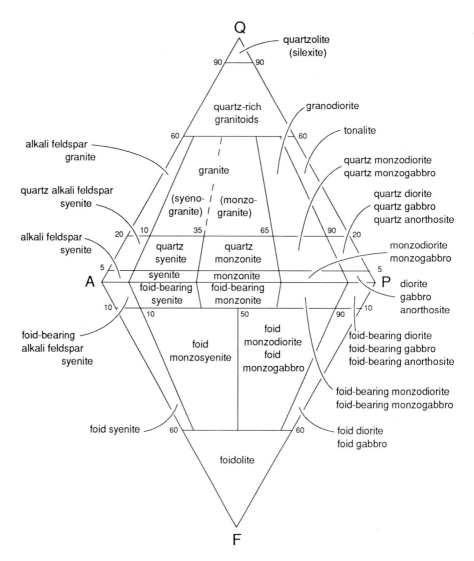

Fig. B.4. Classification and nomenclature of plutonic rocks according to their modal mineral contents using the QAPF diagram (based on Streckeisen, 1976, Fig. 1a). The corners of the double triangle are Q = quartz, A = alkali feldspar, P = plagioclase and F = feldspathoid. However, for more detailed definitions refer to section B.2. This diagram must not be used for rocks in which the mafic mineral content, M, is greater than 90%.

Subcommission suggests that, if there is any uncertainty as to which classification to use, the plutonic root name should be given and prefixed with the term "micro". For example, microsyenite could be used for a rock that was considered to have formed at considerable depth even if many of the individual crystals could not be seen with the naked eye.

The classification is based on modal parameters and is divided into three parts:-

(a) if M is less than 90% the rock is classified according to its felsic minerals, using the now familiar QAPF diagram (Fig. B.4), often simply referred to as the QAPF classification or the QAPF double triangle (section B.10.1).

(b) if M is greater or equal to 90%, the rock is ultramafic and is classified according to its mafic minerals (Fig. B.8), as shown in section B.10.2.

(c) if a mineral mode is not yet available, the "field" classification (Fig. B.9) of section B.10.3 may be provisionally used.

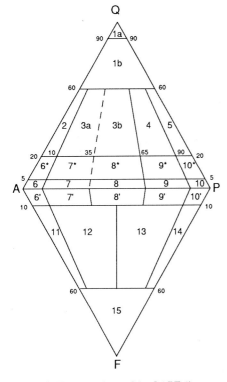

B.10.1. QAPF Classification (M<90%)

The modal classification of plutonic rocks is based on the QAPF diagram and was the first to be completed and recommended by the IUGS Subcommission (Streckeisen 1973 and 1976). The diagram is based on the fundamental work of many earlier petrologists, which is excellently summarised by Streckeisen (1967).

The root names for the classification and the field numbers are given in Fig. B.4 and B.5, respectively.

To use the classification the modal amounts of Q, A, P, and F must be known and recalculated so that their sum is 100%. For example, a rock with Q = 10%, A = 30%, P = 20%, and M = 40% would give recalculated values of Q, A, and P as follows:-

$Q = 100 \times 10 / 60 = 16.7$

$A = 100 \times 30 / 60 = 50.0$

$P = 100 \times 20 / 60 = 33.3$

Although at this stage the rock can then be plotted directly into the triangular diagram, if all that is required is to name the rock it is easier to determine the **plagio-**

Fig. B.5. Field numbers of the QAPF diagram (based on Streckeisen, 1976, Fig. 1a). Fields 6* to 10* are slightly oversaturated variants of fields 6 to 10, respectively, while 6' to 10' are slightly undersaturated variants. The field number 16 is allocated for rocks in which the mafic mineral content, M, is greater than 90%.

clase ratio = $100 \times P / (A + P)$, as the non-horizontal divisions in the QAPF diagram are lines of constant plagioclase ratio. The field into which the rock falls can then easily be determined by inspection.

In the above example rock the plagioclase ratio is 40 so that it can be seen by inspection that the rock falls into QAPF field 8* and is, therefore, a quartz monzonite (Figs. B.4 and B.5).

Similarly, a rock with A = 50%, P = 5%, F = 30%, and M = 15% would recalculate as follows:-

A = $100 \times 50 / 85 = 58.8$

P = $100 \times 5 / 85 = 5.9$

F = $100 \times 30 / 85 = 35.3$

Plagioclase ratio = 9

This rock falls into QAPF field 11 and would, therefore, be called a foid syenite. Furthermore, if the major foid in the rock is nepheline, it would be called a nepheline syenite.

B.10.1.1. *Details of Fields*

Field 2 — rocks in the field of **alkali feldspar granite** have been called alkali granite by many authors. The Subcommission, however, recommends that the term alkali granite be restricted to those rocks that contain alkali amphiboles and/or pyroxenes. The term **alaskite** may be used for a light-coloured (M = < 10) alkali feldspar granite.

Field 3 — the term granite has been used in many senses; in most English and American textbooks it has been restricted to subfield 3a, whereas subfield 3b has contained terms such as adamellite and quartz monzonite. In the European literature, however, granite has been used to cover both subfields, a view adopted by the Subcommission. The Subcommission has also recommended that the term adamellite should no longer be used, as it has been given several meanings, and does not even occur in the Adamello massif as commonly defined (Streckeisen, 1976). Although the term quartz monzonite has also been used with several meanings, the Subcommission decided to retain the term in its original sense, i.e. for rocks in field 8*.

Field 4 — the most widespread rocks in this field are **granodiorites**, commonly containing oligoclase, more rarely andesine. It seems advisable to add the condition that the average An content of the plagioclase should be less than 50% in order to distinguish the common granodiorites from the rare granogabbro in which the An content of the plagioclase is greater than 50%.

Field 5 — the root name **tonalite** should be used whether hornblende is present or not. **Trondhjemite** and **plagiogranite** (of the USSR) may be used for a light-coloured (M = < 10) tonalite.

Fields 6 and 7 — these fields contain the root names **alkali feldspar syenite** and **syenite**, respectively.

Field 8 — the root name is **monzonite**; many so-called "syenites" fall into this field.

Field 9 — the two root names in this field, **monzodiorite** and **monzogabbro**, are separated according to the average composition of their plagioclase; if An is less than 50% the rock is a monzodiorite; if An is greater than 50% the rock is a monzogabbro, and may be further subdivided, if required, as shown below. The terms **syenodiorite** and **syenogabbro** may be used as comprehensive names for rocks between syenite and diorite/gabbro i.e. for monzonites (field 8) and monzodiorite/ monzogabbro, respectively.

Field 10 — the three root names in this field, **diorite**, **gabbro**, and **anorthosite**, are separated according to the average composition of their plagioclase and the colour index; if M is less than 10% the rock is an anorthosite; if An is less than 50% the rock is a diorite; if An is greater than 50% the rock is a gabbro and may be further subdivided, if required, as shown below. Either of the two synonymous terms **dolerite** or **diabase** may be used for medium-grained gabbros rather than the term microgabbro, if required.

Gabbroic rocks — the gabbros (sensu lato) of QAPF field 10, may be further subdivided according to the relative abundances of their orthopyroxene, clinopyroxene, olivine, and hornblende as shown in Fig. B.6. Some of the special terms used are:-

Gabbro (sensu stricto) = plagioclase + clinopyroxene.

Norite = plagioclase + orthopyroxene.

Troctolite = plagioclase + olivine.

Gabbronorite = plagioclase with almost equal amounts of clinopyroxene and orthopyroxene.

Orthopyroxene gabbro = plagioclase + clinopyroxene with minor amounts of orthopyroxene

Clinopyroxene norite = plagioclase + orthopyroxene with minor amounts of clinopyroxene

Hornblende gabbro = plagioclase + hornblende with pyroxene < 5%.

Field 11 — although **foid syenite** is the root name, the most abundant foid present should be used in the name, e.g. **nepheline**

syenite, sodalite syenite, etc. This remark also applies to fields 12 to 15.

Field 12 — the root name **foid monzodiorite** may be replaced by the synonym **foid plagisyenite**. Wherever possible, replace the term foid with the name of the most abundant feldspathoid. **Miaskite**, which contains oligoclase, may also be used.

Field 13 — the two root names in this field, **foid monzodiorite** and **foid monzogabbro**, are separated according to the average composition of their plagioclase, as for rocks in field 9; if An is less than 50% the rock is a foid monzodiorite; if An is greater than 50% the rock is a foid monzogabbro. Wherever possible, replace the term foid with the name of the most abundant feldspathoid. The term **essexite** may be applied to nepheline monzodiorite or nepheline monzogabbro.

Field 14 — the two root names in this field, **foid diorite** and **foid gabbro**, are separated according to the average composition of their plagioclase, as for rocks in field 9; if An is less than 50% the rock is a foid diorite; if An is greater than 50% the rock is a foid gabbro. Wherever possible, replace the term foid with the name of the most abundant feldspathoid. Two special terms may continue to be used, **theralite** for nepheline gabbro and **teschenite** for **analcime gabbro**.

Field 15 — this field contains rocks in which the light-coloured minerals are almost entirely foids and is given the root name **foidolite** to distinguish it from the volcanic equivalent which is called foidite. As these rocks are rather rare the field has not been subdivided, except to note that the most abundant foid should appear in the name, e.g. **nephelinolite**, etc.

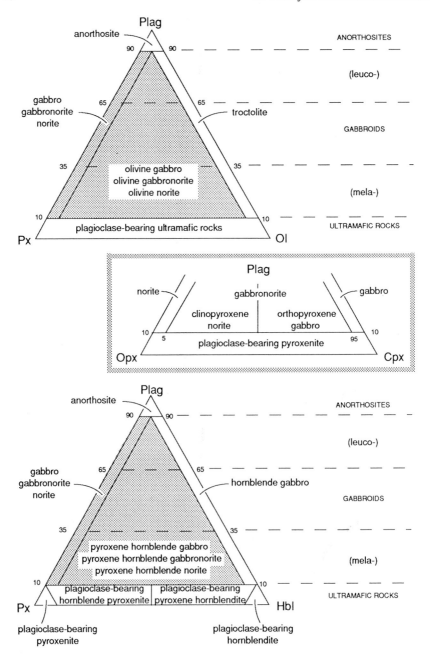

Fig. B.6. Classification and nomenclature of gabbroic rocks based on the proportions of plagioclase (Plag), pyroxene (Px), olivine (Ol), orthopyroxene (Opx), clinopyroxene (Cpx), and hornblende (Hbl) (after Streckeisen, 1976, Fig. 3). Rocks falling in the shaded areas of the triangular diagrams may be further subdivided according to the diagram within the shaded rectangle.

For rocks in the QAPF classification the Subcommission suggests (Streckeisen, 1973, p.30; 1976, p.24) that the prefixes leuco- and mela- may be used to designate the more felsic (lower colour index) and mafic (higher colour index) types within each rock group, when compared with the "normal" types in that group. As the threshold values of M' will obviously vary from rock group to rock group, the limits are shown diagrammatically in Fig. B.7 for the rocks groups to which the terms may be applied. The prefixes should precede the root name, e.g. biotite leucogranite, biotite melasyenite, etc.

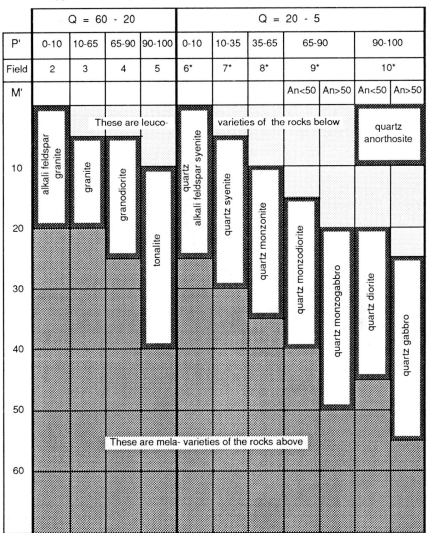

Fig. B.7a. Limits of the use of the terms mela- and leuco- applicable to plutonic rocks classified by the QAPF diagram and with Q greater than 5% (after Streckeisen, 1976, Fig. 5). Abbreviations:- P' = 100*P/(A+P); M' = colour index; An = anorthite content of plagioclase.

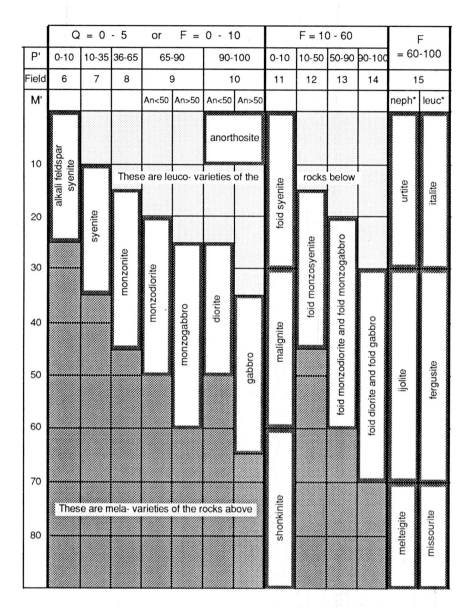

Fig. B.7b. Limits of the use of the terms mela- and leuco- and some special names applicable to plutonic rocks classified by the QAPF diagram and with Q less than 5% or F present (after Streckeisen, 1976, Fig. 5). Abbreviations:- P' = 100*P/(A+P); M' = colour index; An = anorthite content of plagioclase; neph* = nepheline is the predominant foid; leuc* = leucite is the predominant foid.

B.10.2. Ultramafic Rocks (M>90%)

The plutonic ultramafic rocks are classified according to their content of mafic minerals, which consist of olivine, orthopyroxene, clinopyroxene, hornblende, sometimes biotite, and various but usually small amounts of garnet and spinel. Two diagrams are recommended by the Subcommission (Streckeisen, 1973, 1976), one for rocks consisting essentially of olivine, orthopyroxene, and clinopyroxene, and the other for rocks containing hornblende, pyroxenes, and olivine (Fig. B.8.).

Peridotites are distinguished from **pyroxenites** by containing more than 40% olivine. This value, rather than 50%, was chosen because many lherzolites contain up to 60% pyroxene. The peridotites are subdivided into **dunite** (or **olivinite** if the

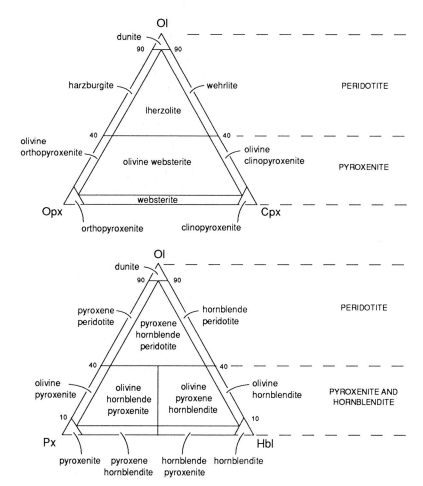

Fig. B.8. Classification and nomenclature of ultramafic rocks based on the proportions of olivine (Ol), orthopyroxene (Opx), clinopyroxene (Cpx), pyroxene (Px) and hornblende (Hbl) (after Streckeisen, 1973, Figs. 2a and 2b).

spinel mineral is magnetite), **harzburgite**, **lherzolite**, and **wehrlite**.

The **pyroxenites** are further subdivided into **orthopyroxenite** (e.g. **bronzitite**), **websterite**, and **clinopyroxenite** (e.g. **diallagite**).

Ultramafic rocks containing garnet or spinel should be qualified in the following manner. If garnet or spinel is less than 5% use garnet-bearing peridotite, chromite-bearing dunite, etc. If garnet or spinel is greater than 5% use garnet peridotite, chromite dunite, etc.

B.10.3. "Field" Classification

The "field" classification of plutonic rocks should be used only as a provisional measure when an accurate mineral mode is not yet available. When available, the plutonic QAPF diagram should be used.

The classification is based on a simplified version of the plutonic QAPF diagram (Streckeisen, 1976) and is shown in Fig. B.9. If the suffix "-oid" is felt to be linguistically awkward then the alternative adjectival form "-ic rock" may be used, i.e. use syenitic rock in place of syenitoid.

B.11. VOLCANIC ROCKS

This classification should be used only if the rock is considered to be volcanic i.e. it is assumed to have been associated with volcanism and has a relatively fine-grained texture in which most of the individual crystals cannot be seen with the naked eye.

The classification of volcanic rocks is divided into three parts:-

(a) if a mineral mode can be determined,

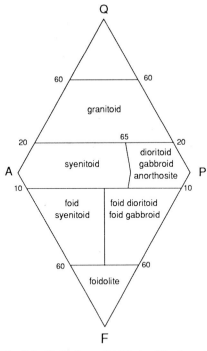

Fig. B.9. Preliminary QAPF classification of plutonic rocks for field use (after Streckeisen, 1976, Fig. 6).

use the QAPF classification (Fig. B.10) of section B.11.1.

(b) if a mineral mode cannot be determined and a chemical analysis is available, use the TAS classification (Fig. B.13) of section B.11.2.

(c) if neither a mineral mode nor chemical analysis is yet available, the "field" classification (Fig. B.17) of section B.11.3 may be provisionally used.

B.11.1. QAPF Classification (M<90%)

This classification should be used only if the rock is considered to be volcanic and if a mineral mode can be determined

(Streckeisen, 1978 and 1979). The root names for the classification are given in Fig. B.10. The numbers of the fields are the same as those for the plutonic rock classification (see Fig. B.5).

B.11.1.1. *Details of Fields*

Field 2 — the root name **alkali feldspar rhyolite,** corresponds with alkali feldspar granite. The term **alkali rhyolite** (= peral-

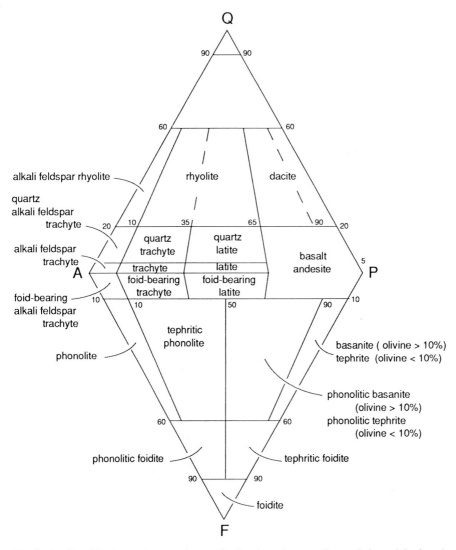

Fig. B.10. Classification and nomenclature of volcanic rocks according to their modal mineral contents using the QAPF diagram (based on Streckeisen, 1978, Fig. 1). The corners of the double triangle are Q = quartz, A = alkali feldspar, P = plagioclase and F = feldspathoid. However, for more detailed definitions refer to section B.2.

kaline rhyolite) can be used when the rock contains alkali pyroxene and/or amphibole. The name rhyolite may be replaced by the synonym **liparite**.

Fields 3a and 3b — in an analogous manner to the granites, **rhyolite (liparite)** covers both fields 3a and 3b. The term **rhyodacite**, which has been used ambiguously for rocks of fields 3b and 4, can be used for transitional rocks between rhyolite and dacite without attributing it to a distinct field.

Fields 4 and 5 — rocks in both these fields are covered by the root name **dacite** in the broad sense. Volcanic rocks of field 5, to which terms such as "plagidacite" and "quartz andesite" have been applied, are frequently also described as dacite, which is the recommended name.

Fields 6, 7, 8 — rocks with root names **alkali feldspar trachyte**, **trachyte** or **latite**, which contain no modal foids but do contain nepheline in the norm, may be qualified with "*ne*-normative" to indicate that they would fall in subfields 6'-8', respectively. **Alkali trachyte** (=peralkaline trachyte) may be used for trachytes containing alkali pyroxene and/or amphibole.

Fields 9 and 10 — these fields contain the large majority of volcanic rocks, including **basalt** and **andesite**, which are tentatively separated using colour index, at a limit of 40 wt% or 35 vol%, and 52% SiO_2 as shown in Fig. B.11. A plagioclase composition (at a limit of An_{50}) is less suitable for the distinction between basalt and andesite, because many andesites commonly contain "phenocrysts" of labradorite or bytownite. Although this may seem rather unsatisfactory, it is unlikely that many of these rocks will be

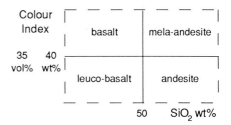

Fig. B.11. Division of QAPF field 9 and 10 rocks into basalt and andesite, using colour index and SiO_2 content (after Streckeisen, 1978, Fig. 2)

classified using the QAPF diagram, as the modes of most basalts and andesites are difficult to accurately determine and so that the TAS classification will have to be used.

Field 11 — the root name **phonolite** is used in the sense of Rosenbusch for rocks consisting essentially of alkali feldspar, any feldspathoid and mafic minerals. The nature of the predominant foid should be added to the root name, e.g. leucite phonolite, analcime phonolite, leucite nepheline phonolite (with nepheline > leucite), etc. Phonolites containing nepheline and/or haüyne as the main foids are commonly described simply as "phonolite".

Field 12 — the root name for these rather rare rocks is **tephritic phonolite**. Although it was originally suggested that the term tephriphonolite was a synonym (Streckeisen, 1978), it is probably better to reserve this term for the root name of TAS field U3, to indicate that the name has been given chemically and may not be identical to those of QAPF field 12.

Field 13 — this field contains the root names **phonolitic basanite** and **phonolitic tephrite** which are separated on the amount of olivine in the CIPW norm. If normative olivine is greater than 10% the rock is called a phonolitic basanite; if less than 10% it is a phonolitic tephrite. Although it was originally suggested that the term phonotephrite was a synonym of phonolitic tephrite (Streckeisen, 1978), it is probably better to reserve this term for the root name of TAS field U2, to indicate that the name has been given chemically and may not be identical to those of QAPF field 13. There is no conflict with the term **phonobasanite** being used as a synonym for phonolitic basanite as the term is not used in TAS.

Field 14 — this field contains the root names **basanite** and **tephrite** which are separated on the amount of olivine in the CIPW norm. If normative olivine is greater than 10% the rock is called a basanite; if less than 10% it is a tephrite. The nature of the dominant foid should be indicated in the name, e.g. nepheline basanite, leucite tephrite, etc.

Field 15 — the general root name of this field is **foidite**, but as these rocks occur relatively frequently the field has been subdivided into 15a, 15b and 15c.

Field 15a — the root name is **phonolitic foidite** but wherever possible, more specific terms, such as phonolitic nephelinite, etc. should be used. Alternatively, the term alkali feldspar foidite could be used as the root name, which would give specific terms such as sanidine nephelinite, etc.

Field 15b — the root names are **tephritic foidite** or **basanitic foidite** and are separated according to olivine content as in field 14. Wherever possible, more specific terms, such as tephritic leucitite, basanitic nephelinite, etc. should be used.

Field 15c — the root name is **foidite** and should be distinguished by the name of the predominant foid, e.g. nephelinite, leucitite, analcimite, etc.

Field 16 — the root name is **ultramafitite**, which should be qualified by indicating the predominant mafic minerals.

B.11.2. The TAS Classification

This classification should be used only if the rock is considered to be volcanic and if a mineral mode cannot be determined, due either to the presence of glass or to the fine-grained nature of the rock, and a chemical analysis of the rock is available.

The main part of the classification is based on the total alkali silica (TAS) diagram. The root names for the classification and the field symbols are given in Fig. B.13 and B.14, respectively (Le Bas et al. 1986). The classification is easy to use as all that is initially required are the values of Na_2O+K_2O and SiO_2 although, if the analysis falls in certain fields, additional calculations, such as the CIPW norm, must be performed in order to arrive at the correct root name.

The TAS classification was originally constructed with the more common rock types in mind using the following principles:-

(a) each field was chosen to accord as closely as possible with the current usage of the root name with the help of data from 24,000 analyses of named fresh volcanic rocks from the CLAIR and PETROS databases (Le Maitre, 1982).

(b) fresh rocks were taken to be those in which $H_2O+ < 2\%$ and $CO_2 < 0.5\%$.

(c) each analysis was recalculated to 100% on an H_2O and CO_2 free basis.

(d) wherever possible, the boundaries were located to minimise overlap between adjacent fields.

(e) the vertical SiO_2 boundaries between the fields of basalt, basaltic andesite, andesite, and dacite, were chosen to be those in common use.

(f) the boundary between the S (for silica Saturated) fields and the U (for silica Undersaturated) fields was chosen to be roughly parallel with the empirically determined contours of 10% normative F in QAPF.

(g) the boundary between the S fields and O (for silica Oversaturated) fields was chosen where there was a density minimum between volcanic rock series that were alkaline and those that were calc-alkaline.

(h) the boundaries between fields S1–S2–S3–T were all made parallel to a pronounced edge found in the distribution of analyses of rocks that had been called trachyte from CLAIR and PETROS.

(i) similarly, the boundaries between fields U1–U2–U3–Ph were also drawn parallel to each other. They are not at right angles to the line separating fields S from U.

However, after the TAS classification was published the Subcommission considered whether or not it was possible to include some of the olivine- and pyroxene-rich ("high-Mg") volcanic rocks e.g. picrites, komatiites, meimechites, and boninites into the scheme. After lengthy discussions this has been done, but only by using some additional parameters not used in TAS. As a result these rocks must be considered first, as they are not the "normal" type of volcanic rocks for which the TAS classification was originally designed.

It is important to note that the TAS diagram published by Le Maitre (1984) has been slightly modified in Le Bas et al. (1986) in the light of further information. Similarly, the TAS classification presented here has been slightly improved in the light of additional information in two ways:-

(a) by "opening" the boundary between fields U1 and F and making it a dashed line. This now allows nephelinites and leucitites to be better classified as they can fall in both fields U1 and F.

(b) by the inclusion of some of the "high-Mg" volcanic rocks such as picrites, komatiites, meimechites, and boninites which has been done after further lengthy discussions. However, as they are not the "normal" type of volcanic rocks for which the TAS classification was originally designed, they must be extracted first, in line with the general logic of the classification scheme.

It must also be stressed that the TAS classification is **purely descriptive**, and that **no genetic significance** is implied. Furthermore, analyses of rocks that are weathered, altered, metasomatised, metamorphosed or have undergone crystal accumulation should be used with caution, as spurious results may be obtained. As a general rule it is, therefore, suggested that only analyses with $H_2O+ < 2\%$ and with $CO_2 < 0.5\%$ should be used, except if the rock is a picrite, komatiite, meimechite or boninite, when this restriction is withdrawn. Note that the application of TAS to altered rocks is discussed by Sabine et al. (1985) who found that many low grade metavolcanic rocks could be satisfactorily classified.

B.11.2.1. *Using TAS*

Before using the classification the following procedures must be adopted:-
(a) all analyses must be recalculated to 100% on an H_2O and CO_2 free basis
(b) if a CIPW norm has to be calculated, in order to determine the correct root name, the amounts of FeO and Fe_2O_3 should be left as determined. If only total iron has been determined, then it is up to the user to justify the method used for partitioning the iron between FeO and Fe_2O_3. One method that can be used to get an estimate of what the FeO and Fe_2O_3 would have been, had they been determined, is that of Le Maitre (1976). Remember, it is the feeling of the Subcommission that rocks should be named according to what they are, and not according to what they might have been.

The analysis must then be checked to see if it is a "high-Mg" volcanic rock, i.e. a picrite, komatiite, meimechite or boninite. This is done as follows (see Fig. B.12):-
(a) **Boninite** — $SiO_2 > 53\%$, MgO > 8%, and $TiO_2 < 0.5\%$
(b) **Picritic rocks** — $SiO_2 < 53\%$, $Na_2O+K_2O < 2.0\%$, and MgO > 18%. These are divided into:-
Picrite — $Na_2O+K_2O > 1\%$
Komatiite — $Na_2O+K_2O < 1\%$ and $TiO_2 < 1\%$
Meimechite — $Na_2O+K_2O < 1\%$ and $TiO_2 > 1\%$

Note that the Subcommission recommends that the term picritic rocks can be used to include the rock names picrite, komatiite, and meimechite.

If the rock is not one of these four ultramafic types it is then classified using the total alkali silica (TAS) diagram as shown in Fig. B.13. Certain of the fields can then be subdivided further as described below.

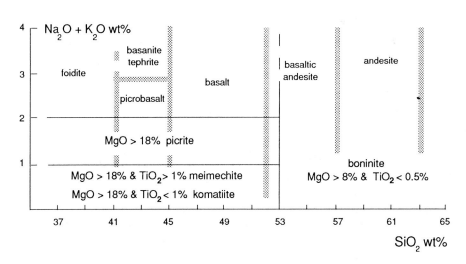

Fig. B.12. Classification and nomenclature of "high-Mg" volcanic rocks (picrite, komatiite, meimechite and boninite) using TAS together with wt% MgO and TiO_2. The thick stippled lines indicate the location of TAS fields.

B.11.2.2. *Details of Fields*

Field B — the root name **basalt** may be divided into **alkali basalt** and **subalkali basalt** according to the state of silica saturation. If the analysis contains normative nepheline, the rock may be called an alkali basalt; if the analysis contains no normative nepheline it may be called a subalkali basalt. It should be noted that in the USSR alkali basalt and subalkali basalt (as defined above) were formerly called subalkali basalt and basalt, respectively; however, it was agreed that in the USSR these two rock types should now be called midalkali basalt and subalkali basalt in order to conform more closely to the IUGS recommendation. Further subdivision was

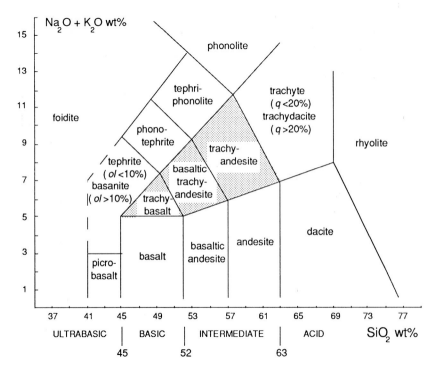

Fig. B.13. Chemical classification and nomenclature of volcanic rocks using the total alkali versus silica (TAS) diagram (after Le Bas et al., 1986, Fig. 2). Rocks falling in the shaded areas may be further subdivided as shown in the table underneath the diagram. The line between the foidite field and the basanite–tephrite field is dashed to indicate that further criteria must be used to separate these types. Abbreviations:- q = normative quartz; ol = normative olivine.

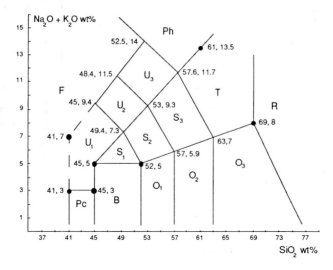

Fig. B.14. Field symbols of the total alkali versus silica (TAS) diagram (after Le Bas et al., 1986, Fig. 1). The pairs of numbers are the coordinates of the intersections of the lines.

not considered desirable at this stage due to the current state of flux and lack of consistency in many of the specialist names used for basalts.

Fields B, O1, O2, O3, R — the root names **basalt** (if SiO2 is greater than 48%), **basaltic andesite, andesite, dacite,** and **rhyolite** may be qualified using the terms **low-K, medium-K,** and **high-K** as shown in Fig. B.15. This is in accord with the concept developed by Peccerillo and Taylor (1976), but the lines have been slightly modified and simplified. It must be stressed that the term **high-K is not**

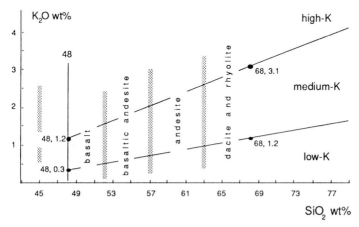

Fig. B.15. Division of basalts (with SiO$_2$ > 48%), basaltic andesites, andesites, dacites and rhyolites in low-K, medium-K and high-K types. Note that high-K is NOT synonymous with potassic. The thick stippled lines indicate the equivalent position of some of the fields in the TAS diagram.

synonymous with potassic, as high-K rocks can have more Na_2O than K_2O.

Field R — the root name **rhyolite** may be further subdivided into **peralkaline rhyolite**, if the peralkaline index, molecular $(Na_2O + K_2O) / Al_2O_3$, is greater than 1.

Field T — this field contains the root names **trachyte** and **trachydacite** which are separated by the amount of CIPW normative Q in Q+an+ab+or (i.e. the normative equivalent of Q and QAPF). If the value is less than 20% the rock is called trachyte; if greater than 20% it is a trachydacite. Trachytes may be further subdivided into **peralkaline trachytes**, if the peralkaline index is greater than 1.

Peralkaline rhyolites and **trachytes** — the Subcommission has considered it useful to further subdivide these rocks into **comenditic rhyolite** (= **comendite**), **comenditic trachyte**, **pantelleritic rhyolite** (= **pantellerite**), and **pantelleritic trachyte** according to the method of Macdonald (1974) which is based on the relative amounts of Al_2O_3 versus total iron as FeO as shown in Fig. B.16.

Field S1 — the root name **trachybasalt** may be divided into **hawaiite** and **potassic trachybasalt** according to the relative amounts of Na_2O and K_2O. If $Na_2O - 2$ is greater than K_2O the rock is considered to be "sodic" and is called hawaiite; if $Na_2O - 2$ is less than K_2O the rock is considered to be "potassic" and is called potassic trachybasalt.

Field S2 — using the same criteria as for field S1, the root name **basaltic trachyandesite** may be divided into **mugearite** ("sodic") and **shoshonite** ("potassic").

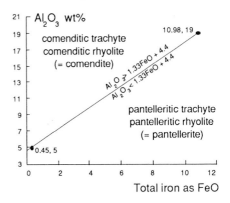

Fig. B.16. Separation of trachytes and rhyolites into comenditic and pantelleritic types using the Al_2O_3 versus total iron as FeO diagram (after Macdonald, 1974).

Field S3 — using the same criteria as for field S1, the root name **trachyandesite** may be divided into **benmoreite** ("sodic") and **latite** ("potassic").

Field U1 and F — field U1 contains the root names **basanite** and **tephrite** while field F contains **foidite**, of which the two main varieties are nephelinite and leucitite. Unlike earlier versions of TAS this boundary is now dashed as, although most named basanites and tephrites do fall in field U1, it has been found that nephelinites and leucitites fall in both fields U1 and F. This means that additional parameters will have to be used to effectively separate the various rock types. This is a problem the Subcommission is currently working on and hopes to come up with a solution in the near future. Until this matter is resolved these names should be assigned with caution.

B.11.3. "Field" Classification

The "field" classification of volcanic rocks should be used only as a provisional measure when neither an accurate mineral mode nor a chemical analysis is yet available. When either become available, the volcanic QAPF diagram or the TAS diagram should be used.

The classification is based on a simplified version of the volcanic QAPF diagram (Streckeisen, 1978 and 1979) and is shown in Fig. B.17. If the suffix "-oid" is felt to be linguistically awkward then the alternative adjectival form "-ic rock" may be used, i.e. use dacitic rock in place of dacitoid.

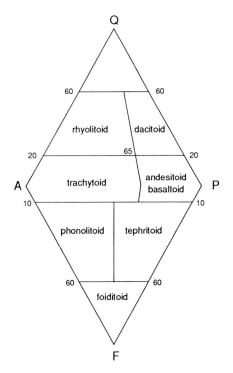

Fig. B.17. Preliminary QAPF classification of volcanic rocks for field use (after Streckeisen, 1979, Fig. 1)

B.12. PUBLICATIONS

These lists of publications includes those important papers that led up to the formation of the Subcommission together with all the official recommendations made on behalf of the Subcommission in English and in other languages. Although not all of these papers have been referred to in this chapter, they are all given for the sake of completeness. Any references cited in this chapter but not found in these lists will be found in section B.13 below.

B.12.1. Prior Work

1964. STRECKEISEN, A. Zur Klassifikation der Eruptivgesteine. *Neues Jahrbuch für Mineralogie. Stuttgart. Monatshefte.* p.195–222.
1965. STRECKEISEN, A. Die Klassifikation der Eruptivgesteine. *Geologische Rundschau. Internationale Zeitschrift für Geologie.* Vol.55, p.478–491.
1967. STRECKEISEN, A. Classification and Nomenclature of Igneous Rocks (Final Report of an Inquiry). *Neues Jahrbuch für Mineralogie. Stuttgart. Abhandlungen.* Vol.107, p.144–240.
†1968, STRECKEISEN, A. Account of Classification and Nomenclature of Igneous Rocks. 11pp. With the following appendices:-
Comments and proposals. 67pp.
Letter of T.F.W. Barth, President of the IUGS.
Report of the Petrographic Committee of the Government of USSR. 10pp.
Report of the Geological Survey of Canada (W.R.A. Baragar & T.N. Irvine). 15pp.
Comments and Proposals by 42 colleagues.

---

†These important documents were circulated to members of the Subcommission prior to the 1968 Prague meeting, which never took place. Although they were never published, they are now housed in the British Museum (Natural History) along with all the Contributions listed in Appendix A.

B.12.2. Official Recommendations

†1973. [STRECKEISEN, A.] Plutonic Rocks. Classification and nomenclature recommended by the IUGS Subcommission on the Systematics of Igneous Rocks. *Geotimes.* Vol.18, No.10, p.26–30.

†1973a. [STRECKEISEN, A.] Commission on Systematics in Petrology. Subcommission on the Systematics of Igneous Rocks. Classification and Nomenclature of Plutonic Rocks. Recommendations. *Geological Newsletter. Quarterly Journal issued by the International Union of Geological Sciences.* Vol. 2, p.110–127.

†1973b. [STRECKEISEN, A.] Classification and Nomenclature of Plutonic Rocks. Recommendations. By the IUGS Subcommission on the Systematics of Igneous Rocks. *Neues Jahrbuch für Mineralogie. Stuttgart. Monatshefte.* p.149–164.

†1974. [STRECKEISEN, A.] Classification and Nomenclature of Plutonic Rocks. Recommendations of the IUGS Subcommission on the Systematics of Igneous Rocks. *Geologische Rundschau. Internationale Zeitschrift für Geologie. Stuttgart.* Vol.63, p.773–785.

1974. STRECKEISEN, A. How should charnockitic rocks be named?. *Centenaire de la Société Géologique de Belgique Géologie des domaines cristallins, Liège.* p.349–360.

1976. STRECKEISEN, A. To each plutonic rock its proper name. *Earth Science Reviews. International Magazine for Geo-Scientists. Amsterdam.* Vol.12, p.1–33.

‡1978. STRECKEISEN, A. IUGS Subcommission on the Systematics of Igneous Rocks. Classification and Nomenclature of Volcanic Rocks, Lamprophyres, Carbonatites and Melilitic Rocks. Recommendations and Suggestions. *Neues Jahrbuch für Mineralogie. Stuttgart. Abhandlungen.* Vol.134, p.1–14.

‡1979. STRECKEISEN, A. Classification and nomenclature of volcanic rocks, lampro-

phyres, carbonatites, and melilitic rocks: Recommendations and suggestions of the IUGS Subcommission on the Systematics of Igneous Rocks. *Geology. The Geological Society of America. Boulder.* Vol.7, p.331–335.

‡1980. STRECKEISEN, A. Classification and Nomenclature of Volcanic Rocks, Lamprophyres, Carbonatites and Melilitic Rocks. IUGS Subcommission on the Systematics of Igneous Rocks. Recommendations and Suggestions. *Geologische Rundschau. Internationale Zeitschrift für Geologie.* Vol.69, p.194–207.

§1981. SCHMID, R. Descriptive nomenclature and classification of pyroclastic deposits and fragments: Recommendations of the IUGS Subcommission on the Systematics of Igneous Rocks. *Geology. The Geological Society of America. Boulder.* Vol.9, p.41–43.

§1981a. SCHMID, R. Descriptive nomenclature and classification of pyroclastic deposits and fragments: Recommendations by R. Schmid on behalf of the IUGS Subcommission on the Systematics of Igneous Rocks. *Neues Jahrbuch für Mineralogie. Stuttgart. Monatshefte.* p.190-196.

§1981b. SCHMID, R. Descriptive Nomenclature and Classification of Pyroclastic Deposits and Fragments. *Geologische Rundschau. Internationale Zeitschrift für Geologie.* Vol.70, p.794-799

1984. LE MAITRE, R.W. A proposal by the IUGS Subcommission on the Systematics of Igneous Rocks for a chemical classification of volcanic rocks based on the total alkali silica (TAS) diagram. *Australian Journal of Earth Science, Melbourne.* Vol.31, p.243–255.

1984. ZANETTIN, B. Proposed new chemical classification of volcanic rocks. *Episodes*, Vol.7, p.19–20.

1986. LE BAS, M.J., LE MAITRE, R.W., STRECKEISEN, A. & ZANETTIN, B. A Chemical Classification of Volcanic Rocks Based on the Total Alkali – Silica Diagram. *Journal of Petrology. Oxford.* Vol.27, p.745–750.

---

† The contents of these papers are virtually the same and were published without the author's name.

‡ The contents of these papers are virtually the same.

---

§ The contents of these papers are virtually the same.

B.12.3. In Chinese

LE BAS, M.J., LE MAITRE, R.W., STRECKEISEN, A. & ZANETTIN, B., 1988. A Chemical Classification of Volcanic Rocks Based on the Total Alkali – Silica Diagram [transliterated from Chinese]. *Dizhi Keji Dongtai*, Vol.9, p.25-29.

B.12.4. In Danish

SØRENSEN, H., 1981. Hvad skal bjergarterne hedde? *Dansk geologisk Forening, Årsskrift for 1980*, side 39–46, København.
SØRENSEN, H., 1986. Hvad skal bjergarterne hedde? II. Vulkanske bjergater, pyroclaster m.v. *Dansk geologisk Forening, Årsskrift for 1985*, side 59–65, København.

B.12.5. In Russian

Akademiya Nauk SSSR, Petrograficheskii Komitet, 1975. Klassifikatsiya i nomenklatura plutonicheskikh (intrusivn'ykh) gorn'ykh porod [transliterated from Russian]. *"Nedra"*, *Moskva*, 25pp.

B.12.6. In Slovakian

KAMENICKÝ, J., 1975. Návrh IUGS na klasifikáciu a nomenklatúru plutonických hornín. *Mineralia Slovaca, Bratislava*. Vol.7, p.1-12.

B.12.7. In Spanish

STRECKEISEN, A., 1974 (translated by C. Pineda & A. Tobar). Rocas Plutonicas. Clasificacion y nomenclatura recomendada por la Union Internacional de Ciencias Geologicas (I.U.G.S.) Subcomision en Sistematica de Rochas Igneas. *Instituto de Investigaciones Geologicas, Universidad de Chile, Santiago,*

Chile. 10pp.
TERUGGI, M.E., 1980. La Clasificacion de las Rochas Igneas: segun la Subcomision de Sistematica de las Rocas Igneas de la Union Internacional de Ciencias Geologicas (IUGS). Con Glosario Ingles – Castellano y Castellano – Ingles de los Terminos Usados. Coleccion Ciencias de la Tierra. No.1. *Ediciones Científicas Argentinas Librart (ECAL), Buenos Aires.* 34pp.

B.13. ADDITIONAL REFERENCES

These are some additional references that have been cited in this chapter. Any reference that cannot be found in this list will be found in section B.12 above.

BERGMAN, S.C., 1987. Lamproites and other potassium-rich igneous rocks: a review of their occurrence, mineralogy and geochemistry. In: J.G. Fitton & B.G.J. Upton (Eds.), Alkaline Igneous Rocks. *Geological Society of London Special Publication*, No.30, p.103-190.
LE MAITRE, R.W., 1976. Some problems of the projection of chemical data into mineralogical classifications. *Contributions to Mineralogy and Petrology*, Vol.56, p.181–189.
LE MAITRE, R.W., 1982. Numerical Petrology. *Elsevier, Amsterdam*, 281pp.
MACDONALD, R., 1974. Nomenclature and Petrochemistry of the Peralkaline Oversaturated Extrusive Rocks. *Bulletin Volcanologique*. Vol. 38, p.498–516.
MITCHELL, R.H., 1985. A review of the mineralogy of lamproites. *Transactions and Proceedings of the Geological Society of South Africa*, Vol.88, p.411-437.
MITCHELL, R.H., 1986. Kimberlites. Mineralogy, Geochemistry and Petrology. *Plenum Press, New York*. 442pp.
PECCERILLO, A. & TAYLOR, S.R., 1976. Geochemistry of the Eocene calc-alkaline volcanic rocks from the Kastamonu area, northern Turkey. *Contributions to Mineralogy and Petrology*, Vol.58, p.63–81.

Rock, N.M.S., 1987. The nature and origin of lamprophyres: an overview. In: J.G. Fitton & B.G.J. Upton (Eds.), Alkaline Igneous Rocks. *Geological Society of London Special Publication*, No.30, p.191-226.

Sabine, P.A., Harrison, R.K., & Lawson, R.I., 1985 Classification of volcanic rocks of the British isles on the total alkali oxide–silica diagram, and the significance of alteration. *British Geological Survey Report*, Vol 17., No.4, p.1–9.

Woolley, A.R. & Kempe D.R.C., 1989 (In press). Carbonatites: nomenclature, average chemical compositions and element distribution. In: K. Bell (Ed.), Carbonatite Genesis and Evolution. *Unwin Hyman, London.*

B.14. TEXT BOOKS

The following textbooks have wholly or partially adopted the IUGS Subcommission classifications. In most cases the classifications has been adopted as originally recommended, but some have made minor modifications of their own.

Best, M.G., 1982. Igneous and Metamorphic Petrology. *Freeman, San Francisco.* 630pp.

Cox, K.G., Bell, J.D. & Pankhurst, R.J., 1979. The Interpretation of the Igneous Rocks. *George Allen & Unwin, London.* 450pp.

Efremova, S.V. & Stafeev, K.G., 1985. Petrokhmicheskie metody issledovaniva goriykh porod [transliterated from Russian].

"Nedra", Moscow. 511pp.

Ehlers, E.G. & Blatt, H., 1982. Petrology. Igneous, Sedimentary and Metamorphic. *Freeman, San Francisco.* 732pp.

Hovorka, D. & Suk, M., 1987. Geochémia a genéza eruptivn¥ch a metamorfovan¥ch hornín. *Universita Komenského, Bratislava.* 163pp.

Hyndman, D.W., 1972. Petrology of Igneous and Metamorphic Rocks. *McGraw-Hill, New York.* 533pp.

Lameyre, J., 1986. Roches et Minéraux. *Doin Éditeurs, Paris.* 350pp.

Middlemost, E.A.K., 1985. Magmas and Magmatic Rocks. *Longman, London.* 266pp.

Murray, J.W., 1981. A Guide to Classification in Geology. *Ellis Horwood, Chichester.* 112pp.

Raymond, L.A., 1984. Petrography Laboratory Manual. Vol.1. Handspecimen Petrography. *Geology Services International, Boone, N.Carolina.*

Streckeisen, A., (in press). Systematik der Eruptivgesteine. *Springer, Heidelberg.*

Thorpe, R.S & Brown, G.C., 1985. The Field Description of Igneous Rocks. *Geological Society of London Handbook, Open Universituy Press, Milton Keynes.* 154pp.

Wells, M.K. & Wells, A.K., (in press). Petrology of the Igneous Rocks. *Unwin Hyman, London.* 14th. Edition

Williams, H., Turner, F.J. & Gilbert, C.M., 1982. Petrography. *Freeman, San Francisco.* 626pp.

Wimmenauer, W., 1985. Petrographie der magmatischen und metamorphen Gesteine. *Enke, Stuttgart.* 382pp.

C. Glossary of Terms

In order to make the glossary a standard reference for the future, every effort has been made to make the list of rock names as complete as possible. The initial list included all the names found in standard reference texts and glossaries such as Johannsen (1931, 1932, 1937, 1938 & 1939), Tröger (1935 & 1938), Sørensen (1974) and Tomkeieff et al. (1983), except for varietal names as outlined in section C.1.1. This list was later extensively edited and modified by suggestions from contributors and particularly when the references were checked. The final list contains 1586 terms which, as far as the Subcommission is aware, includes the largest number of igneous rock names ever published.

C.1. DETAILS

Most of the glossary entries consist of five types of information:-
(a) the term, with any alternative spellings in brackets
(b) a brief petrological description or comments on the term
(c) the author(s), year and page number of the source reference
(d) the origin of the term
(e) the location of the term in three standard texts.
Where given the last three are printed in italics and enclosed in brackets. Each of these five types of information is described in further detail below.

C.1.1. Choice of Terms

The principles with which names have been included in the glossary have been somewhat subjective. In general, varietal names have not been included where they are self explanatory e.g. hornblende granite, biotite granite. However, varietal names have been included where they have assumed a special meaning (e.g. nepheline syenite, quartz monzonite) or are not self evident (e.g. Oslo-essexite, Puys-andesite) or are in common usage (e.g. mid-ocean ridge basalt, I-type granite).

Terms in parentheses are alternative spellings, usually due to different transliterations into English, often from German or Russian. Terms separated only by commas are discrete names but with the same derivations e.g. pallioessexite, palliogranite.

As the publication is in English, foreign accents have not been used in the names, except where they have traditionally been used e.g. haüynite.

The glossary of terms also contains all the root names recommended by the Subcommission together with some general, adjectival and chemical terms that are necessary for their definition and understanding. All these terms are listed in **BOLD CAPITALS** as they form part of the Subcommission classification. A total of 297 entries (or 19%) fall into this category. No attempt has been made to include all igneous textural terms as some petrological (as well as mineralogical) knowledge is expected of the reader.

Any of the others terms may also be used if they are felt to serve a useful purpose, although it is hoped that some of the terms described as "obsolete" would not be revived. Note, however, that the Subcommission has not defined any of these terms.

C.1.2. Petrological Description

The petrological descriptions were designed to be concise and informative. Wherever possible the names have been described in terms of the root name that would have been used if the rock had been classified using the IUGS Subcommission recommendations as set out earlier in the book. In most cases this has been achieved by using the phrase "**a variety of**" followed by the root name. A total of 447 entries (or 28%) fall into this category.

The term "**obsolete**" is used when, to the best knowledge of members of the Subcommission, the term has not been used in publications for a long time. In many cases the terms were only used once in the original publication. It is hoped that these terms will not be revived. A total of 406 entries (or 26%) fall into this category.

The term "**local**" is used when, to the best knowledge of members of the Subcommission, the name has only been used for rocks from the particular local region or area where it was named. It is suggested that such names should not be used outside their local areas. A total of 316 entries (or 20%) fall into this category.

An attempt has also been made to ensure that the mineral names used in the descriptions conform to those recommended by the International Mineralogical Associa-

tion (IMA), except where a mineral name was used as part of the original definition or description, e.g. barkevikite.

C.1.3. Source Reference

After the petrological description the author(s) and date of the original source reference where the term was first used is given. This differs somewhat from many of the references given in some of the other standard texts where the source reference cited often refers to where the rock was first collected but not necessarily named. This explains some of the apparent discrepancies that will be found between these and some of those in the standard texts, in particular Tröger and Tomkeieff.

The page number given with the reference is where the name is first used or where it is defined and described in detail.

As many as possible of the references have been seen and checked by members of the Subcommission. Only 15 out of 787 original publications, or 1.9%, have not been seen and checked. These references are all preceded with an "*".

C.1.4. Origin of Name

Wherever possible the source reference is followed by information concerning the origin of the name, unless the derivation is obvious e.g. pyroxenite. Most rocks are named after a geographical locality or region, a Greek, Latin or other linguistic root or a person. In some cases where the derivation is obvious (i.e. mineralogical) the type locality is given instead. A total of 866 entries (or 55%) have this information.

Table C.1. List of countries and linguistic roots, etc. used 10 or more times in the origin of a rock name. The accumulative percentage is based on the 866 glossary entries for which there is any information. A total of 109 different types of origin were found to be present.

	Number	Acc%
USSR	89	10.3
Italy	69	18.2
USA	64	25.6
Norway	62	32.8
from Greek	62	40.0
named after people	53	46.1
West Germany	46	51.4
UK	39	55.9
France	33	59.7
Canada	31	63.3
Czechoslovakia	23	65.9
Sweden	22	68.5
from Latin	16	70.3
Madagascar	13	71.8
Portugal	12	73.2
Australia	11	74.5
South Africa	10	75.6

All the geographical localities, except the very local, are given in their English spelling and conform to the style found in "The Times Atlas of the World", Comprehensive Edition, 1977.

The frequency with which some countries and linguistic roots have been used is given in Table C.1. As can be seen 33% of the rocks come from or are named after only 4 countries — the USSR, Italy, USA and Norway. Similarly, 10% of the rocks have been named after Greek, Latin or German words.

C.1.5. Location in Standard Texts

The last piece of information given is the location of the term in either Johannsen (1931, 1932, 1937, 1938 or 1939), Tröger (1935 or 1938) or Tomkeieff et al., (1983). This is done in the following manner. For example:-

Trög. 508 indicates that the name occurs as rock Number 508 in Tröger (1935). *Trög(38) 775⅔* indicates that the name occurs as rock Number 775⅔ in Tröger (1938). Note that both of these references can be found in Tröger (1969)[†], a special reprinted facsimile edition.

Joh. v.4, p.165 indicates that the name can be found in Johannsen volume 4 on page 165. *Joh. v.1(2nd.Ed.), p.238* indicates that the name can be found in Johannsen volume 1, 2nd. edition on page 165. The dates of the various volumes are v.1 = 1931; v.1 (2nd.Ed.) = 1939; v.2 = 1932; v.3 = 1937; v.4 = 1938.

Tom. 261 indicates that the name can be found on page 261 in Tomkeieff et al. (1983).

The contribution of these texts has been considerable, although many discrepancies and errors have been found. Of the 1586 terms listed in this glossary, 834, or 53%, can be found in Tröger, 615 (39%) in Johannsen, and 1179 (74%) in Tomkeieff. Only the relatively small number of 499, or 31%, of the terms can be found in all three texts. Comparing the texts with each other, Tomkeieff contains 339 names that are not in Tröger or Johannsen; Tröger contains 40 names that are not in Johannsen or Tomkeieff while Johannsen only contains 16 names that are not in Tröger or

---

† Spezielle Petrographie der Eruptivgesteine. Ein Nomenklatur-Kompendium. mit 1. Nachtrag. Eruptivgesteinsnamen. *Verlag der Deutschen Mineralogischen Gesellschaft. Schweizerbart, Stuttgart*, 360pp. + 80pp.

Tomkeieff. Finally, a total of 324, or 20%, of the names in this glossary are not contained in Tröger, Johannsen or Tomkeieff.

C.2. HISTORICAL PERSPECTIVE

For those interested in the historical perspective of igneous petrology, it is interesting to look at the distribution of new rock names with time. For example, as can be seen in Table C.2, of the 1534 rock names for which the year in which the name was first used is known, only 37 are from the period prior to 1800 while 523 were defined in the period 1800–99 and a further 974 in the period 1900–88.

Similarly, Table C.3 shows the "best" and "worst" 2-, 5- and 10-year periods since 1800. Of the 121 rock names produced in the period 1934–1938, about half are from Tröger and Johannsen.

Table C.4 gives a list of some of the more prolific individual years for the production of new rock names and references containing new rock names. The largest number of new rock names produced in one year was 54 in 1973 which was mostly the work of the Subcommission – many of these names, however, were newly defined terms which had previously been

Table C.2. Distribution of new rock names and references containing new rock names by centuries.

Period	Rocks		References	
	Number	Acc%	Number	Acc%
pre 1800	37	2.4	22	2.8
1800–99	523	36.5	287	39.3
1900–88	974	100.0	478	100.0
Totals	1534		787	

Table C.3. The "best" and "worst" 2-, 5- and 10-year periods since 1800 for the production of new rock names and references containing new rock names.

	Rocks			References	
2-year periods					
Best	1937–38	71		1912–13	31
5-year periods					
Best	1909–13	121		1912–16	62
	1934–38	121		1913–19	62
Worst	1800–05	0		1800–05	0
10-year periods					
Best	1911–20	204		1893–02	114
Worst	1827–36	8		1800–09	4

Table C.4. List of years in which 20 or more new rock names and 10 or more references containing new rock names were published. The accumulative percentages are based on the 1534 rock names for which the source references are known.

	Rocks			References	
Year	Num	Acc%	Year	Num	Acc%
1973	54	3.5	1913	18	2.3
1938	50	6.8	1896	14	4.1
1920	37	9.2	1900	14	5.9
1911	36	11.5	1912	13	7.5
1913	32	13.6	1917	13	9.2
1935	30	15.6	1899	13	10.8
1898	28	17.4	1893	13	12.5
1983	27	19.2	1928	12	14.0
1917	26	20.9	1898	12	15.5
1978	24	22.4	1895	12	17.0
1811	24	24.0	1902	11	18.4
1931	23	25.5	1937	11	19.8
1926	22	26.9	1906	11	21.2
1901	22	28.4	1935	11	22.6
1921	22	29.8	1916	11	24.0
1900	21	31.2	1905	10	25.3
1912	21	32.6	1901	10	26.6
1937	21	33.9	1904	10	27.9
1896	20	35.2	1911	10	29.1
1933	20	36.5	1907	10	30.4
1895	20	37.8	1915	10	31.7
			1914	10	33.0
			1910	10	34.2

used as varietal names. The next three most prolific years, 1938, 1920 and 1911 are mainly due to the efforts of Johannsen.

The largest number of references with new rocks names produced in one year was 18 in 1913, due to suprisingly to 16 different authors.

Only 11 or about 50% of the years are common to both lists indicating that a large number of publications does not automatically correspond to a large number of rock names.

Since 1800 there have only been 18 years when no new rock names were published. They were 1800–05, 1812, 1818, 1828–30, 1856, 1867, 1871, 1950–51, 1987–88.

Figure C.1 presents histograms of the frequency with which new rocks names have appeared in the literature for 10 year periods since 1760 i.e. over the last 228 years. In order to see the influence of different "schools" of petrology, the contribution of new rock names from publications written in English (Fig. C.1b), German (Fig. C.1c) and French (Fig. C.1d) are given as well.

The most prolific period for new rock names (Fig. C.1a) was from the 1890's to the 1930's, with minor peaks in the 1810's, mainly due to Brongniart, Cordier and Pinkerton, and in the 1970's, due to the Subcommission's work. The French school

Fig. C.1 (next column). Histograms of ten year intervals from 1760 to the present for:- (a) all new rock names; (b) new rock names published in English; (c) new rock names published in German; (d) new rock names published in French; (e) the number of references containing new rock names; (f) the number of new rock names per reference. The year values along the horizontal axis are the lower value for each class.

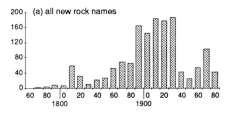

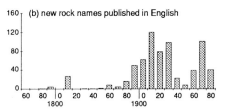

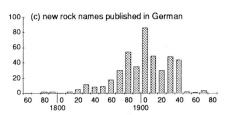

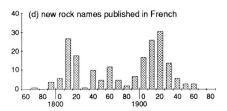

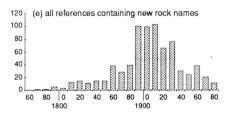

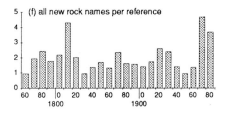

has two distinct peaks of activity, the earlier due to Brongniart and Cordier, while the later peak is due mainly to Lacroix. The major activity of the German school slightly preceded that of the English and the second peak of the French.

A slightly different picture is presented by the distribution of the number of references containing new rock names (Fig. C.1e) where the early and late peaks shown in the distribution of new rock names are missing.

Finally, the histograms of the number of new rock names per reference (Fig. C.1f) is interesting. On average each publication produced about 2 new rock names. However, two distinct peaks are present, one reflecting the the pioneering work of Brongniart, Cordier and Pinkerton in the 1810's, the other the work of the Subcommission in the 1970's. In this respect, the Subcommission's work can be said to be one of the most "productive" (to use a hackneyed political term) in the last 170 years. Whether we should be congratulated or not, only time will tell!

C.3. THE GLOSSARY

A-TYPE GRANITE. A general term for granitic rocks typically occurring in rift zones and in the interiors of stable continental plates. They are usually mildly alkaline granites with low CaO and Al_2O_3, high Fe/Fe+Mg, high K_2O/Na_2O and K_2O values and consist of quartz, K-feldspar, minor plagioclase and Fe-rich biotite, and sometimes alkali amphibole. The prefix A stands for anorogenic. *(Loiselle & Wones, 1979, p.468)*

ABESSEDITE. An obsolete local term for a variety of peridotite composed of olivine, hornblende and phlogopite. *(Cotelo Neiva, 1947a, p.105; Abessédo Mine, Bragança district, Portugal; Tomk. p.3)*

ABSAROKITE. A collective term, related to banakite and shoshonite, for a trachyandesitic rock, containing phenocrysts of olivine and augite in a groundmass of augite, calcic plagioclase, alkali feldspar and sometimes leucite. Later defined chemically in several ways in terms of K_2O and SiO_2. *(Iddings, 1895a, p.938; Absaroka Range, Yellowstone National Park, Wyoming, USA; Trög. 271; Joh. v.4, p.44; Tomk. p.4)*

ABYSSAL THOLEIITE. A term defined as a variety of tholeiitic basalt with $K_2O < 0.4\%$ occurring on the ocean floor; synonymous with mid-ocean ridge basalt. *(Myiashiro & Shido, 1975, p.268)*

ACHNAHAITE. An obsolete name used as a magma type for a biotite-bearing eucrite or variety of gabbro. *(Niggli, 1936, p.360; Achnaha, Ardnamurchan, Scotland; Trög(38). 362; Tomk. p.6)*

ACHNELITH. A term for pyroclastic fragments formed from solidified lava spray and whose external features are controlled by surface tension and not by fracturing. *(Walker & Croasdale, 1971, p.308; from the Greek achne = spray)*

ACID. A commonly used chemical term now defined in the TAS classification (Fig. B.13) as rocks containing more than 63% SiO_2. See also intermediate, basic and ultrabasic. *(Abich, 1841, p.12; Tomk. p.6)*

ACIDITE. An obsolete general term applied to all acid rocks. *(Cotta, 1864a, p.824; Trög. 778; Tomk. p.6)*

ADAM-GABBRO, ADAM-DIORITE, ADAM-TONALITE. A series of obsolete terms suggested for rocks intermediate between an adamellite and gabbro, diorite and tonalite, respectively. *(Johannsen, 1920b, p.168; Tomk. p.7)*

ADAMELLITE. A term originally used for orthoclase-bearing tonalite of the Adamello massif, but later used for granites with about equal amounts of alkali feldspar and plagioclase, which do not occur in the Adamello. The term should be avoided because of ambiguity and it is recommended that such rocks should be called monzogranite of QAPF field 3b (Fig. B.4). *(Cathrein, 1890, p.74; Mt. Adamello, Alto Adige, Italy; Trög. 779; Joh. v.2, p.308; Tomk. p.6)*

AEGIAPITE. A mnemonic name suggested for a variety of pyroxenite consisting essentially of *aegi*rine and *ap*atite. *(Belyankin & Vlodavets, 1932, p.63; Cape Turij, Kola Peninsula, USSR; Trög(38). 775⅔; Tomk. p.8)*

AEGINEITE. A mnemonic name suggested for a rock consisting essentially of *aegi*rine and *ne*pheline. *(Belyankin, 1929, p.22)*

AEGIRINITE. An intrusive rock consisting essentially of aegirine-augite with minor nepheline and albite. *(Polkanov, 1940, p.303; Gremiakha-Vyrmes pluton, Kola peninsula, USSR; Tomk. p.8)*

AEGIRINOLITH (AIGIRINOLITH, EGIRINOLITH). An obsolete name for a variety of alkali clinopyroxenite containing aegirine-augite, titanite, and magnetite. The original spelling was aigirinolith. *(Kretschmer, 1917, p.201; Trög. 780; Tomk. p.8)*

AEGISODITE. A mnemonic name suggested for a rock consisting of *aegi*rine and *sod*alite. *(Belyankin, 1929, p.22)*

AETNA-BASALT. A term from an obsolete chemical classification, based on feldspar composition rather than SiO_2 alone, for basaltic rocks in which $CaO : Na_2O : K_2O \approx 8 : 2.5 : 1$ and $SiO_2 \approx 50\%$. *(Lang, 1891, p.236; Mt. Etna, Sicily, Italy; Tomk. p.9)*

AFRIKANDITE (AFRICANDITE). A variety of melilitolite consisting of pyroxene, olivine, mica, perovskite, Ti-magnetite, and melilite. *(Chirvinskii et al., 1940, p.31; Railway station Africanda, Kola Peninsula, USSR; Tomk. p.9)*

AGGLOMERATE. Now defined in the pyroclastic classification (Table B.1) as a consolidated or unconsolidated coarse pyroclast material in which rounded pyroclasts predominate. *(Leonhard, 1823b, p.685; from the Latin agglomero = join; Joh. v.2, p.289; Tomk. p.9)*

AGGLUTINATE. A variety of agglomerate in which the ejecta were plastic when emplaced and are cemented together by thin skins of glass. *(Tyrrell, 1931, p.66; from the Latin gluten = glue; Tomk. p.9)*

AGPAITE (AGPAITIC). A general term for nepheline syenites characterised by the molecular ratio of $Na_2O + K_2O / Al_2O_3 > 1$ (originally >

1.2), high contents of Na, Fe, Cl and Zr with low Mg and Ca. Agpaite is now restricted (Sørensen, 1960) to peralkaline nepheline syenites characterised by complex Zr and Ti minerals, such as eudialyte and mosandrite, rather than simple minerals such as zircon and ilmenite. Agpaitic has commonly, but incorrectly, been used as a synonym for peralkaline. Cf. miaskitic. *(Ussing, 1912, p.341; Agpat, Ilimaussaq, Greenland; Trög. 781; Joh. v.1(2nd.Ed.), p.238; Tomk. p.10)*

AIGIRINOLITH. See aegirinolith.

AILLIKITE. An ultramafic carbonate-rich lamprophyre consisting of various phenocrysts including olivine, diopsidic pyroxene, amphiboles and phlogopite in a matrix of similar minerals with at least partly primary carbonate and minor perovskite but no melilite. Now defined modally in the lamprophyre classification (Table B.3). *(Kranck, 1939, p.75; Aillik, Labrador, Canada; Tomk. p.10)*

AILSYTE (AILSITE). A local name for a variety of microgranite or quartz microsyenite consisting of orthoclase, quartz and riebeckite. Cf. paisanite. *(Heddle, 1897, p.266; Ailsa Craig, Firth of Clyde, Scotland; Trög. 32; Tomk. p.10)*

AIOUNITE. A local name for a variety of melteigite containing Ti-augite and biotite in a cryptocrystalline groundmass of uncertain composition. *(Duparc, 1926, p.b119; El Aïoun, Morocco; Trög(38). 404½; Tomk. p.10)*

AKENOBEITE. A local name for an aplitic dyke rock which is associated with granodiorite and composed of oligoclase, orthoclase, quartz, chloritised biotite and garnet. *(Kato, 1920, p.17; Higashiyama, Akénobé district, Tajima, Japan; Trög. 112; Joh. v.2, p.360; Tomk. p.11)*

AKERITE. A collective term for varieties of microsyenite and micromonzonite consisting of alkali feldspar, with more or less oligoclase, biotite, pyroxene and often quartz. They are characterised by rectangular oligoclase. *(Brögger, 1890, p.43; Vestre Åker, Oslo district, Norway; Trög. 782; Joh. v.3, p.62; Tomk. p.11)*

AKOAFIMITE. A leucocratic variety of hornblende-bearing quartz norite of QAPF field 10* (Fig. B.4) belonging to the charnockitic rock series. It is suggested (Streckeisen, 1974, p.357) that this term should be abandoned. *(Schüller, 1949, p.574; Akoafim, Cameroon)*

ALABRADORITE. An obsolete name applied to rocks which contain alkali feldspar but no labradorite. *(Senft, 1857, Table II; Tomk. p.11)*

ALASKITE. A leucocratic variety of alkali feldspar granite consisting almost entirely of quartz and alkali feldspar. Now defined as a synonym for leucocratic alkali feldspar granite of QAPF field 2 (Fig. B.4). *(Spurr, 1900a, p.189; named after Alaska, USA; Trög. 14; Joh. v.2, p.106; Tomk. p.11)*

ALBANITE. A local name for a variety of leucitite composed essentially of leucite and clinopyroxene in about equal proportions with subordinate plagioclase. Minor amounts of alkali feldspar, nepheline and olivine may be present. The name was previously used for a bituminous material from Albania. *(Washington, 1920, p.47; Lake Albano, Alban Hills, near Rome, Italy; Trög. 784; Tomk. p.11)*

ALBASALT. An obsolete name proposed for a basalt with excess alumina and alkalis. *(Belyankin, 1931, p.30; from an abbreviation of aluminate basalt; Tomk. p.11)*

ALBITITE. A variety of alkali feldspar syenite consisting almost entirely of albite. *(Turner, 1896, p.728; Meadow Valley, Plumas County, California, USA; Trög. 169; Joh. v.3, p.138; Tomk. p.12)*

ALBITOPHYRE. An obsolete term for a porphyritic rock containing albite phenocrysts in a feldspathic groundmass. A variety of keratophyre. *(Coquand, 1857, p.78; Trög. 787; Joh. v.3, p.139; Tomk. p.12)*

ALBORANITE. An obsolete local name for a variety of basalt containing phenocrysts of hypersthene, augite and calcic plagioclase in a groundmass of plagioclase, augite and glass. *(Becke, 1899, p.553; Alboran Island, near Cabo de Gata, Spain; Trög. 127; Joh. v.3, p.284; Tomk. p.13)*

ALENTEGITE. An obsolete name suggested for a group of mica quartz "diorites" containing 16% to 33% of quartz. Erroneously spelt alentecite by Tomkeieff. *(Marcet Riba, 1925, p.293; Alemtejo, Portugal; Tomk. p.13)*

ALEUTITE. A term proposed for a variety of

porphyritic basaltic andesite with a fine-grained groundmass whose feldspars are intermediate in composition between those found in basalt and andesite. *(Spurr, 1900a, p.190; Katmai region, Aleutian Peninsula, Alaska, USA; Trög. 342; Joh. v.3, p.179; Tomk. p.13)*

ALEXOITE. An obsolete local name for a variety of dunite composed of olivine, pyrrhotite and smaller amounts of magnetite and pentlandite. *(Walker, 1931, p.5; Alexo Mine, Dundonald Township, Ontario, Canada; Trög. 775; Joh. v.4, p.407; Tomk. p.13)*

ALGARVITE. A name proposed for a variety of biotite melteigite. *(Lacroix, 1922, p.646; Navete, Caldas de Monchique, Algarve, Portugal; Trög. 610; Joh. v.4, p.330; Tomk. p.14)*

ALKALI. A prefix given to rocks which contain either 1) modal foids and/or alkali amphiboles or pyroxenes or 2) normative foids or acmite. *(Iddings, 1895b, p.183; Tomk. p.14)*

ALKALI ANDESITE. A term used as a synonym for trachyandesite. *(Original reference uncertain; Trög. 1003; Tomk. p.14)*

ALKALI BASALT. A term originally used for basalts containing accessory foids. Such rocks generally contain a Ti-augite and olivine as their main ferromagnesian phases. Now defined chemically as a variety of basalt in TAS field B (Fig. B.13) which contains normative nepheline. Cf. subalkali basalt. *(Hibsch, 1910, p.402; Böhmisches Mittelgebirge (now České Středohoří), N. Bohemia, Czechoslovakia; Trög. 381; Joh. v.4, p.69; Tomk. p.14)*

ALKALI DIORITE. A variety of diorite with alkaline character due to the presence of such dark constituents as barkevikite, kaersutite, etc. *(Niggli, 1931, p.323)*

ALKALI FELDSPAR CHARNOCKITE. A member of the charnockitic rock series equivalent to hypersthene alkali feldspar granite QAPF field 2 (Fig. B.4). *(Streckeisen, 1974, p.355)*

ALKALI FELDSPAR FOIDITE. A term suggested as an alternative to phonolitic foidite. *(Streckeisen, 1978, p.7)*

ALKALI FELDSPAR GRANITE. A special term for a variety of granite in which plagioclase is less than 10% of the total feld-spar. Now defined modally in QAPF field 2 (Fig. B.4). *(Streckeisen, 1973, p.26)*

ALKALI FELDSPAR RHYOLITE. A special term for a variety of rhyolite in which plagioclase is less than 10% of the total feld-spar. Now defined modally in QAPF field 2 (Fig. B.10). *(Streckeisen, 1978, p.4)*

ALKALI FELDSPAR SYENITE. A special term for a variety of syenite in which plagioclase is less than 10% of the total feldspar. Now defined modally in QAPF field 6 (Fig. B.4). *(Streckeisen, 1973, p.26)*

ALKALI FELDSPAR TRACHYTE. A special term for a variety of trachyte in which plagioclase is less than 10% of the total feld-spar. Now defined modally in QAPF field 6 (Fig. B.10). *(Streckeisen, 1978, p.4)*

ALKALI GABBRO. A variety of gabbro in QAPF field 10', with alkaline character due to the presence of analcime or nepheline and such dark constituents as barkevikite, kaersutite, Ti-augite, etc. *(Niggli, 1931, p.326; Trög. 789; Tomk. p.15)*

ALKALI GRANITE. A term used, but not recommended, as a synonym for peralkaline granite i.e. a granite containing alkali amphibole or pyroxene. It should not be used as a synonym for alkali feldspar granite. *(Rosenbusch, 1896, p.56; Trög. 56; Tomk. p.15)*

ALKALI RHYOLITE. A term used, but not recommended, as a synonym for peralkaline rhyolite i.e. a rhyolite containing alkali amphibole or pyroxene. It should not be used as a synonym for alkali feldspar rhyolite. *(Original reference uncertain)*

ALKALI SYENITE. A term used, but not recommended, as a synonym for peralkaline syenite i.e. a syenite containing alkali amphibole or pyroxene. It should not be used as a synonym for alkali feldspar syenite. *(Rosenbusch, 1907, p.141; Trög. 176; Tomk. p.16)*

ALKALI TRACHYTE. A term used, but not recommended, as a synonym for peralkaline trachyte i.e. a trachyte containing alkali amphibole or pyroxene. It should not be used as a synonym for alkali feldspar trachyte. *(Rosenbusch, 1908, p.916; Trög. 207 – 217; Tomk. p.16)*

ALKALIPLETE. An obsolete term for a melano-cratic rock in which $(Na_2O + K_2O) > CaO$. *(Brögger, 1898, p.265; Tomk. p.16)*

ALKALIPTOCHE. An obsolete term for igneous rocks poor in alkalis. *(Loewinson-Lessing, 1900b, p.241; Tomk. p.16)*

ALKORTHOSITE. An obsolete name for a variety of syenite consisting predominatly of Na-orthoclase. *(Eckermann, 1942, p.403; Tomk. p.16)*

ALLALINITE. A local name for a saussuritised gabbro, frequently with smaragdite, in which the original textures are preserved. *(Rosenbusch, 1896, p.328; Allalin, Zermatt, Switzerland; Joh. v.3, p.229; Tomk. p.17)*

ALLGOVITE. An obsolete name temporarily proposed for a group of basaltic rocks which could not be classified as melaphyres or trapps as their age was unknown. *(Winkler, 1859, p.669; Allgovia, Allgäuer Alps, Bavaria, West Germany; Trög. 790; Joh. v.3, p.299; Tomk. p.17)*

ALLIVALITE. A variety of troctolite composed of olivine and highly calcic plagioclase. *(Harker, 1908, p.71; Allival, now Hallival, Island of Rhum, Scotland; Trög. 364; Joh. v.3, p.348; Tomk. p.17)*

ALLOCHETITE. A porphyritic fine-grained variety of nepheline monzosyenite containing phenocrysts of labradorite, orthoclase, Ti-augite, and nepheline in a felted ground-mass of augite, biotite, hornblende, nepheline, and orthoclase. *(Ippen, 1903, p.133; Allochet Valley, Mt. Monzoni, Alto Adige, Italy; Trög.518; Joh. v.4, p.176; Tomk. p.17)*

ALLOITE. An obsolete term for a tuff consisting of fragments of feldspathic glass and some crystals. *(Cordier, 1816, p.366; Tomk. p.18)*

ALNÖITE. An ultramafic lamprophyre with phenocrysts of phlogopite-biotite, olivine and augite in a groundmass of melilite (often altered to calcite), augite and/or biotite with minor perovskite, garnet and calcite. Now defined in the lamprophyre classification (Table B.3). *(Rosenbusch, 1887, p.805; Alnö Island, Westnorrland, Sweden; Trög. 746; Joh. v.4, p.385; Tomk. p.19)*

ALPHA GRANITE. A possible term suggested for granites falling into QAPF field 3a. *(Streckeisen, 1973, p.28)*

ALSBACHITE. A local name for a porphyritic

dyke rock containing phenocrysts of quartz, feldspar and garnet in a fine-grained groundmass of quartz, feldspar and colourless mica. *(Chelius, 1892, p.2; Melibocus, Alsbach, Odenwald, West Germany; Trög. 111; Joh. v.2, p.359; Tomk. p.20)*

ALVIKITE. A special term in the carbonatite classification for the medium- to fine-grained variety of calcite carbonatite consisting principally of calcite. *(Eckermann, 1942, p.403; Alvik, Alnö Island, Westnorrland, Sweden; Tomk. p.21)*

AMBONITE. A local collective name for a series of andesites and dacites containing cordierite possibly formed by assimilation of sillimanite-cordierite gneiss. *(Verbeek, 1905, p.100; Ambon Island, Moluccas, Indonesia; Trög. 791; Joh. v.3, p.176; Tomk. p.21)*

AMHERSTITE. A leucocratic variety of quartz monzodiorite of QAPF field 9* (Fig. B.4) belonging to the charnockitic rock series. The rock consists essentially of andesine-microcline antiperthite with minor quartz and hypersthene. It is suggested (Streckeisen, 1974, p.357) that this term should be abandoned. *(although attributed to Watson & Taber, 1913 the reference does not contain the name although it does describe the type rock; Amherst Co., Virginia, USA; Trög. 294; Tomk. p.22)*

AMIATITE. A term from an obsolete chemical classification, based on feldspar composition rather than SiO_2 alone, for a class of rocks in which $CaO : Na_2O : K_2O \approx 1.1 : 1 : 1.8$ and $SiO_2 \approx 63\%$. *(Lang, 1891, p.226; Mt. Amiata, Tuscany, Italy; Trög. 792; Tomk. p.22)*

AMNEITE. A mnemonic name suggested for a rock consisting of *am*phibole and *ne*pheline. *(Belyankin, 1929, p.22)*

AMPASIMENITE. A local name for a porphyritic variety of ijolite or nephelinite with a glassy matrix. *(Lacroix, 1922, p.647; Ampasimena, Madagascar; Trög. 793; Tomk. p.22)*

AMPHIBOLDITE. An erroneous spelling of amphibololite. *(Tomkeieff et al., 1983, p.23)*

AMPHIBOLEID. An obsolete field term for a coarse-grained igneous rock consisting almost entirely of amphibole. *(Johannsen, 1911, p.320; Tomk. p.22)*

AMPHIBOLIDE. A revised spelling recommended to replace the field term amphiboleid. Now obsolete. *(Johannsen, 1926, p.182; Joh. v.1, p.57; Tomk. p.22)*

AMPHIBOLITE. A term originally used for any rock with a groundmass of hornblende in which other minerals are disseminated. Now solely used for metamorphic rocks consisting of hornblende and plagioclase. *(Brongniart, 1813, p.40; Trög. 700; Joh. v.4, p.442; Tomk. p.22)*

AMPHIBOLOLITE. A general term used in France for medium- to coarse-grained igneous rocks composed almost entirely of amphibole. *(Lacroix, 1894, p.270; Trög. 700; Tomk. p.23)*

AMPHIGENITE. An obsolete term originally used for leucite-tephrites but later used for intrusive rock composed of more than 90% modal leucite. *(Cordier, 1842, Vol.1, p.388; from the French amphigène = leucite; Trög. 794; Joh. v.4, p.337; Tomk. p.23)*

AMYGDALITE. An old term for a coarse-grained amygdaloidal basalt with "nodules or kernals of chalcedony, agate, calcareous spar". Cf. mandelstein. *(Pinkerton, 1811a, p.89; Tomk. p.24)*

AMYGDALOID. A term proposed for a rock containing amygdales. *(Cronstedt, 1758, p.228; Tomk. p.24)*

AMYGDALOPHYRE. An obsolete term for a porphyritic rock containing amygdales. *(Jenzsch, 1853, p.395; Trög. 795; Tomk. p.24)*

ANABOHITSITE. A local name for an variety of websterite composed of hypersthene, augite, ilmenite and magnetite with minor olivine and hornblende. *(Lacroix, 1914, p.419; Anabohitsy, Madagascar; Trög. 680; Joh. v.4, p.464; Tomk. p.24)*

ANALCIME BASALT. A term used for an alkaline mafic volcanic rock consisting of olivine, Ti-augite and analcime with minor feldspars. The name should not be used as the term basalt is now restricted to a rock containing essential plagioclase. As the rock is a variety of foidite it should be given the appropriate name, e.g. olivine analcimite. *(Lindgren, 1886, p.727; Highwood Mts., Montana, USA; Trög. 655; Joh. v.4, p.345; Tomk. p.25)*

ANALCIME BASANITE. Now defined in QAPF field 14 (Fig. B.10) as a variety of basanite in which analcime is the most abundant foid. *(although attributed to Hibsch, 1920, p.69 he does not use the name, but only describes it in the group of nepheline basanites; Trög. 599)*

ANALCIME DIORITE. Now defined in QAPF field 14 (Fig. B.4) as a variety of foid diorite in which analcime is the most abundant foid.

ANALCIME GABBRO. Now defined in QAPF field 14 (Fig. B.4) as a variety of foid gabbro in which analcime is the most abundant foid. The special term teschenite may be used as an alternative.

ANALCIME MONZODIORITE. Now defined in QAPF field 13 (Fig. B.4) as a variety of foid monzodiorite in which analcime is the most abundant foid.

ANALCIME MONZOGABBRO. Now defined in QAPF field 13 (Fig. B.4) as a variety of foid monzogabbro in which analcime is the most abundant foid.

ANALCIME MONZOSYENITE. Now defined in QAPF field 12 (Fig. B.4) as a variety of foid monzosyenite in which analcime is the most abundant foid. The term is synonymous with analcime plagisyenite.

ANALCIME PHONOLITE. Now defined in QAPF field 11 (Fig. B.10) as a variety of phonolite in which analcime is the most abundant foid. *(Pelikan, 1906, p.118; Radzein (now Radejčín), between Lovosice and Teplice, N. Bohemia, Czechoslovakia; Joh. v.4, p.132)*

ANALCIME PLAGISYENITE. Now defined in QAPF field 12 (Fig. B.4) as a variety of foid plagisyenite in which analcime is the most abundant foid. The term is synonymous with analcime monzosyenite.

ANALCIME SYENITE. Now defined in QAPF field 11 (Fig. B.4) as a variety of foid syenite in which analcime is the most abundant foid. *(Hibsch, 1899, p.72; Schlossberg, near Velké Březno, Böhmisches Mittelgebirge (now České Středohoří), N. Bohemia, Czechoslovakia; Joh. v.4, p.114; Tomk. p.24)*

ANALCIMITE. An alkaline volcanic rock composed essentially of analcime, Ti-augite, opaques and only minor olivine. A variety of foidite in QAPF field 15c (Fig. B.10) in which analcime is the most abundant foid.

(Gemmellaro, 1845, p.318; the reference cited by Johannsen and Tomkeieff is erroneous; Cyclope Island, Catania, Sicily, Italy; Trög. 654; Joh. v.4, p.336; Tomk. p.24)

ANALCIMOLITH. An obsolete term proposed for a monomineralic volcanic rock consisting of analcime i.e. a leucocratic analcimite. *(Johannsen, 1938, p.336; Tomk. p.25)*

ANALCITITE. As analcime is the mineral spelling recommended by the International Mineralogical Association analcimite should be used in preference to analcitite. *(Pirsson, 1896, p.690; Joh. v.4, p.337)*

ANAM-AEGISODITE. A mnemonic name suggested for a rock consisting of *an*alcime, *am*phibole, *aegi*rine and *sod*alite. *(Belyankin, 1929, p.22)*

ANAMESEID. An obsolete field term for dark coloured, non-porphyritic aphanitic igneous rock. Also called melano-aphaneid. *(Johannsen, 1911, p.321; Tomk. p.25)*

ANAMESEID PORPHYRY. An obsolete field term for a dark coloured porphyritic igneous rock with an aphanitic groundmass. Also called melanophyreid. *(Johannsen, 1911, p.321)*

ANAMESITE. An obsolete term for basaltic rocks which are between basalt and dolerite in texture. *(Leonhard, 1832, p.150; from the Greek anamesos = in the middle; Trög. 796; Joh. v.3, p.291; Tomk. p.25)*

ANATECTITE (ANATEXITE). An igneous rock produced by the remelting of crustal rocks. Cf. prototectite and syntectite. *(Loewinson-Lessing, 1934, p.7; Tomk. p.26)*

ANCHORITE. An obsolete local name for a diorite containing occasional syenite veins and scattered mafic segregation patches. *(Lapworth, 1898, p.419; Anchor Inn, Nuneaton, Warwickshire, England; Trög. 797; Tomk. p.26)*

ANDELATITE. An obsolete term suggested for an extrusive rock intermediate between andesite and latite. *(Johannsen, 1920b, p.174; Trög. 798; Tomk. p.27)*

ANDENDIORITE. A local name suggested for diorites which are younger than the usual ones. *(Stelzner, 1885, p.201; Trög. 799)*

ANDENGRANITE. A local name suggested for granitic rocks which are younger than the usual ones. *(Stelzner, 1885, p.201; Trög. 799)*

ANDENNORITE. A local name suggested for norites which are younger than the usual ones. *(Wolff, 1899, p.482; Trög. 799; Tomk. p.27)*

ANDERSONITE. An obsolete name suggested for a group of amphibole tonalites. *(Marcet Riba, 1925, p.293; named after W. Anderson)*

ANDESIBASALT. A term used in the USSR as a synonym for basaltic andesite.

ANDESILABRADORITE. A name used in France for rocks transitional between andesite and labradorite (an old French group name which includes basalt). *(Tröger, 1935, p.322; Trög. 890; Tomk. p.27)*

ANDESINE BASALT. A term for a dark coloured olivine-bearing andesite. *(Iddings, 1913, p.191; Lava of 1910, Mt. Etna, Sicily, Italy; Trög. 327; Tomk. p.27)*

ANDESINITE. A term proposed for a coarse-grained rock consisting essentially of andesine. *(Turner, 1900, p.110; Trög. 293; Joh. v.3, p.146; Tomk. p.27)*

ANDESITE. An intermediate volcanic rock, usually porphyritic, consisting of plagioclase (frequently zoned from labradorite to oligoclase), pyroxene, hornblende and/or biotite. Now defined modally in QAPF fields 9 and 10 (Figs. B.10 and B.11) and, if modes are not available, chemically in TAS field O2 (Fig. B.13). *(Buch, 1836, p.190; Andes Mts, South America; Trög. 324; Joh. v.3, p.160; Tomk. p.27)*

ANDESITE-BASALT. An obsolete term originally used as a synonym for basanite and later for rocks intermediate in composition between andesite and basalt. *(Bořický, 1874, p.43; Trög. 801; Tomk. p.27)*

ANDESITE-TEPHRITE. An obsolete term for a foid-bearing variety of trachyandesite composed of plagioclase, sanidine, augite and haüyne. Olivine and nepheline may also be present. *(Colony & Sinclair, 1928, p.307; Trög. 267; Tomk. p.27)*

ANDESITOID. A variety of andesite containing considerable amounts of sanidine in the groundmass. Now used as a general term in the provisional "field" classification (Fig. B.17) for rocks tentatively identified as andesite. *(Sigmund, 1902, p.282; Trög. 802; Tomk. p.28)*

ANGARITE. A local name for Siberian basalts and dolerites. *(Loewinson-Lessing et al., 1932, p.71; River Angara, Siberia, USSR; Tomk. p.28)*

ANKARAMITE. A porphyritic melanocratic basanite with abundant phenocrysts of pyroxene and olivine. *(Lacroix, 1916a, p.182; Ankaramy, Ampasindava, Madagascar; Trög. 408; Joh. v.3, p.338; Tomk. p.29)*

ANKARANANDITE. A group name for varieties of hypersthene alkali feldspar syenite to hypersthene syenite of QAPF fields 6 – 6* to 7 – 7* (Fig. B.4) belonging to the charnockitic rock series. It is suggested (Streckeisen, 1974, p.357) that this term should be abandoned. *(Giraud, 1964, p.49; Betafo-Ankaranando, Madagascar)*

ANKARATRITE. A melanocratic variety of olivine nephelinite containing biotite. *(Lacroix, 1916b, p.256; Ankaratra, Madagascar; Trög. 623; Joh. v.4, p.366; Tomk. p.29)*

ANORTHITISSITE. An obsolete name given to a melanocratic variety of gabbro consisting of 70% hornblende and 22% anorthite. *(Tröger, 1935, p.175; Koswinski Mts., Urals, USSR; Trög. 403; Joh. v.3, p.350; Tomk. p.30)*

ANORTHITITE. A term proposed for a variety of anorthosite consisting essentially of anorthite. In the original description the composition of the plagioclase was probably incorrectly determined. *(Turner, 1900, p.110; Trög. 300; Joh. v.3, p.338; Tomk. p.30)*

ANORTHOBASE. According to Tomkeieff et al. (1983) this rock is a diabase containing basic plagioclase. However, the term does not appear in the reference cited. The same incorrect reference is also cited by Loewinson-Lessing (1932). *(Belyankin, 1911, p.363; Tomk. p.30)*

ANORTHOCLASE BASALT. A variety of trachybasalt containing phenocrysts of anorthoclase, augite and enstatite in a groundmass of these minerals, labradorite and opaques. *(Skeats & Summers, 1912, p.33; Sugarloaf Hill, Mt. Macedon, Victoria, Australia; Trög. 273; Tomk. p.30)*

ANORTHOCLASITE. A variety of alkali feldspar syenite composed almost entirely of anorthoclase. *(Loewinson-Lessing, 1901, p.114; Trög. 805; Joh. v.3, p.5; Tomk. p.30)*

ANORTHOLITE. An erroneous term stated to be synonymous with anorthosite. However, the reference cited (Hunt, 1864) does not contain the term, only the term anorthosite. *(Tomkeieff et al., 1983, p.30)*

ANORTHOPHYRE. An obsolete name for a porphyritic anorthoclase syenite. Cf. pilandite. *(Loewinson-Lessing, 1900a, p.174; Trög. 804; Tomk. p.30)*

ANORTHOSITE. A leucocratic plutonic rock consisting essentially of plagioclase often with small amounts of pyroxene. Now defined modally in QAPF field 10 (Fig. B.4). The term is synonymous with plagioclasite. *(Hunt, 1862, p.62; Laurentian Mts., Quebec, Canada; Trög. 291; Joh. v.3, p.196; Tomk. p.31)*

ANORTHOSYENITE. An obsolete term for a variety of syenite with phenocrysts of anorthoclase. *(Loewinson-Lessing, 1900a, p.174; Trög. 805; Tomk. p.31)*

ANOTERITE. An obsolete name for a variety of rapakivi granite with euhedral quartz thought to have crystallised at a high level. *(Sederholm, 1891, p.21; from the Greek anoteros = higher up; Tomk. p.31)*

ANTHRAPHYRE. An obsolete term proposed for igneous rocks intruding anthracite-rich formations. *(Ebray, 1875, p.291)*

ANTIFENITEPEGMATITE. A sodic syenite pegmatite composed mainly of antiperthite with minor biotite and opaques. *(Barth, 1927, p.97; Island of Seiland, Finnmark, Norway; Trög. 173)*

ANTIRAPAKIVI. A term originally used for a texture in which ovoids of plagioclase are mantled by orthoclase. Tomkeieff et al. use it as a variety of rapakivi granite with this texture. *(Vakar, 1931, p.1026; Umylymnan Mt., Kolyma region, Siberia, USSR; Tomk. p.32)*

ANTSOHITE. A local name for a lamprophyric dyke rock consisting of phenocrysts of biotite in a groundmass of biotite, hornblende and interstitial quartz. *(Lacroix, 1922, p.431; Antsohy, Tsaratanana, Madagascar; Trög. 147; Tomk. p.32)*

APACHITE. A variety of peralkaline phonolite rich in sodic amphiboles and containing sodic pyroxene and aenigmatite. *(Osann, 1896,*

p.402; Apache (now Davis) Mts., Texas, USA; Trög. 466; Joh. v.4, p.129; Tomk. p.33)

APANEITE. A mnemonic name from *apa*tite and *ne*pheline, used for alkaline intrusive rocks consisting mainly of apatite with variable amounts of nepheline and minor aegirine and biotite i.e. a nepheline-bearing apatitolite. *(Vlodavets, 1930, p.34; Khibina complex, Kola Peninsula, USSR; Trög(38). 775½; Tomk. p.33)*

APATITOLITE. A rock of magmatic origin composed essentially of apatite. *(Loewinson-Lessing, 1936, p.989; Khibina complex, Kola Peninsula, USSR; Tomk. p.33)*

APHANEID. An obsolete field term for non-porphyritic rocks which contain a megascopically indeterminable component. *(Johannsen, 1911, p.318; from the Greek aphanes = invisible; Tomk. p.33)*

APHANIDE. A revised spelling recommended to replace the field term aphaneid. Now obsolete. *(Johannsen, 1926, p.182; from the Greek aphanes = invisible; Joh. v.1, p.56; Tomk. p.33)*

APHANITE. A term originally suggested by Haüy for fine-grained rocks, including trapp, ophite and variolite, of dioritic composition. Now generally applied to all fine-grained igneous rocks. *(D'Aubuisson de Voisins, 1819, p.147; Trög. 806; Tomk. p.33)*

APLITE. A term used both for fine-grained granitic rocks, consisting only of feldspar and quartz, and as a group name for any leucocratic fine-grained to aphanitic dyke rock. *(Leonhard, 1823a, p.51; from the Greek aploos = simple; Trög. 807; Joh. v.2, p.91; Tomk. p.34)*

APLO-. A prefix used for light coloured rocks with simple mineralogy and few ferromagnesian minerals e.g. aplodiorite (Trög. 105), aplogranite (Trög. 808). Cf. haplo-. *(Bailey & Maufe, 1916, p.160; Tomk. p.34)*

APLOID. An obsolete term for a nepheline-bearing aplite. *(Shand, 1910, p.377; Trög. 809; Tomk. p.34)*

APLOSYENITE. An obsolete term used as a family name for syenites with less than 5% mafics. *(Tröger, 1935, p.79; Trög. 163–165)*

APO-. A prefix used to denote that one rock has been derived from another by a specific alteration process. This can include devitrification e.g. aporhyolite (Trög. 810), apoperlite and apoobsidian would be acid volcanic rocks whose structure was once glassy. *(Bascom, 1893, p.828; from the Greek apo = from, off; Tomk. p.34)*

APOGRANITE. An albitised or greisenised granite located in the apices of intrusions and frequently mineralised in Sn, W, Be, Nb–Ta, Li, and B. *(Beus et al., 1962, p.5)*

APOTROCTOLITE. A term used for a coarse-grained rock consisting essentially of alkali feldspar and olivine, with minor pyroxene, biotite and chlorite. *(Barth, 1944, p.57; Tomk. p.35)*

APPINITE. A local, general term for medium- to coarse-grained, meso- to melanocratic rocks with conspicuous hornblende in a base of oligoclase-andesine and/or orthoclase with or without quartz. The plutonic equivalent of hornblende vogesite and spessartite. *(Bailey & Maufe, 1916, p.167; Appin district, near Ballachulish, Scotland; Trög. 811; Tomk. p.35)*

ARAPAHITE. A local name for a variety of basalt containing over 50% magnetite. *(Washington & Larsen, 1913, p.452; named after Arapaho Indians, Colorado, USA; Trög. 777; Joh. v.3, p.303; Tomk. p.36)*

ARENDALITE. A comprehensive term for rocks of the charnockitic series found in the Arendal area. It is suggested (Streckeisen, 1974, p.357) that this term should be abandoned. *(Bugge, 1940, p.81; Arendal, Norway)*

ARGEINITE. A local name for a variety of olivine hornblendite composed of hornblende and olivine. *(Lacroix, 1933, p.194; Argeine, Pyrénées, France; Trög. 707; Tomk. p.37)*

ARIEGITE. A local group name for a variety of websterite composed of clinopyroxene, orthopyroxene abundant spinel and often pyrope-rich garnet and hornblende. *(Lacroix, 1901a, p.360; Lac de Lherz, now Lhers, Ariège, Pyrénées, France; Trög. 684; Joh. v.4, p.461; Tomk. p.38)*

ARIZONITE. A local name for a dyke rock consisting of 80% quartz with 18% orthoclase and 2% muscovite. *(Spurr & Washington, 1917, p.34; Helvitia, Arizona, USA; Trög. 7; Joh. v.2, p.32; Tomk. p.38)*

ARKESINE. An obsolete name for hornblende biotite granodiorites and granites in the Dent Blanche nappe of the Swiss and Italian Alps. *(Jurine, 1806, p.373; Mont Blanc, France; Tomk. p.39)*

ARKITE. A local name for a leucocratic variety of nepheline fergusite consisting of leucite or pseudoleucite in a matrix of nepheline, melanite garnet, biotite, sodic pyroxene, amphibole and minor alkali feldspar. *(Washington, 1901, p.617; named after the usual abbreviation of Arkansas, USA; Trög. 629; Joh. v.4, p.274; Tomk. p.39)*

ARSOITE. A local name for a variety of trachyte consisting of phenocrysts of sanidine, diopside, andesine, and a little olivine in a groundmass of sanidine, oligoclase and diopside with minor amounts of leucite. The same rock had previously been called ciminite by Washington. *(Reinisch, 1912, p.121; Arso flow of 1302, Epomeo, Ischia, Italy; Trög. 252; Joh. v.4, p.36; Tomk. p.40)*

ASCHAFFITE. A local name for a variety of kersantite (a lamprophyre) with abundant biotite in a matrix of plagioclase and quartz. *(Gümbel, 1865, p.206; Aschaffenburg, Bavaria, West Germany; Trög. 812; Joh. v.3, p.190; Tomk. p.40)*

ASCLERINE. An obsolete term for a tuff consisting of fragments of altered feldspathic glass and some crystals. *(Cordier, 1816, p.372; Tomk. p.40)*

ASH TUFF. Now defined in the pyroclastic classification (Table B.1) as a pyroclastic rock in which the average pyroclast size is between 2 mm. and 1/16 mm. *(Hibsch, 1896, p.234; Tomk. p.41)*

ASH, ASH GRAIN. Now defined in the pyroclastic classification (Table B.1) as a pyroclast of any shape but with mean diameter between 2 mm. and 1/16 mm. *(Schmid, 1981, p.42; Tomk. p.41)*

ASH-STONE. A term used for a lithified volcanic ash. *(Williams, 1941, p.279; Tomk. p.40)*

ASO LAVA. Aso lava is the same as ash-stone in Japan. *(Williams, 1941, p.279; Aso volcano, Japan; Tomk. p.41)*

ASPERITE. An obsolete field term for a variety of andesite with trachytic character. *(Becker,*

1888, p.151; from the Latin asper = rough; Trög. 813)

ASPRONE. An alternative name for sperone. *(*Gmelin, 1814; Tomk. p.41)*

ASSYNTITE. A local name for a variety of nepheline syenite composed of abundant orthoclase, smaller amounts of sodalite, and nepheline with aegirine-augite, biotite, and large crystals of titanite. *(Shand, 1910, p.403; Borralan complex, Assynt, Scotland; Trög. 439; Joh. v.4, p.101; Tomk. p.42)*

ASTRIDITE. A green ultrabasic rock composed mainly of chromo-jadeite and picotite. *(*Willems, 1934, p.120; named after Astrid, Queen of Belgium; Tomk. p.42)*

ATATSCHITE. An obsolete term originally used for a glassy orthophyre containing sillimanite and cordierite, but later shown to be contact metamorphosed tuffs and agglomerates with porphyry boulders. The correct transliteration should be Atachite. *(Morozewicz, 1901, p.16; Atatch Ridge, Magnitnaya Mt., S. Urals, USSR; Trög. 258; Tomk. p.42)*

ATLANTITE. An obsolete term for a melanocratic variety of basanite. *(Lehmann, 1924, p.118; named after islands of Atlantic Ocean; Trög. 577; Joh. v.4, p.240; Tomk. p.42)*

AUGANITE. A collective name for augite andesites composed essentially of augite and plagioclase. *(Winchell, 1912, p.657; Trög. 814; Joh. v.3, p.280; Tomk. p.44)*

AUGITITE. A volcanic rock composed essentially of augite and opaque phenocrysts in an indeterminate dark coloured matrix which may be analcime. *(Rosenbusch, 1883, p.404; Trög. 580; Joh. v.4, p.21; Tomk. p.45)*

AUGITOPHYRE. An obsolete name for an augite porphyry. *(Palmieri & Scacchi, 1852, p.67; Joh. v.3, p.322; Tomk. p.45)*

AVEZACITE. A local name for a variety of pyroxene hornblendite which occurs as dykes and consists of phenocrysts of hornblende in a groundmass of hornblende, augite and abundant ilmenite. *(Lacroix, 1901b, p.826; Avezac-Prat, SW Lannemezan, Pyrénées, France; Trög. 714; Joh. v.4, p.449; Tomk. p.48)*

BAHIAITE. An obsolete local name for a variety of hornblende orthopyroxenite consisting of hypersthene, hornblende and small amounts of olivine and spinel. *(Washington, 1914a, p.86; near Maracas, Bahia, Brazil; Trög. 679; Joh. v.4, p.432; Tomk. p.50)*

BAJAITE. Originally used as "bajaite series rocks" the term is now used as a rock name for a variety of boninite with $MgO \approx 8\%$, $SiO_2 \approx 56\%$ and characterised by high contents of Sr (>1000 ppm) and high K/Rb ratios (>1000). It was originally described as a magnesian andesite. *(Rogers et al., 1985, p.392; Baja California, Mexico)*

BALDITE. A variety of analcimite composed essentially of analcime and Ti-augite with minor olivine and opaques. The rock was originally called analcime basalt. *(Johannsen, 1938, p.393; Big Baldy Mt., Little Belt Mts., Montana, USA; Trög(38). 640½; Tomk. p.51)*

BALGARITE. A local name for a variety of trachyte with marked spheroidal structure composed essentially of perthite with minor amounts of biotite and aegirine-augite. *(Borisov, 1963, p.225; Balgarovo, Bulgaria)*

BALTORITE. A local comprehensive term for potassic vogesites and potassic minettes. *(Comucci, 1937, p.737; Alto Baltoro, Karakorum, Himalaya)*

BANAKITE. A collective term, related to absarokite and shoshonite, for a trachyandesitic rock containing phenocrysts of augite and sometimes olivine in a groundmass of sanidine mantling labradorite-andesine, augite, biotite, analcime and opaques. It is similar to absarokite but contains less olivine and augite. *(Iddings, 1895a, p.947; named after Bannock Indians, Yellowstone National Park, Wyoming, USA; Trög. 527; Joh. v.4, p.44; Tomk. p.51)*

BANATITE. A term used for a series of rocks ranging from granite to diorite (but mainly granodiorite) that were intruded in Upper Cretaceous time in the Banat and adjacent areas of Hungary and Yugoslavia. *(Cotta, 1864b, p.13; Banat district, Transylvania, Romania; Trög. 816; Joh. v.2, p.348; Tomk. p.51)*

BANDAITE. A variety of glassy andesite composed of labradorite, sanidine, hypersthene and augite in a base of glass containing potential quartz and feldspar. *(Iddings, 1913, p.111; Bandai San, Iwasciro province, Japan; Trög.*

158; Joh. v.2, p.416; Tomk. p.52)

BARAMITE. Altered ultramafic plutonic rock composed of serpentine, magnesite and opal. *(Hume et al., 1935, p.26; Baramia mine, Upper Egypt, Egypt; Trög(38). 677½; Tomk. p.52)*

BARNEITE. A mnemonic name suggested for a rock consisting of *bar*kevikite and *ne*pheline. *(Belyankin, 1929, p.22)*

BARSHAWITE. A melanocratic variety of analcime nepheline monzosyenite consisting of kaersutite, Ti-augite, alkali feldspar, andesine, nepheline and analcime. *(Johannsen, 1938, p.283; Barshaw, Paisley, Scotland; Trög(38). 564½; Tomk. p.53)*

BASALATITE. A term suggested for an extrusive rock intermediate between basalt and latite. *(Johannsen, 1920c, p.212; Trög. 1004)*

BASALT. A volcanic rock consisting essentially of calcic plagioclase and pyroxene. Olivine and minor foids or minor interstitial quartz may also be present. Now defined modally in QAPF fields 9 and 10 (Figs. B.10 and B.11) and, if modes are not available, chemically in TAS field B (Fig. B.13). For some other varieties of basalt see also alkali basalt, high-alumina basalt, island arc basalt, mid-ocean ridge basalt (MORB), olivine basalt, olivine tholeiite, subalkali basalt, tholeiite, tholeiitic basalt, transitional basalt. *(a term of great antiquity, probably Egyptian in origin, usually attributed to Pliny[†], 77AD – see Johannsen for further discussion; Trög. 378; Joh. v.3, p.246; Tomk. p.54)*

BASALT-TRACHYTE. An obsolete name for a trachyte containing augite and hornblende. *(Vogelsang, 1872, p.541; Tomk. p.55)*

BASALT-WACKE. An obsolete term for an altered basalt. *(Cotta, 1855, p.44; Tomk. p.55)*

BASALTIC ANDESITE. A term for a volcanic rock which has the feldspar composition expected in andesites with the ferromagnesian minerals more commonly found in basalts e.g. olivine. Now defined chemically in TAS field O1 (Fig. B.13). *(Anderson, 1941,*

p.389; one of the Modoc basalts, Medicine Lake, California, USA)

BASALTIC KOMATIITE (KOMATIITIC BASALT). A member of the komatiite series with MgO in the range 5% to 15% typically displaying spinifex texture and evidence of rapid quenching of high temperature magma. Chemically they are intermediate between tholeiitic basalt and boninite and are thought to have been komatiite magmas contaminated by crustal material. *(Viljoen & Viljoen, 1969, p.61; Komati river, Barberton, Transvaal, South Africa)*

BASALTIC TRACHYANDESITE. A group term introduced for rocks intermediate between trachyandesite and trachybasalt, i.e. an analogous name to basaltic andesite in the oversaturated rocks. Now defined chemically in TAS field S2 (Fig. B.13). *(Le Bas et al., 1986, p.747)*

BASALTIN. An obsolete term for a fine-grained variety of basalt. *(Pinkerton, 1811a, p.32; Tomk. p.54)*

BASALTINE. An obsolete term for a porphyritic or fine-grained basalt. *(Born, 1790, p.395; Tomk. p.54)*

BASALTITE. An obsolete term originally used for a variety of melaphyre, but later used in several other senses ranging from specific types of basalt to a group name for rocks consisting of labradorite, augite, nepheline and leucite. *(Raumer, 1819, p.101; Trög. 817; Tomk. p.54)*

BASALTOID. The term was originally used as the name for a black Egyptian basalt, but later as a collective term for black basic volcanic rocks. Now used as a general term in the provisional "field" classification (Fig. B.17) for rocks tentatively identified as basalt. *(Haüy, 1822, p.542; Tomk. p.55)*

BASALTON. An obsolete term for a coarse-grained basalt e.g. dolerite. *(Pinkerton, 1811a, p.72; Tomk. p.55)*

BASANITE. A term originally used for a porphyritic basalt containing pyroxene phenocrysts, but later used as a group name for rocks composed of clinopyroxene, plagioclase, essential foids and olivine. Now defined modally in QAPF field 14 (Fig. B.10) and, if modes are not available, chemically in

---

[†]There is considerable confusion with the derivation of the word basalt as some versions of Pliny contain the phrase "quem vocant basalten" while others contain "quem vocant basanites".

TAS field U1 (Fig. B.13). *(*a term of great antiquity usually attributed to Theophrastus, 320BC - see Johannsen for further discussion; from the Greek basanos = touchstone; Tróg. 818; Joh. v.4, p.230; Tomk. p.55)*

BASANITIC FOIDITE. A collective term for alkaline volcanic rocks consisting of foids with some plagioclase as defined modally in QAPF field 15b (Fig. B.10). It is distinguished from tephritic foidite by having more than 10% modal olivine. If possible the most abundant foid should be used in the name e.g. basanitic nephelinite, basanitic leucitite, etc. *(Streckeisen, 1978, p.7)*

BASANITOID. A term originally used for volcanic rocks intermediate between olivine basalt and basanite. Has also been used for rocks with the composition of basanite but without modal nepheline. *(Bücking, 1881, p.154; Joh. v.4, p.69; Tomk. p.55)*

BASIC. A commonly used chemical term now defined in the TAS classification (Fig. B.13) as a rock with SiO_2 between 45% and 52%. See also ultrabasic, intermediate and acid. *(Abich, 1841, p.12; Tomk. p.56)*

BASITE. An obsolete group name for all basic igneous rocks. *(Cotta, 1864a, p.824; Tróg. 819; Tomk. p.56)*

BASITE-PORPHYRY. An obsolete term for a porphyritic nepheline and leucite rock. *(Vogelsang, 1872, p.542; Tomk. p.56)*

BATUKITE. A melanocratic variety of leucitite essentially composed of clinopyroxene with small amounts of olivine and leucite. *(Iddings & Morley, 1917, p.595; Batuku, Sulawesi, Indonesia; Tróg.647; Joh. v.4, p.369; Tomk. p.57)*

BAUCHITE. A local name for varieties of fayalite-bearing and orthopyroxene granite and quartz monzonite of QAPF fields 3, 7* and 8* (Fig. B.4). It is suggested (Streckeisen, 1974, p.357) that this term should be abandoned. *(Oyawoye, 1965, p.689; Bauchi, Jos plateau, Nigeria; Tomk. p.57)*

BAULITE. An obsolete term for a variety of rhyolite. *(Bunsen, 1851, p.199; Mt. Baula, Faeroe Islands; Tomk. p.57)*

BEBEDOURITE. A local name for a variety of biotite pyroxenite composed essentially of aegirine-augite and biotite with perovskite and opaques. *(Tróger, 1928, p.202; Bebedouro, Minas Geraes, Brazil; Tróg.692; Joh. v.4, p.452; Tomk. p.58)*

BEFORSITE. A special term in the carbonatite classification for the medium- to fine-grained variety of dolomite carbonatite which often occurs as dykes and consists principally of dolomite. *(Eckermann, 1928a, p.386; Bergeforsen, near Alnö Island, Westnorrland, Sweden; Tróg. 759; Tomk. p.59)*

BEKINKINITE. A melanocratic variety of theralite consisting of abundant brown amphibole and lesser Ti-augite with minor olivine, nepheline, labradorite and alkali feldspar. *(Rosenbusch, 1907, p.441; Mt. Bekinkina, Madagascar; Tróg. 515; Joh. v.4, p.332; Tomk. p.59)*

BELOEILITE. A local name for a variety of sodalitolite consisting of about 70% sodalite with nepheline, alkali feldspar, oligoclase and a little aegirine, originally described as a feldspathic variety of tawite. *(Johannsen, 1920b, p.163; Beloeil, now Mt. St. Hilaire, Quebec, Canada; Tróg. 506; Joh. v.4, p.282; Tomk. p.60)*

BELUGITE. An obsolete term for a plutonic rock whose feldspars are intermediate in composition between those typical of diorite and gabbro. *(Spurr, 1900a, p.189; Beluga River, Alaska, USA; Tróg. 330; Joh. v.3, p.159; Tomk. p.60)*

BENMOREITE. Originally described as a member of the sodic volcanic rock series falling in Daly's compositional gap between mugearite and trachyte, with a differentiation index between 65 and 75. It consists essentially of anorthoclase or sodic sanidine, Fe-olivine and ferroaugite. Now defined chemically as the sodic variety of trachyandesite in TAS field S3 (Fig. B.13). *(Tilley & Muir, 1963, p.439; Ben More, Island of Mull, Scotland)*

BENTONITE. An altered and devitrified ash or tuff, composed largely of smectite clay, but commonly with shard and phenocryst outlines preserved. *(Knight, 1898, p.491; Fort Benton Formation, Wyoming, USA)*

BERESHITE. See bjerezite, which is an earlier transliteration. *(Zavaritskii, 1955, p.375; River Beresh, West Siberia, USSR; Tomk.*

p.61)

BERESITE. A miners term for a kaolinised dyke rock which was originally a quartz porphyry but now looks like a greisen. *(Rose, 1837, p.186; Beresovsk, Urals, USSR; Trög. 9; Joh. v.2, p.23; Tomk. p.61)*

BERGALITE. A local name for an ultramafic lamprophyre similar to alnöite except that the matrix contains essential feldspathoids as well as melilite. The rock could be described as a haüyne micromelilitolite. The term is essentially synonymous with polzenite. *(Söllner, 1913, p.415; Oberbergen, Kaiserstuhl, Baden, West Germany; Trög. 662; Joh. v.4, p.379; Tomk. p.61)*

BERGMANITE. An obsolete term for a variety of serpentinite. *(Pinkerton, 1811b, p.53; named after T. Bergman, 1780; Tomk. p.61)*

BERINGITE. An local name for a melanocratic variety of trachyte containing nearly 50% of barkevikite. *(Starzynski, 1912, p.671; Bering Island, Bering Sea, USSR; Trög. 231; Joh. v.3, p.106; Tomk. p.61)*

BERMUDITE. A local name for a lamprophyric rock consisting of biotite and minor Ti-augite in a groundmass of analcime (probably after nepheline). Possibly equivalent in composition to ouachitite. *(Pirsson, 1914, p.340; named after Bermuda; Trög. 621; Joh. v.4, p.345; Tomk. p.61)*

BERONDRITE. A melanocratic variety of theralite with kaersutite, Ti-augite and labradorite rimmed with alkali feldspar and small amounts of nepheline. *(Lacroix, 1920, p.21; Beronda Valley, Bezavona Massif, Madagascar; Trög. 545; Joh. v.4, p.198; Tomk. p.62)*

BESCHTAUITE. A local name for a variety of porphyritic rhyolite that contains phenocrysts of sanidine and oligoclase. *(*Bayan, 1866; Mt. Beschtau, Pyatigorsk, N. Caucasus, USSR; Trög. 95; Tomk. p.62)*

BETA GRANITE. A possible term suggested for granites falling into QAPF field 3b. *(Streckeisen, 1973, p.28)*

BIELENITE. An obsolete local name for a variety of olivine websterite consisting of diallage, enstatite and olivine. *(Kretschmer, 1917, p.153; Biele River (now Bělá), Moravia, Czechoslovakia; Trög. 820; Joh. v.4, p.437; Tomk. p.62)*

BIGWOODITE. A variety of alkali feldspar syenite consisting essentially of orthoclase, microcline and albite with traces of foids and sodic pyroxene and/or amphibole. *(Quirke, 1936, p.179; Bigwood Township, Ontario, Canada; Trög(38). 163½; Joh. v.4, p.31; Tomk. p.62)*

BIMSSTEIN (BYNFTEIN, BYMSFTEIN). An old German term for pumice. *(Boetius de Boot, 1636, p.576; Trög. 821; Joh. v.2, p.282)*

BINARY GRANITE. A term applied to a variety of granite containing both muscovite and biotite. *(Keyes, 1895, p.714; Joh. v.2, p.225; Tomk. p.63)*

BINEITE. A mnemonic name suggested for a rock consisting of *bi*otite and *ne*pheline. *(Belyankin, 1929, p.22)*

BINEMELITE. A mnemonic name suggested for a rock consisting of *bi*otite, *ne*pheline and *me*lilite. *(Belyankin, 1929, p.22)*

BIOTITITE. A rock consisting almost entirely of biotite. *(Washington, 1927, p.187; Villa Senni, near Rome, Italy; Trög. 720; Joh. v.4, p.441; Tomk. p.66)*

BIQUAHORORTHANDITE. An unwieldy name constructed by Johannsen to illustrate some of the possibilties of the mnemonic classification of Belinakin (1929) for a granitic rock consisting of *bi*otite, *qua*rtz, *hor*nblende, *ortho*clase and *and*esine. Cf. hobiquandorthite and topatourbiolilepiquorthite. *(Johannsen, 1931, p.125)*

BIRKREMITE (BJERKREIMITE). A member of the charnockitic rock series, this term was originally used for a hypersthene alkali feldspar granite in QAPF field 2 (Fig. B.4), but as it was later shown to contain mesoperthite, it became a mesoperthite charnockite of QAPF field 3. It is suggested (Streckeisen, 1974, p.358) that this term should be abandoned. *(Kolderup, 1903, p.117; Birkrem, Egersund district, Norway; Trög. 15; Joh. v.2, p.46; Tomk. p.66)*

BISTAGITE. An obsolete local name for a variety of clinopyroxenite composed essentially of diopside. *(Yachevskii, 1909, p.46; Bis-Tag Range, Yenisey province, Siberia, USSR; Trög(38). 821½; Tomk. p.67)*

BIZARDITE. A local name applied to an ultramafic lamprophyre or aillikite. *(Stansfield, 1923b, p.551; Bizard Island, Montreal,*

Quebec, Canada; Trög. 665; Tomk. p.68)

BIZEUL. According to Tomkeieff a local French name for diabase or diorite. *(Tomkeieff et al., 1983, p.68)*

BJEREZITE. A term for a porphyritic micromonzonite or monzogabbro rich in nepheline phenocrysts and containing analcime, Ti-augite rimmed by aegirine. Later transliterated as Bereshite. *(Erdmannsdörffer, 1928, p.88; Bjerez River, Yenisey River, Siberia, USSR; Trög. 562; Joh. v.4, p.292; Tomk. p.68)*

BJERKREIMITE. An alternative spelling of birkremite.

BJÖRNSJÖITE. A local name for a variety of trachyte consisting essentially of albite and small amounts of aegirine-augite. *(Brögger, 1932, p.34; Lake Björnsjö, Oslo district, Norway; Trög. 205; Tomk. p.68)*

BLACOLITE. An obsolete term for a variety of serpentinite. *(Pinkerton, 1811b, p.53; named after J. Black, 1760; Tomk. p.69)*

BLAIRMORITE. A leucocratic variety of analcimite with phenocrysts of analcime in a groundmass of analcime and aegirine-augite with minor nepheline, sanidine and melanite garnet. *(Knight, 1905, p.275; Blairmore, Alberta, Canada; Trög. 653; Joh. v.4, p.256; Tomk. p.70)*

BLATTERSTEIN. A local German name for a variolite. *(Leonhard, 1823a, p.108; Trög. 822; Tomk. p.71)*

BLOCK. Now defined in the pyroclastic classification (Table B.1) as a pyroclast with an average diameter exceeding 64 mm. and whose generally angular shape indicates that it was solid when formed. Cf. bomb. *(Wentworth, 1935, p.227; Tomk. p.71)*

BODERITE. A term proposed for a melanocratic variety of nosean phonolite, originally called a nosean sanidine biotite augitite, with more alkali feldspar than nosean, biotite and augite. Cf. riedenite and rodderite. *(Taylor et al., 1967, p.409 and Frechen, 1971, p.41 for the description; Boder Wald, near Rieden, Eifel district, near Koblenz, West Germany)*

BOGUEIRITE. An obsolete local name for a variety of hornblende orthopyroxenite composed of enstatite and hornblende. *(Cotelo Neiva, 1947b, p.118; Bogueiro Hill, Bra-*

gança district, Portugal; Tomk. p.73)

BOGUSITE. A term synonymous with teschenite or analcime gabbro. *(Johannsen, 1938, p.220; Boguschowitz, Silesia, Poland; Trög(38). 565½; Tomk. p.73)*

BOJITE. A plutonic rock originally described as consisting of labradorite, brown hornblende, minor augite and a little biotite i.e. a variety of gabbro. Tröger (1935), however, showed the plagioclase to be andesine and the rock to be a diorite. The term was later adopted by Johannsen (1938) in its original sense as an alternative to hornblende gabbro. *(Weinschenk, 1899, p.541; named after Boii, a celtic tribe, Pfaffenreuth district, Bavaria; Trög. 309; Joh. v.3, p.226; Tomk. p.73)*

BOLSENITE. A term from an obsolete chemical classification, based on feldspar composition rather than SiO_2 alone, for a class of rocks in which $CaO : Na_2O : K_2O \approx 1.9 : 1 : 4.8$ and $SiO_2 \approx 60\%$. *(Lang, 1891, p.221; Lake Bolsena, Umbria, Italy; Tomk. p.74)*

BOMB. Now defined in the pyroclastic classification (Table B.1) as a pyroclast with an average diameter exceeding 64 mm. and whose shape or surface indicates that it was wholly or partially molten during its formation and subsequent transport. Cf. block. *(Wentworth & Williams, 1932, p.46; Tomk. p.74)*

BONINITE. A high magnesium relatively siliceous glassy lava consisting of phenocrysts of protoenstatite (which inverts to clinoenstatite), orthopyroxene, clinopyroxene and olivine in a glassy base full of crystallites. The rock exhibits textures characteristic of rapid growth and was originally described as an hyaloandesite. Now defined chemically in the TAS classification (Fig. B.12). *(Petersen, 1890a, p.25 and Petersen, 1890b for the description; Bonin Islands, now Ogaswaragunto Islands, Japan; Trög. 160; Joh. v.3, p.174; Tomk. p.74)*

BORENGITE. A dyke rock consisting essentially of K-feldspar with minor fluorite. *(Eckermann, 1960, p.519; Bäräng, Alnö Island, Westnorrland, Sweden)*

BORNITE. An obsolete term for an altered porphyry, predating its use as a mineral name. *(Pinkerton, 1811b, p.239; named after Baron von Born; Tomk. p.75)*

BOROLANITE. A local name for a coarse-grained variety of nepheline syenite consisting of alkali feldspar, nepheline (or its alteration products), melanite and biotite, and sometimes characterised by the presence of pseudoleucite. *(Horne & Teall, 1892, p.163; Loch Borolan, now Borralan, Assynt, Scotland; Trög. 420; Joh. v.4, p.139; Tomk. p.75)*

BORZOLITE. An obsolete term for a hornblende-bearing rock with carbonate amygdales associated with serpentinite. *(Issel, 1880, p.187; Borzoli, Genoa, Italy; Tomk. p.75)*

BOSTONITE. A variety of fine-grained leucocratic alkali feldspar syenite consisting almost entirely of alkali feldspar and characterised by a bostonitic texture i.e. irregular, subparallel laths of feldspar arranged in a divergent manner. *(Hunter & Rosenbusch, 1890, p.447; Boston, Massachusetts, USA; Trög. 171; Joh. v.4, p.22; Tomk. p.75)*

BOTTLEITE. A local Irish name for a tachylyte. *(Kinahan, 1875, p.426; Tomk. p.75)*

BOWRALITE. A variety of peralkaline syenite pegmatite composed of idiomorphic feldspars with interstitial arfvedsonite and aegirine-augite. *(Mawson, 1906, p.606; Bowral, New South Wales, Australia; Trög. 195; Joh. v.3, p.32; Tomk. p.76)*

BOXITE. A name given to a hypothetical set of data used to illustrate the problems of the statistical interpretation of rock compositions. *(Aitchison, 1984, p.534)*

BRACCIANITE. An aphyric variety of tephritic leucitite essentially composed of leucite and clinopyroxene in about equal proportions with small amounts of plagioclase. Minor olivine and nepheline may be present. *(Lacroix, 1917e, p.1030; Lake Bracciano, Sabatini volcanoes, near Rome, Italy; Trög. 583; Joh. v.4, p.307; Tomk. p.77)*

BRAGANÇAITE. A local name for a variety of hornblende harzburgite composed of olivine, enstatite and hornblende. *(Cotelo Neiva, 1947a, p.112; Bragança district, Portugal; Tomk. p.77)*

BRANDBERGITE. A local name for a peralkaline granite containing biotite and small amounts of arfvedsonite. *(Chudoba, 1930, p.389;*

Brandberg, Erongo Mts., Namibia; Trög. 27; Joh. v.2, p.99; Tomk. p.77)

BRECCIOLE. A term originally used for piperino-like basaltic tuffs. *(Brongniart, 1824, p.3; Val Nera, Chiampo Valley, near Arzignano, Italy; Tomk. p.77)*

BRONZITITE. A variety of orthopyroxenite composed almost entirely of bronzite. *(Williams, 1890, p.47; Trög. 675; Joh. v.4, p.459; Tomk. p.78)*

BUCHNERITE. An obsolete name for lherzolite. *(Wadsworth, 1884, p.85; named after Dr. Otto Buchner for writings on meteorites; Trög. 823)*

BUCHONITE. A local name for a variety of melanocratic tephrite containing labradorite, hornblende, Ti-augite, biotite, nepheline, and minor orthoclase. *(Sandberger, 1872, p.208; Buchonia, now Rhön Mts., West Germany; Trög. 579; Joh. v.4, p.202; Tomk. p.79)*

BUGITE. A term applied to a series of dark fine-grained rocks, belonging to the charnockitic rock series, with 50% to 75% SiO_2 and consisting mainly of plagioclase, hypersthene and quartz. The series is characterised by the absence of orthoclase and a complete transition in mineral composition from the basic to acid end members. It is suggested (Streckeisen, 1974, p.358) that this term should be abandoned. *(Bezborod'ko, 1931, p.129; Bug River, Podolia, Ukraine, USSR; Trög(38). 131½; Tomk. p.80)*

BURGASITE. A local name for a potassic volcanic rock with marked spheroidal structure composed of almost equal amounts of alkali feldspar and various zeolites with minor augite, biotite and aegirine-augite. It may be a metasomatised andesitic rock. *(Borisov, 1963, p.225; Burgas, Bulgaria)*

BUSONITE. An erroneous spelling of busorite. *(Tomkeieff et al., 1983, p.80)*

BUSORITE. A local name for a coarse-grained cancrinite syenite with primary calcite, alkali feldspar, lepidomelane and aegirine. *(Béthune, 1956, p.394; Busori, Lueshe, Kivu, Zaire; Tomk. p.80)*

BYTOWNITITE. A variety of anorthosite composed mainly of highly calcic plagioclase. Cf. allivalite. *(Johannsen, 1920a, p.60; Trög. 299; Joh. v.3, p.196; Tomk. p.80)*

CABYTAUITE. A mnemonic name from *c*alcite, *byt*ownite and *augite*, given to an altered dolerite composed mainly of highly calcic plagioclase, minor augite, calcite and biotite with secondary iron oxides. *(Vlodavets, 1952, p.659; Tomk. p.81)*

CALC-ALKALI BASALT. A basalt not defined by its mineralogy but by its association with rocks of the basalt-andesite-dacite suite of the orogenic belts and island arcs. *(Original reference uncertain)*

CALC-TONALITE. An obsolete term for a tonalite in which the alkali feldspar to plagioclase ratio is less than 1 to 6.5. *(Bailey & Maufe, 1916, p.160; Tomk. p.85)*

CALCICLASITE. A term for varieties of anorthosite containing calcic plagioclase. *(Johannsen, 1937, p.338; Trög. 300; Tomk. p.83)*

CALCIOCARBONATITE. A chemically defined variety of carbonatite (Fig. B.2) in which wt% $CaO/(CaO+MgO+FeO+Fe_2O_3 +MnO) > 0.8$. *(Woolley & Kempe, in press)*

CALCIOPLETE. An obsolete term for a melanocratic rock in which $CaO > (Na_2O + K_2O)$. *(Brögger, 1898, p.265; Tomk. p.83)*

CALCIPTOCHE. An obsolete adjectival term for igneous rocks poor in calcium. *(Loewinson-Lessing, 1901, p.118; Tomk. p.84)*

CALCITE CARBONATITE. A variety of carbonatite, defined modally in the carbonatitic rock classification, in which more than 90% of the carbonate is calcite. *(Streckeisen, 1978, p.12)*

CALCITFELS. An obsolete term for a calcite-rich member of the ijolite-melteigite series. *(Brögger, 1921, p.136; Tomk. p.84)*

CALCITITE. A term proposed for a calcareous carbonatite or calcite carbonatite. *(Belyankin & Vlodavets, 1932, p.53; Tomk. p.84)*

CALCOGRANITONE. An obsolete term for a gabbroic rock impregnated with calcite. *(Mazzuoli & Issel, 1881, p.326; Tomk. p.85)*

CALTONITE. An obsolete local name for a variety of analcime basanite consisting of olivine, Ti-augite, labradorite and analcime. *(Johannsen, 1938, p.242; Calton Hill, Derbyshire, England; Trög(38). 599½; Tomk. p.86)*

CAMPANITE. A variety of leucite tephrite with large leucite crystals associated with essential plagioclase and clinopyroxene with minor sanidine, nepheline and haüyne. *(Lacroix, 1917b, p.207; named after Campania, Italy; Mt. Somma, Naples, Italy; Trög. 573; Joh. v.4, p.190; Tomk. p.86)*

CAMPTONITE. A variety of lamprophyre composed of phenocrysts of combinations of olivine, kaersutite, Ti-augite, and Ti-biotite in a matrix of the same minerals (minus olivine) with plagioclase and sometimes subordinate alkali feldspar and feldspathoids. A volatile-rich basanite or alkali basalt. Now defined in the lamprophyre classification (Table B.3). *(Rosenbusch, 1887, p.333; Campton Falls, New Hampshire, USA; Trög. 322; Joh. v.4, p.63; Tomk. p.86)*

CAMPTOSPESSARTITE. A local term used for a basic spessartite containing Ti-augite. *(Tröger, 1931, p.139; Golenz, SW Bautzen, Lausitz, Saxony, East Germany; Trög. 322; Joh. v.3, p.193; Tomk. p.86)*

CAMPTOVOGESITE. A name suggested for a basic variety of vogesite containing basaltic hornblende. *(Loewinson-Lessing, 1905a, p.15; Tagil river, Ural Mts., USSR; Trög. 250)*

CANADITE. A coarse-grained variety of nepheline syenite in which the feldspar is essentially albite or sodic plagioclase, with abundant mafic minerals, notably biotite and amphibole. *(Quensel, 1913, p.177; Monmouth Township, Ontario, Canada; Trög. 484; Joh. v.4, p.185; Tomk. p.86)*

CANCALITE. A variety of potassic lamproite containing normative olivine but no normative leucite. *(Fuster et al., 1967, p.35; Cancarix and Calasparra, Spain)*

CANCARIXITE. A local name for a variety of peralkaline quartz syenite consisting essentially of sanidine and lesser amounts of aegirine-augite and quartz. *(Parga-Pondal, 1935, p.66; Cancarix, Sierra de las Cabras, Spain; Trög(38). 233½; Tomk. p.87)*

CANCRINITE DIORITE. Now defined in QAPF field 14 (Fig. B.4) as a variety of foid diorite in which cancrinite is the most abundant foid. *(Streckeisen, 1976, p.21)*

CANCRINITE GABBRO. Now defined in QAPF field 14 (Fig. B.4) as a variety of foid gabbro in which cancrinite is the most abundant foid. *(Streckeisen, 1976, p.21)*

CANCRINITE MONZODIORITE. Now defined in QAPF field 13 (Fig. B.4) as a variety of foid monzodiorite in which cancrinite is the most abundant foid. *(Streckeisen, 1976, p.21)*

CANCRINITE MONZOGABBRO. Now defined in QAPF field 13 (Fig. B.4) as a variety of foid monzogabbro in which cancrinite is the most abundant foid. *(Streckeisen, 1976, p.21)*

CANCRINITE MONZOSYENITE. Now defined in QAPF field 12 (Fig. B.4) as a variety of foid monzosyenite in which cancrinite is the most abundant foid. The term is synonymous with cancrinite plagisyenite. *(Streckeisen, 1976, p.21)*

CANCRINITE PLAGISYENITE. Now defined in QAPF field 12 (Fig. B.4) as a variety of foid plagisyenite in which cancrinite is the most abundant foid. The term is synonymous with cancrinite monzosyenite. *(Streckeisen, 1976, p.21)*

CANCRINITE SYENITE. Now defined in QAPF field 11 (Fig. B.4) as a variety of foid syenite in which cancrinite is the most abundant foid. *(Törnebohm, 1883, p.399; Trög. 431; Joh. v.4, p.108)*

CANTALITE. An obsolete term used in several ways. It was originally defined as an extremely silica-rich rock but later redefined for two compositions of pitchstone and a tephritic trachyte. *(Leonhard, 1821, p.122; Verrières, Plomb du Cantal, Auvergne, France; Trög. 49; Tomk. p.87)*

CARBONATITE. A collective term for igneous rocks largely composed of primary carbonate minerals. *(Brögger, 1921, p.350; Fen complex, Telemark, Norway; Trög. 752; Tomk. p.88)*

CARBOPHYRE. An obsolete term proposed for igneous rocks intruding Carboniferous formations. *(Ebray, 1875, p.291; Tomk. p.89)*

CARMELOITE. A local name for an iddingsite-bearing basalt. *(Lawson, 1893, p.8; Carmelo Bay, California, USA; Trög. 326; Joh. v.3, p.282; Tomk. p.89)*

CARROCKITE. An obsolete term for devitrified tachylyte veins in granophyre. *(Groom, 1889, p.43; Carrock Fell, Cumbria, England; Tomk. p.89)*

CARVOEIRA. A local Brazilian name for a quartz-tourmaline rock. *(Eschwege, 1832, p.178; Villa Rica, Minas Gerais, Brazil; Joh. v.2, p.22; Tomk. p.90)*

CASCADITE. A local name for a sodic melanocratic lamprophyre having phenocrysts of biotite, olivine and augite in a matrix essentially of alkali feldspar with patches of what could have been leucite. Now considered to be a lamproite. *(Pirsson, 1905, p.55; Cascade Creek, Highwood Mts, Montana, USA; Trög. 229; Joh. v.4, p.30; Tomk. p.90)*

CAUCASITE (KAUKASITE). A local name for a variety of granite containing anorthoclase or sanidine. *(Belyankin, 1924, p.422; Caucasus Mts., Central Ridge, USSR; Trög. 880; Tomk. p.91)*

CAVALORITE. An obsolete local name proposed in place of the term oligoclasite for a leucocratic plutonic rock consisting of orthoclase and lesser oligoclase with hornblende. *(Capellini, 1878, p.124; Mt. Cavaloro, Bologna, Italy; Trög(38). 824½; Joh. v.3, p.145; Tomk. p.92)*

CECILITE. An aphyric variety of melilite leucitite consisting of essential leucite and clinopyroxene with small amounts of plagioclase, melilite and nepheline. *(Cordier, 1868, p.117; Tomb of Cecilia Metella, Capo di Bove, Rome, Italy; Trög. 671; Joh. v.4, p.306; Tomk. p.92)*

CEDRICITE. A melanocratic leucite diopside lamproite essentially composed of leucite and diopside with minor olivine pseudomorphs, amphibole, phlogopite and rutile. *(Wade & Prider, 1940, p.64; Mt. Cedric, West Kimberley district, West Australia; Tomk. p.92)*

CERAMICITE. A spelling of keramikite attributed to Koto (1916) but not found in the reference cited. *(Tomkeieff et al., 1983, p.94)*

CHARNO-ENBERBITE. A member of the charnockitic rock series equivalent to hypersthene granodiorite of QAPF field 4 (Table B.4). The term is synonymous with opdalite. *(Tobi, 1972)*

CHARNOCKITE. A term now applied to any hypersthene granite of QAPF field 3 (Table B.4). The term was originally defined as a specific type of hypersthene granite found associated with other members of what was called the charnockitic rock series, ranging from granitic and intermediate members through to noritic and hypersthene peridotites and pyroxenites. *(Holland, 1900, p.134; Tombstone of Job Charnock, St. John's Churchyard, Calcutta, India; Trög. 16; Joh. v.2, p.234; Tomk. p.96)*

CHIBINITE. See khibinite.

CHLOROPHYRE. An obsolete term originally applied to green porphyries from Egypt. *(Rozière, 1826, p.303; Trög. 825; Tomk. p.100)*

CHRISTIANITE. A term from an obsolete chemical classification, based on feldspar composition rather than SiO_2 alone, for granitic rocks in which $CaO : Na_2O : K_2O \approx 1 : 2.3 : 2.4$ and $SiO_2 \approx 70\%$. Later (Brögger, 1921) the spelling was changed to kristianite and the term was used as a local name for a biotite granite with approximately equal amount of Na_2O and K_2O. *(Lang, 1891, p.226; Christiania, Oslo district, Norway; Trög. 826; Tomk. p.100)*

CHROMITITE. An ultramafic rock consisting essentially of chromite. *(Johannsen, 1920c, p.225; Trög. 771; Joh. v.4, p.466; Tomk. p.100)*

CIMINITE. A variety of trachyte characterised by phenocrysts of sanidine, augite, labradorite and olivine in a trachytic groundmass of orthoclase, labradorite, augite, olivine and magnetite. The same rock was later called arsoite by Reinisch. *(Washington, 1896b, p.838; Fontana di Fiescoli, Cimino volcano, near Viterbo, Italy; Trög. 254; Joh. v.3, p.90; Tomk. p.101)*

CINERITE. A basic volcanic ash consisting of glass and pyroxene fragments. Later extended to include all types of formations derived from volcanic ash. *(Cordier, 1816, p.385; from the Latin cinera = ashes; Tomk. p.101)*

CIZLAKITE. A melanocratic variety of quartz monzodiorite or monzogabbro consisting of abundant augite with smaller amounts of hornblende and plagioclase with minor alkali feldspar, quartz and biotite. *(Nikitin & Klemen, 1937, p.172; Cizlak, N. Slovenia, Yugoslavia)*

CLINKSTONE. See klingstein.

CLINOPYROXENE NORITE. A basic plutonic rock consisting mainly of calcic plagioclase, orthopyroxene and minor clinopyroxene. Now defined modally in the gabbroic rock classification (Fig. B.6). *(Johannsen, 1937, p.238)*

CLINOPYROXENITE. An ultramafic plutonic rock consisting almost entirely of clinopyroxene. Now defined modally in the ultramafic rock classification (Fig. B.8). *(Wyllie, 1967, p.2)*

COCCITE. See kokkite.

COCCOLITE. An obsolete name applied to granular pyroxene rocks from Sweden. *(Cordier, 1842, Vol.4, p.45; from the Greek kokkos = grain or kernal; Tomk. p.108)*

COCITE. A melanocratic leucite-rich lamprophyre largely composed of clinopyroxene and subordinate olivine in a matrix of alkali feldspar, leucite and clinopyroxene with minor biotite. Similar to tjosite but with leucite instead of nepheline. *(Lacroix, 1933, p.195; Coc Pia, Upper Tonkin, Vietnam; Trög. 500; Joh. v.4, p.279; Tomk. p.108)*

COLORADOITE. A variety of quartz latite that corresponds to the opdalitic magma-type of Niggli. *(Niggli, 1923, p.118; San Cristobel Quadrangle, Colorado, USA; Trög. 99; Tomk. p.111)*

COLOUR INDEX (COLOUR RATIO). The volume percentage or ratio of dark coloured or mafic minerals to light coloured or felsic minerals in a rock. In terms of the QAPF terminolgy the colour index is denoted by the parameter M' (section B.3.(i)). *(Shand, 1916, p.403; Tomk. p.111)*

COLUMBRETITE. A variety of tephritic leucite phonolite essentially composed of alkali feldspar and plagioclase with subordinate leucite/analcime, clinopyroxene and amphibole. *(Johannsen, 1938, p.168; Bauzá, Columbrete Islands, Spain; Trög(38). 533; Tomk. p.111)*

COMENDITE. A variety of porphyritic leucocratic peralkaline rhyolite containing phenocrysts of quartz, alkali feldspar, aegirine, arfvedsonite or riebeckite and minor biotite. It is now defined and distinguished from pantellerite as a variety of peralkaline rhyolite of TAS field R (Fig. B.10) in which $Al_2O_3 > 1.33$ x total iron as $FeO + 4.4$ (Fig.

B.16). *(Bertolio, 1895, p.50; Comende, San Pietro Island, Sardinia; Trög. 48; Joh. v.2, p.65; Tomk. p.111)*

COMMENDITE. An erroneous spelling of comendite. *(Irvine & Baragar, 1971, p.530)*

CONCHAITE. An obsolete name for a variety of hornblende orthopyroxenite composed of bronzite and hornblende. *(Cotelo Neiva, 1947b, p.122; Concha Farm, San Antonio, Portugal; Tomk. p.114)*

CONGRESSITE. A local name for a coarse-grained variety of urtite consisting of about 70% nepheline with minor alkali feldspar, biotite, occasional sodalite and accessory minerals. *(Adams & Barlow, 1913, p.90; Congress Bluff, Craigmont Hill, Ontario, Canada; Trög. 605; Joh. v.4, p.280; Tomk. p.116)*

COPPAELITE. A local term for a variety of biotite melilitite consisting of almost equal amounts of pyroxene and melilite with variable amounts of phlogopite and no olivine. *(Sabatini, 1903, p.378; Coppaeli di Sotto, near Rieti, Umbria, Italy; Trög.751; Joh. v.4, p.371; Tomk. p.119)*

CORCOVADITE. An obsolete local name for the hypabyssal equivalent of andengranite (granitic rocks which are younger than the usual ones) consisting of phenocrysts of plagioclase, hornblende, minor quartz, and biotite in a fine-grained groundmass of plagioclase, quartz, and orthoclase. *(Scheibe, 1933, p.160; Marmato, Cauca River, Colombia; Trög. 114; Tomk. p.120)*

CORONITE. A term for any igneous rock containing coronas. *(Shand, 1945, p.251; Tomk. p.121)*

CORSILITE. An old term for Corsite. *(Pinkerton, 1811b, p.78; Tomk. p.122)*

CORSITE. An orbicular variety of gabbro composed essentially of calcic plagioclase with hornblende, minor hypersthene and quartz. Spheroids are of variable composition but the matrix is richer in hornblende. This rock has also been called kugeldiorite, miagite and napoleonite. *(Zirkel, 1866b, p.133; named after Corsica; Trög. 361; Joh. v.3, p.230; Tomk. p.122)*

CORTLANDTITE. A variety of pyroxene olivine hornblendite composed of large crystals of hornblende poikilitically enclosing olivine, hypersthene and augite. The rock had previously been called hudsonite, which was objected to as the name was already in use as a mineral name. *(Williams, 1886, p.30; named after Cortlandt, New York, USA; Trög. 708; Joh. v.4, p.425; Tomk. p.123)*

COSCHINOLITE. An obsolete term for a variety of brownish borzolite (hornblende-bearing rock) with cavities indicating former amygdales. *(Issel, 1880, p.187; from the Latin coccineus = scarlet-red or the Greek "honeycomb"; Tomk. p.123)*

COVITE. A local name for a melanocratic nepheline syenite with sodic pyroxene and amphibole. *(Washington, 1901, p.615; Magnet Cove, Arkansas, USA; Trög. 485; Joh. v.4, p.103; Tomk. p.124)*

COXITE. A name given to a hypothetical set of data used to illustrate the problems of the statistical interpretation of rock compositions. *(Aitchison, 1984, p.534)*

CRAIGMONTITE. A coarse-grained leucocratic variety of nepheline diorite consisting of approximately ⅔ nepheline and ⅓ oligoclase with small amounts of corundum. *(Adams & Barlow, 1910, p.313; Craigmont Hill, Ontario, Canada; Trög. 548; Joh. v.4, p.299; Tomk. p.125)*

CRAIGNURITE. A local Hebridian name for a series of volcanic rocks occurring as cone sheets and ranging in composition from intermediate to acid. They are characterised by acicular and skeletal crystals of plagioclase, augite and hornblende in a felsic groundmass. *(Thomas & Bailey, 1924, p.224; Craignure, Island of Mull, Scotland; Trög. 98; Tomk. p.125)*

CRINANITE. A variety of olivine analcime dolerite or gabbro composed of olivine, Ti-augite, and labradorite with minor analcime. Although it has less analcime and more olivine than teschenite the two names have been used interchangeably. *(Flett, 1911, p.117; Loch Crinan, Sound of Jura, Scotland; Trög. 566; Joh. v.4, p.70; Tomk. p.127)*

CRISTULITE. An obsolete name for a porphyritic rock containing phenocrysts of quartz and feldspar. *(Cordier, 1868, p.85; Tomk. p.127)*

CROMALTITE. A local name for an alkali pyroxenite consisting of aegirine-augite, melanite, biotite, iron ore and apatite. *(Shand, 1910, p.394; Ledmore Lodge, near Cromalt Hills, Assynt, Scotland; Trög. 694; Joh. v.4, p.454; Tomk. p.128)*

CRYPTODACITE. A term for a dacite in which the silica phase is present in a glassy groundmass. *(Belyankin, 1923, p.100; Tomk. p.130)*

CRYSTAL TUFF. Now defined in the pyroclastic classification (Fig. B.1) as a tuff in which crystal fragments are more abundant than either lithic or vitric fragments. *(Pirsson, 1915, p.193)*

CUMBERLANDITE. A variety of dunite composed of olivine, ilmenite, magnetite and small amounts of labradorite and spinel. *(Wadsworth, 1884, p.75; Cumberland, Rhode Island, USA; Trög. 766; Joh. v.3, p.335; Tomk. p.134)*

CUMBRAITE. An obsolete local name for a dyke rock composed of phenocrysts of calcic plagioclase, enstatite and augite in a groundmass of glass. *(Tyrrell, 1917, p.306; Great Cumbrae, Firth of Clyde, Scotland; Trög. 125; Tomk. p.134)*

CUMULATE. A general term proposed for igneous rocks in which most of the crystals are thought to have accumulated by gravitational settling from a magma. *(Wager et al., 1960, p.73; from the Latin cumulus = heap; Tomk. p.134)*

CUMULOPHYRE. An obsolete term for a porphyritic rock in which the clusters of phenocrysts resemble cumulus clouds! *(Iddings, 1909, p.224; Tomk. p.134)*

CUSELITE (KUSELITE). A local name for an altered lamprophyric rock containing phenocrysts of plagioclase and uralitised diopside in a granular groundmass of plagioclase, orthoclase, chloritised biotite, hornblende, diopside and interstitial quartz. *(Rosenbusch, 1887, p.503; Cusel (now Kusel), Rheinland-Pfalz, West Germany; Trög. 218; Joh. v.3, p.190; Tomk. p.135)*

CUYAMITE. A local name for an olivine-free or olivine-poor variety of teschenite. *(Johannsen, 1938, p.243; Cuyamas Valley, California, USA; Trög(38). 565⅔; Tomk. p.136)*

DACITE. A volcanic rock composed of quartz and sodic plagioclase with minor amounts of biotite and/or hornblende and/or pyroxene. The volcanic equivalent of granodiorite and tonalite. Now defined modally in QAPF fields 4 and 5 (Fig. B.10) and, if modes are not available, chemically in TAS field O3 (Fig. B.13). *(Hauer & Stache, 1863, p.72; Dacia, a Roman province, Transylvania, Romania; Trög. 148; Joh. v.2, p.390; Tomk. p.137)*

DACITE-ANDESITE. An obsolete term for an andesitic rock containing olivine and quartz. *(Dannenberg, 1900, p.239; Caucasus, USSR; Tomk. p.137)*

DACITOID. A collective name for acidic volcanic rocks having the chemical composition of dacite but containing no quartz. Now used as a general term in the provisional "field" classification (Fig. B.17) for rocks tentatively identified as dacite. *(Lacroix, 1919, p.297; Tomk. p.137)*

DAHAMITE. An obsolete local term for a fine-grained variety of peralkaline granite containing phenocrysts of albite in a groundmass of equal amounts of alkali feldspar, albite and quartz. Minor riebeckite is present. *(Pelikan, 1902, p.397; Dahamis, Socotra Island, Indian Ocean; Trög. 30; Tomk. p.137)*

DAMKJERNITE (DAMTJERNITE). A melanocratic variety of nepheline lamprophyre containing phenocrysts of biotite and Ti-augite in a groundmass of the same minerals with some nepheline, calcite, and orthoclase. As pointed out by Sæther (1957) the name was actually misspelt by Brögger in the belief that the locality was named Damkjern instead of Damtjern. The strictly correct spelling should have been damtjernite but damkjernite is in common usage. *(Brögger, 1921, p.276; Damkjern, Fen complex, Telemark, Norway; Trög. 499; Joh. v.4, p.277; Tomk. p.138)*

DANCALITE. A variety of analcime tephrite with phenocrysts of oligoclase, augite with aegirine rims and rare amphibole in a groundmass of plagioclase laths with interstitial analcime. *(De-Angelis, 1925, p.82; Dankalia, Ethiopia; Trög. 570; Joh. v.4, p.189; Tomk. p.138)*

DANUBITE. An obsolete term for an amphibole hypersthene andesite in which the labradorite phenocrysts were thought to be nepheline. *(Krenner, 1910, p.130; Trög. 828)*

DAVAINITE. An obsolete term for a granular variety of hornblendite, probably metamorphic, consisting essentially of brown hornblende with cores of pyroxene. Hypersthene and feldspar are also present but in small amounts. *(Wyllie & Scott, 1913, p.501; Garabal Hill, Loch Beinn Damhain or Ben Davain, Scotland; Trög. 702; Joh. v.4, p.448)*

DELDORADITE (DELDORADOITE). A leucocratic variety of cancrinite syenite containing biotite and aegirine. *(Johannsen, 1938, p.77; Deldorado creek, Uncompahgre quadrangle, Colorado, USA; Trög(38). 412½)*

DELLENITE. A name proposed for a rock intermediate between dacite and rhyolite. Cf. toscanite. *(Brögger, 1895, p.59; Lake Dellen, Helsingland, Sweden; Trög. 96; Joh. v.2, p.309; Tomk. p.140)*

DELLENITOID. A rock which is chemically a dellenite but contains no visible orthoclase. *(Lacroix, 1923, p.9)*

DEVONITE. A temporary name suggested by A. Johannsen for a variety of porphyritic diabase. *(Clarke, 1910, p.40; Mount Devon, Missouri, USA; Trög. 315; Joh. v.3, p.320)*

DIABASE. A term, that has been used in two distinct ways, for medium-grained rocks of basaltic composition. British usage implied heavy alteration, but French, German and American usage implied an ophitic texture. The original definition included a transitional texture between that of basalts and that of coarse-grained rocks. Now regarded as being synonymous with dolerite and an approved synonym for microgabbro of QAPF field 10 (Fig. B.4). *(Brongniart, 1807, p.456; from the Greek diabasis = crossing over; Trög. 390; Joh. v.3, p.311; Tomk. p.144)*

DIABASITE. A term proposed, but never used, for a diabase aplite. *(Polenov, 1899, p.464; Joh. v.3, p.61)*

DIALLAGITE. Although originally suggested for a rock composed essentially of labradorite and diallage (pyroxene), it was more commonly used for rocks consisting mainly of diallage. *(Cloizeaux, 1863, p.108; Trög. 681; Joh. v.4, p.457)*

DIASCHISTITE. A term for a rock from a minor intrusion which has a different composition from the main intrusion. *(Johannsen, 1931, p.5; from the Greek diaschistos = cleaved; Tomk. p.147)*

DIATECTITE (DIATEXITE). A term used for a rock formed by diatexis i.e. complete or almost complete melting of pre-existing rocks. *(Fiedler, 1936, p.493; Tomk. p.148)*

DIMELITE. A mnemonic name suggested for a rock consisting essentially of *di*opside and *me*lilite. *(Belyankin, 1929, p.22)*

DIOPSIDITE. A variety of pyroxenite consisting almost entirely of diopside (or chrome-diopside). *(Lacroix, 1895, p.752; Lhers, Ariége, Pyrénées, France; Trög. 682; Joh. v.4, p.455)*

DIOREID. An obsolete field term for a coarse-grained igneous rock containing more than 50% amphibole; feldspar is subordinate. *(Johannsen, 1911, p.320; Tomk. p.150)*

DIORIDE. A revised spelling recommended to replace the field term dioreid. Now obsolete. *(Johannsen, 1926, p.181; Joh. v.1, p.57; Tomk. p.150)*

DIORITE. A plutonic rock consisting of sodic plagioclase, commonly hornblende and often with biotite or augite. Now defined modally in QAPF field 10 (Fig. B.4). *(D'Aubuisson de Voisins, 1819, p.146; from the Greek diorizein = to distinguish; Trög. 308; Joh. v.3, p.146)*

DIORITITE. An obsolete name for a variety of diorite or diorite-aplite which occurs as dykes. *(Polenov, 1899, p.464; Joh. v.3, p.61; Tomk. p.150)*

DIORITOID. A group name for igneous rocks consisting mainly of plagioclase, amphibole and biotite. Now used as a general term in the provisional "field" classification (Fig. B.9) for rocks tentatively identified as diorite or monzodiorite. *(Gümbel, 1888, p.87; Tomk. p.150)*

DIORITOPHYRITE. An obsolete term for a porphyritic diorite aplite. *(Polenov, 1899, p.464; Tomk. p.150)*

DIOROID. An obsolete mnemonic name for a *dior*ite containing feldspath*oid*s. *(Shand, 1910, p.378; Tomk. p.150)*

DISSOGENITE. A term applied to igneous rocks containing two different sets of minerals, such as might be created by assimilation. *(Lacroix, 1922, p.374; from the Greek dissos = twofold, genos = family; Trög. 832; Tomk. p.153)*

DITRO-ESSEXITE. A comprehensive term for various plutonic rocks of essexitic and theralitic chemistry from the alkaline complex of Ditro, Romania. The term includes such types as diorite, monzonite, essexite with large amounts of barkevikite and kaersutite. *(Streckeisen, 1952, p.274 and Steckeisen, 1954, p.394 for the description; Ditro (now Ditrau), Transylvania, Romania)*

DITROITE. A biotite-bearing variety of nepheline syenite with cancrinite and primary calcite; sodalite penetrates along fractures and intergranular boundaries. Used by Brögger (1890) as a general term for nepheline syenites of granular texture. *(Zirkel, 1866a, p.595; Ditro (now Ditrau), Transylvania, Romania; Trög. 427; Joh. v.4, p.98; Tomk. p.153)*

DOLEREID. An obsolete field term for a rock containing more than 50% of ferromagnesian minerals when the species cannot be determined megascopically. Cf. gabbreid and dioried. *(Johannsen, 1911, p.320; Tomk. p.154)*

DOLERINE. An obsolete name for a rock with a feldspathic groundmass containing chlorite. *(Jurine, 1806, p.374; Mont Blanc, France; Tomk. p.154)*

DOLERITE. A rock intermediate in grain-size between basalt and gabbro and composed essentially of plagioclase, pyroxene and opaque minerals; often with ophitic texture. If olivine is present the rock may be called olivine dolerite – if quartz, quartz dolerite. Now regarded as being synonymous with diabase and an approved synonym for microgabbro of QAPF field 10 (Fig. B.4). *(D'Aubuisson de Voisins, 1819, p.556; from the Greek doleros = deceptive; Trög. 833; Joh. v.3, p.291; Tomk. p.154)*

DOLEROID. An obsolete mnemonic name for a dolerite containing foids. *(Shand, 1910, p.378; Tomk. p.154)*

DOLOMITE CARBONATITE. A variety of carbonatite, defined modally in the carbonatitic rock classification, in which more than 90% of the carbonate is dolomite. Cf.

rauhaugite. *(Tröger, 1935, p.303; Trög. 758)*

DOMITE. A local name for a variety of trachyte that contains a few phenocrysts of oligoclase and biotite in a potentially quartz-bearing glassy groundmass. *(Buch, 1809, p.244; Puy de Dôme, Auvergne, France; Trög. 103; Joh. v.3, p.66; Tomk. p.156)*

DOREITE. A local name for a variety of trachyandesite characterised by microphenocrysts of andesine and augite. *(Lacroix, 1923, p.328; Mont Dore, Auvergne, France; Trög. 268; Tomk. p.157)*

DORGALITE. An obsolete local name for a variety of olivine basalt containing phenocrysts of olivine only. *(Amstutz, 1925, p.300; Mt. Pirische, near Dorgali, Sardinia; Trög. 328; Joh. v.3, p.282; Tomk. p.157)*

DRAKONTITE (DRAKONITE). A local name for a variety of trachyte containing biotite and alkali amphibole. *(Reinisch, 1912, p.121; Drachenfels, Siebengebirge, near Bonn, West Germany; Trög. 834; Joh. v.4, p.35; Tomk. p.157)*

DRUSITE. An obsolete name for a gabbroic rock containing concentric coronas of minerals (drusy structure). *(*Fedorov, 1896, p.168; Tomk. p.158)*

DUCKSTEIN. A local German name for a non-stratified tuff (trass). *(although attributed to Wolff, 1914, p.404 it is not the original reference; Tomk. p.159)*

DUMALITE. A local name for a variety of trachyandesite characterised by intersertal texture and a glassy groundmass of nepheline composition. *(Loewinson-Lessing, 1905b, p.276; Dumala Gorge, N. Caucasus, USSR; Trög. 556; Joh. v.4, p.190; Tomk. p.159)*

DUNGANNONITE. A local name for a variety of nepheline-bearing diorite containing corundum and scapolite. *(Adams & Barlow, 1908, p.67; Dungannon, Renfrew County, Ontario, Canada; Trög. 550; Joh. v.4, p.59)*

DUNITE. An ultramafic plutonic rock consisting essentially of olivine. Now defined modally in the ultramafic rock classification (Fig. B.8). *(Hochstetter, 1859, p.275; Dun Mountain, Nelson, New Zealand; Trög. 724; Joh. v.4, p.405; Tomk. p.159)*

DURBACHITE. A fine- to medium-grained melanocratic variety of syenite consisting of large flakes of biotite with hornblende and megacrysts of orthoclase in a groundmass of oligoclase and a little quartz. *(Sauer, 1892, p.247; Durbach, near Offenburg, Black Forest, West Germany; Trög. 243; Joh. v.3, p.86; Tomk. p.160)*

DUST GRAIN. Now defined in the pyroclastic classification (Table B.1) as a pyroclast with mean diameter less than 1/16 mm. *(Schmid, 1981, p.42)*

DUST TUFF. Now defined in the pyroclastic classification (Table B.1) as a pyroclastic rock in which the average pyroclast size is less than 1/16 mm. *(Schmid, 1981, p.43)*

EGERINOLITH. An erroneous spelling of aigirinolith or aegirinolith. *(Tomkeieff et al., 1983, p.165)*

EHRWALDITE. A lamprophyric rock of basanitic composition composed of Ti-augite, biotite, serpentinised olivine and glass. *(Pichler, 1875, p.927; Wetterschroffen, Ehrwald, Tyrol, Austria; Trög. 347; Tomk. p.165)*

EICHSTÄDTITE. An obsolete name suggested for a group of quartz norites. *(Marcet Riba, 1925, p.293; named after F. Eichstädt)*

EKERITE. A local name for a peralkaline granite containing anorthoclase microperthite and small amounts of arfvedsonite and aegirine. *(Brögger, 1906, p.136; Eker, Oslo district, Norway; Trög. 18; Tomk. p.166)*

ELKHORNITE. A variety of alkali feldspar syenite containing some labradorite. *(Johannsen, 1937, p.92; Elkhorn, Montana, USA; Trög(38). 244½; Tomk. p.167)*

ELVAN. A local term for dyke rocks of granitic composition containing phenocrysts of quartz and orthoclase with tourmaline occurring as isolated crystals and in radiating groups. *(Conybeare, 1817, p.401; from the Celtic el = rock and van = white; Trög. 113; Joh. v.2, p.300; Tomk. p.168)*

ELVANITE. An alternative term for elvan. *(Cotta, 1866, p.214; Joh. v.2, p.300; Tomk. p.168)*

ENDERBITE. A member of the charnockitic rock series consisting essentially of quartz, antiperthite, hypersthene and magnetite. It is equivalent to hypersthene tonalite of QAPF field 5 (Table B.4). *(Tilley, 1936, p.312; named after Enderby Land, Antarctica; Joh. v.1(2nd.Ed.), p.250; Tomk. p.169)*

ENGADINITE. A term for granites with low quartz content, that correspond to the engadinitic magma-type of Niggli (1923, p.110). *(Tröger, 1935, p.20; Platta mala granite, near Ramosch, Engadine, Switzerland; Trög. 13; Tomk. p.171)*

ENGELBURGITE. A variety of hybrid biotite granodiorite containing leucocratic spots consisting of alkali feldspar and abundant titanite. *(Frentzel, 1911, p.145; Engelburg, Bavaria, West Germany; Trög(38). 107½; Tomk. p.171)*

ENSTATITITE. A variety of orthopyroxenite composed almost entirely of enstatite. *(Streng, 1864, p.260; Trög. 674; Tomk. p.171)*

ENSTATOLITE. An obsolete term for a variety of orthopyroxenite composed almost entirely of enstatite. *(Pratt & Lewis, 1905, p.30; Trög. 836; Joh. v.4, p.459; Tomk. p.171)*

EORHYOLITE. An old name proposed for a devitrified acid igneous rock of Archean age. *(Nordenskjöld, 1893, p.153; Tomk. p.172)*

EPIBASITE. An obsolete term suggested for non-porphyritic lamprophyres – somewhat a contradiction of terms. *(Sederholm, 1926, p.14; Tomk. p.173)*

EPIBUGITE. A leucocratic variety of enderbite of QAPF field 5 (Fig. B.4). It is a member of the bugite series with 68% to 72% SiO_2, 0% to 8% hypersthene, oligoclase-andesine antiperthite and quartz. It is suggested (Streckeisen, 1974, p.358) that this term should be abandoned. *(Bezborod'ko, 1931, p.137; Bug River, Podolia, Ukraine, USSR; Trög(38). 129½; Tomk. p.173)*

EPIDIABASE. An obsolete name for epidiorite. *(Issel, 1892, p.324; Trög. 395; Tomk. p.173)*

EPIDIORITE. A term originally applied to a diorite in which the pyroxene had been altered to amphibole. Current usage as a field term implies rocks of basaltic composition recrystallised by regional metamorphism. *(Gümbel, 1874, p.10; Trög. 395; Tomk. p.174)*

ERCALITE. An obsolete local name for a variety of granophyre composed of clusters of quartz and sodic plagioclase surrounded by a graphic intergrowth of quartz and feldspar. *(Watts, 1925, p.327; Ercall, near Wrekin, Shropshire, England; Tomk. p.177)*

ESBOITE. A local name for a variety of diorite,

often orbicular, consisting of abundant sodic plagioclase with small amounts of orthoclase and biotite. *(Sederholm, 1928, p.72; Esbo, now Espoo, near Helsinki, Finland; Trög. 306; Tomk. p.179)*

ESMERALDITE. A local name for a granular variety of quartzolite consisting predominatnly of quartz with some muscovite. It is thought to be an extreme differentiate but, unlike greisen, not produced by pneumatolytic action. Cf. northfieldite. *(Spurr, 1906, p.382; Esmeralda County, Nevada, USA; Trög. 5; Joh. v.2, p.17; Tomk. p.179)*

ESPICHELLITE. A local name for a variety of camptonite (a lamprophyre) containing analcime. *(Souza-Brandão, 1907, p.2; Cape Espichel, Lisbon, Portugal; Trög. 559; Joh. v.4, p.207; Tomk. p.179)*

ESSEXITE. A variety of nepheline monzogabbro or nepheline monzodiorite containing Ti-augite, kaersutite and/or biotite with labradorite, lesser alkali feldspar and nepheline. Now defined as a synonym for nepheline monzogabbro or nepheline monzodiorite of QAPF field 13 (Fig. B.4). *(Sears, 1891, p.146; Salem Neck, Essex County, Massachusetts, USA; Trög. 542; Joh. v.4, p.191; Tomk. p.179)*

ESSEXITE-AKERITE. A term suggested for a dark plagioclase-rich akerite. *(Brögger, 1933, p.35; Lortlauptal, W Vikersund, Oslo district, Norway; Trög. 276)*

ESSEXITE-BASALT. An obsolete name for a volcanic rock consisting of serpentinised olivine phenocrysts in a groundmass of olivine, augite, labradorite, sanidine, biotite, ore and interstitial nepheline. Johannsen (1938) renamed it westerwaldite. *(Lehmann, 1924, p.175; Utanjilua, Patagwa, Malawi; Trög. 382; Joh. v.4, p.203; Tomk. p.180)*

ESSEXITE-DIABASE. An obsolete name for a variety of nepheline tephrite or essexite containing small amounts of orthoclase. *(Erdmannsdörffer, 1907, p.22; Trög. 601; Tomk. p.180)*

ESSEXITE-DIORITE. An obsolete term for a variety of diorite containing Ti-augite similar to rongstockite and kauaiite. *(Tröger, 1935, p.124; Trög. 282, 284)*

ESSEXITE-FOYAITE. An obsolete term for a variety of nepheline monzosyenite, synony-

mous with husebyite. *(Tröger, 1935, p.214; Trög. 508)*

ESSEXITE-GABBRO. An obsolete term for a variety of essexite rich in plagioclase and poor in feldspathoids. *(Lacroix, 1909, p.543; Jordanne Valley, Cantal, Auvergne, France; Trög. 283; Tomk. p.180)*

ESSEXITE-MELAPHYRE. An obsolete term for a geologically old altered essexite-basalt. *(Brögger, 1906, p.120; Holmestrand, Oslo district, Norway; Trög. 389; Tomk. p.180)*

ESSEXITE-PORPHYRITE. A variety of essexite containing plagioclase phenocrysts. *(Brögger, 1906, p.121; Trög. 557; Tomk. p.180)*

ESTERELLITE. A local name for a porphyritic variety of quartz diorite consisting of zoned plagioclase and hornblende in a groundmass of quartz, orthoclase and plagioclase. *(Michel-Levy, 1897, p.39; Esterel, Provence, France; Trög. 143; Joh. v.2, p.400; Tomk. p.180)*

ETINDITE. A variety of leucite nephelinite essentially composed of nepheline, clinopyroxene and leucite. *(Lacroix, 1923, p.65; Etinde volcano, Cameroon; Trög. 619; Joh. v.4, p.367; Tomk. p.180)*

ETNAITE. A local term used for trachyandesites from Mt. Etna. *(Rittmann, 1960, p.117; Lava of 1669, Catania, Mt. Etna, Sicily, Italy)*

EUCRITE (EUKRITE, EUCRYTE). A term originally and properly applied to meteorites composed of anorthite and augite (Rose,1863), it was extended to include varieties of gabbro in which the plagioclase was bytownite. The term should no longer be used as a rock name. *(Cotta, 1866, p.148; from the Greek eukritos = clear, distinct; Trög. 358; Joh. v.3, p.347; Tomk. p.181)*

EUDIALYTITE. A melanocratic plutonic rock consisting essentially of eudialyte with minor variable amounts of microcline, arfvedsonite and aegirine. *(Eliseev, 1937, p.1100; Lovozero complex, Kola Peninsula, USSR; Tomk. p.181)*

EUKTOLITE. A name given by Rosenbusch to a melanocratic volcanic rock largely composed of melilite, leucite and olivine, when he was unaware that the rock had already been called venanzite. *(Rosenbusch, 1899, p.110; from the Greek euktos = desired, San Venanzo, Umbria, Italy; Trög. 838; Joh. v.4, p.361)*

EULYSITE. An obsolete term for a variety of peridotite composed essentially of olivine, diopside, and opaques with anthophyllite. *(Erdmann, 1849, p.837; from the Greek eulytos= easy to dissolve or break; Trög. 726; Joh. v.4, p.412; Tomk. p.181)*

EUPHOTIDE. An obsolete term originally suggested by Haüy for a saussuritised gabbro. *(Brongniart, 1813, p.42; Trög. 839; Joh. v.3, p.229; Tomk. p.181)*

EURITE. A name originally suggested by d'Aubuisson for a compact felsitic rock and later extended to cover all aphanitic rocks of granitic composition. Cf. felsite. Although this name appears in Johannsen's index it was not found in the text. *(Brongniart, 1813, p.43; from the Greek eurys = broad; Rocher de Sanadoire, Auvergne, France; Trög. 840; Tomk. p.182)*

EURYNITE. An obsolete term for a porphyritic eurite (felsitic rock). *(Cordier, 1868, p.81; Tomk. p.182)*

EUSTRATITE. A local name for a lamprophyric dyke rock with rare phenocrysts of olivine, corroded hornblende, and occasional augite in a groundmass of augite, Ti-magnetite, feldspar, and glass. *(Kténas, 1928, p.1632; Haghios Eustratios, now Ayios Evstrátios Island, Greece; Trög. 521; Joh. v.4, p.178; Tomk. p.182)*

EUTECTITE. According to Tomkeieff an igneous rock formed by the crystallisation of residual liquids. However, the reference cited (Bowen, 1914) does not contain the name. *(Tomkeieff et al., 1983, p.183)*

EUTECTOFELSITE (EUTEKTOFELSITE, EUTECTOPHYRE). An obsolete term for a variety of quartz porphyry with a whitish earthy tuff-like appearance which cleaves into imperfect tablets. The description given by Tomkeieff appears to be erroneous. *(Kotô, 1909, p.189; Trög. 841; Tomk. p.183)*

EVERGREENITE. A local name for a variety of alkali feldspar granite containing wollastonite, which was originally identified as enstatite. *(Ritter, 1908, p.751; Evergreen Mine, Apex, Oregon, USA; Trög. 71; Tomk. p.184)*

EVISITE. A local comprehensive term for peralkaline granites and syenites with aegirine and/or riebeckite corresponding to the evisitic magma-type of Niggli (1923, p.148). *(Tröger, 1935, p.317; Evisa, Corsica; Trög. 842; Tomk. p.184)*

FARRISITE. A local name for a melanocratic variety of melilite lamprophyre containing augite, barkevikite, biotite and olivine but no feldspar. *(Brögger, 1898, p.64; Lake Farris, Norway; Trög. 522; Joh. v.4, p.389; Tomk. p.189)*

FARSUNDITE. A member of the charnockitic rock series equivalent to hypersthene monzogranite of QAPF field 3b (Table B.4), consisting of oligoclase, microcline, quartz, hornblende, hypersthene and a little clinopyroxene. The term is also frequently used as a comprehensive name for granitic rocks of the Farsund area. It is suggested (Streckeisen, 1974, p.354) that this term should be abandoned, to avoid ambiguity. *(Kolderup, 1903, p.110; Farsund, Egersund, Norway; Trög. 106; Tomk. p.189)*

FASIBITIKITE. A fine-grained variety of peralkaline granite, similar to rockallite, containing large amounts of aegirine and riebeckite and characterised by the presence of eucolite. *(Lacroix, 1915, p.257; Ampasibitika, Madagascar; Trög. 64; Tomk. p.190)*

FASINITE. A variety of melteigite consisting of 60% – 70% Ti-augite and some nepheline, with minor olivine, biotite, and orthoclase. *(Lacroix, 1916b, p.257; from the Madagascan fasina = sand, refering to long sand beaches; Trög. 611; Joh. v.4, p.252; Tomk. p.190)*

FELDSPAR-BASALT. An obsolete name proposed for a fine-grained rock consisting essentially of plagioclase and augite to distinguish it from leucite- and nepheline-basalt. *(Zirkel, 1870, p.108; Trög. 888; Tomk. p.190)*

FELDSPATHIDOLITE. An obsolete term for an igneous rock composed mainly of feldspathoids. *(Loewinson-Lessing, 1901, p.114; Tomk. p.191)*

FELDSPATHINE. An obsolete name for a feldspar rock. *(Cordier, 1868, p.74; Tomk. p.191)*

FELDSPATHITE. An obsolete collective name for monzonite and quartz monzonite aplites. *(Kolenec, 1904, p.162; Trög. 844)*

FELDSPATHOIDITE. A general term for extrusive rocks composed essentially of feldspathoids, now superseded by the term foidite. *(Johannsen, 1938, p.336; Tomk. p.192)*

FELDSPATHOLITE. An obsolete term for an igneous rock composed mainly of feldspar. *(Loewinson-Lessing, 1901, p.114; Tomk. p.192)*

FELDSPATTAVITE. An obsolete term originally used for a feldspathic variety of tawite (a sodalite aegirine rock). *(Tröger, 1935, p.318; Trög. 846; Tomk. p.192)*

FELSEID. An obsolete field term for an aphanitic, non-porphyritic light-coloured igneous rock. Also called leuco-aphaneid. *(Johannsen, 1911, p.321; Tomk. p.192)*

FELSEID PORPHYRY. An obsolete field term for a porphyritic light-coloured igneous rock with an aphanitic groundmass. Also called leucophyreid. *(Johannsen, 1911, p.321)*

FELSIC. A collective term for modal quartz, feldspars and feldspathoids which was introduced to stop the normative term salic incorrectly being used for that purpose. See also mafic and femic. *(Cross et al., 1912, p.561)*

FELSIDE. A revised spelling recommended to replace the field term felseid. Now obsolete. *(Johannsen, 1926, p.182; Joh. v.1, p.57; Tomk. p.192)*

FELSITE (FELSITIC). As a rock term initially used for the microcrystalline groundmass of porphyries but now more commonly used for microcrystalline rocks of granitic composition. The name had previously been used by Kirwan (1794) for a microcrystalline variety of feldspar. *(Gerhard, 1815, p.18; from feldspar; Trög. 847; Tomk. p.192)*

FELSOANDESITE. An obsolete term used for andesites in which the groundmass is cryptocrystalline. *(Johannsen, 1937, p.170; Tomk. p.193)*

FELSOGRANOPHYRE. An obsolete name for a porphyry in which the texture of the groundmass is between granitic and felsitic. *(Zirkel, 1894a, p.168; Tomk. p.193)*

FELSOPHYRE. An obsolete group name for a porphyritic rock with a microcrystalline groundmass containing some glass. *(Vogelsang, 1872, p.534; from the German fels = rock and phyre = porphyritic; Trög. 848; Joh. v.2, p.275; Tomk. p.193)*

FELSOPHYRITE. A term originally proposed for non-porphyritic rocks with microcrystalline texture and some glass. Later used for effusive rocks of diorite composition. Now obsolete. *(Vogelsang, 1872, p.534; Tomk. p.194)*

FELSOVITROPHYRE. A term proposed for a porphyry in which the texture of the groundmass is between felsitic and glassy. *(Vogelsang, 1875, p.160; Tomk. p.194)*

FELSPARITE. An obsolete term proposed for a granite rich in feldspar. *(Boase, 1834, p.17; Tomk. p.194)*

FELSTONE. An obsolete term for a felsite. *(Leonhard, 1823a, p.210; Tomk. p.194)*

FEMIC. A name used in the CIPW normative classification for one of two major groups of normative minerals which includes the Fe and Mg silicates such as olivine and pyroxene as well as the Fe and Ti oxides, apatite and fluorite. The other group is named saliB. Cf. mafic. *(Cross et al., 1902, p.573; Joh. v.1, p.173; Tomk. p.194)*

FENITE. A metasomatic rock, normally associated with carbonatites or ijolites and occasionally with nepheline syenites and peralkaline granites, composed of alkali feldspar, sodic pyroxene and/or alkali amphibole. Some varieties are monomineralic alkali feldspar rocks. *(Brögger, 1921, p.156; Fen complex, Telemark, Norway; Trög. 187; Joh. v.4, p.32; Tomk. p.194)*

FERGANITE. An obsolete name for a variety of clinopyroxenite consisting essentially of clinopyroxene with minor olivine, plagioclase and magnetite. *(Lebedev & Vakhrushev, 1953, p.119; Fergana, ESE of Tashkent, USSR; Tomk. p.195)*

FERGUSITE. A plutonic rock consisting of roughly 70% pseudoleucite (alkali feldspar, nepheline, kalsilite and minor analcime) and 30% pyroxene. Now defined modally as a variety of foidite in QAPF field 15 (Figs. B.4 and B.7). *(Pirsson, 1905, p.74; Fergus County, Highwood Mts., Montana, USA; Trög. 628; Joh. v.4, p.325; Tomk. p.195)*

FERRILITE. An obsolete local name for a dolerite. *(Kirwan, 1794, p.229; Rowley Regis, Staffordshire, England; Tomk. p.195)*

FERROCARBONATITE. A chemically defined variety of carbonatite (Fig. B.2) in which wt% MgO < (FeO+Fe$_2$O$_3$+MnO) and CaO / (CaO+MgO+FeO+Fe$_2$O$_3$+MnO) < 0.8. *(Le Bas, 1977, p.37)*

FERRODIORITE. A term for rocks of gabbroic appearance containing Fe-rich pyroxenes and olivines but with plagioclase more sodic than An$_{50}$. Many such rocks were previously called ferrogabbro (e.g. from Skaergaard). *(Wager & Vincent, 1962, p.26)*

FERROGABBRO. A variety of gabbro in which the pyroxenes and olivines are Fe-rich varieties. Since the plagioclase of gabbros should be more calcic than An$_{50}$, most of the rocks originally given this name (e.g. from Skaergaard) have been renamed as ferrodiorites. *(Wager & Deer, 1939, p.98; Tomk. p.196)*

FERROLITE. An obsolete term originally proposed for iron ore rocks, but later restricted to iron ore of magmatic origin. *(*Wadsworth, 1893, p.91; Tomk. p.196)*

FERROPLETE. An obsolete term for an igneous rock rich in iron oxides. *(Brögger, 1898, p.266; Tomk. p.196)*

FIASCONITE. A local name for a variety of leucite basanite in which the plagioclase is rich in anorthite. *(Johannsen, 1938, p.307; Mt. Fiascone, Vulsini district, near Viterbo, Italy; Trög(38). 595¾; Tomk. p.196)*

FINANDRANITE. A K-rich variety of alkali feldspar syenite consisting essentially of microcline and the amphibole torendrikite, with some biotite and ilmenite. *(Lacroix, 1922, p.378; Ambatofinandrahana, Madagascar; Trög. 178; Joh. v.3, p.13; Tomk. p.197)*

FINLANDITE. A name suggested for a group of gabbros "sensu stricto". *(Marcet Riba, 1925, p.293; named after Finland)*

FIRMICITE. An obsolete term for rocks consisting of hornblende and feldspar. *(Pinkerton, 1811b, p.42; named after Firmicus, who first mentioned alchemy; Tomk. p.197)*

FITZROYITE. A local name for a variety of lamproite consisting essentially of leucite and phlogopite in a chlorite-rich groundmass. Erroneously spelt fitzroydite by Sørensen (1974). *(Wade & Prider, 1940, p.59; Fitzroy River basin, Kimberley district, West Australia; Tomk. p.198)*

FLOOD BASALT. A general term for basaltic lavas occurring in continental regions as vast accumulations of sub-horizontal flows that have been erupted, probably from fissures, in rapid succession on a regional scale. Many of the lavas are tholeiitic basalts, but transitional basalts are also common. The term is synonymous with plateau basalt. *(Tyrrell, 1937, p.94; Tomk. p.201)*

FLORIANITE. An obsolete term for a variety of granite containing red plagioclase, white orthoclase, quartz and pinite. However, the source reference cited by Tomkeieff does not contain the name although it does mention granites from the Florian Hills. *(Tomkeieff et al., 1983, p.201; Florian Hills (Swabian Alb), West Germany)*

FLORINITE. A local name for a melanocratic variety of monchiquite (a lamprophyre) with phenocrysts of olivine and augite and smaller augites and biotite in an altered matrix. *(Lacroix, 1933, p.197; Sainte-Florine, Brassac, Auvergne, France; Trög. 406; Tomk. p.201)*

FLUOLITE. An obsolete term, originally used by Lampadius, for a variety of pitchstone from Iceland. *(Glocker, 1831, p.720; Tomk. p.204)*

FOID DIORITE. A collective term for alkaline plutonic rocks consisting of feldspathoids (10-60% of the felsic minerals), acid plagioclase and large amounts of mafic minerals. Now defined modally in QAPF field 14 (Fig. B.4). *(Streckeisen, 1973, p.26)*

FOID DIORITOID. Used as a general term in the provisional "field" classification (Fig. B.9) for rocks thought to contain essential foids and in which plagioclase is thought to be more abundant than alkali feldspar. *(Streckeisen, 1973, p.28)*

FOID GABBRO. A collective term for basic alkaline plutonic rocks consisting of feldspathoids (10-60% of the felsic minerals), calcic plagioclase and large amounts of mafic minerals. Now defined modally in QAPF field 14 (Fig. B.4). *(Streckeisen, 1973, p.26)*

FOID GABBROID. Used as a general term in the provisional "field" classification (Fig. B.9) for rocks thought to contain essential foids and in which plagioclase is thought to be more abundant than alkali feldspar. *(Streckeisen, 1973, p.28)*

FOID MONZODIORITE. A collective term for alkaline plutonic rocks consisting of feldspathoids (10-60% of the felsic minerals), acid plagioclase with subordinate alkali feldspar and large amounts of mafic minerals. Now defined modally in QAPF field 13 (Fig. B.4). *(Streckeisen, 1973, p.26)*

FOID MONZOGABBRO. A collective term for basic alkaline plutonic rocks consisting of feldspathoids (10-60% of the felsic minerals), calcic plagioclase with subordinate alkali feldspar and large amounts of mafic minerals. Now defined modally in QAPF field 13 (Fig. B.4). *(Streckeisen, 1973, p.26)*

FOID MONZOSYENITE. A collective term for rare alkaline plutonic rocks consisting of feldspathoids, alkali feldspar, plagioclase and mafic minerals. Now defined modally in QAPF field 12 (Fig. B.4). The term is synonymous with foid plagisyenite. *(Streckeisen, 1973, p.26)*

FOID PLAGISYENITE. Now defined as a synonym for foid monzosyenite. *(Streckeisen, 1973, p.26)*

FOID SYENITE. A collective term for leucocratic alkaline plutonic rocks consisting of feldspathoids (10-60% of the felsic minerals), alkali feldspar and mafic minerals. Now defined modally in QAPF field 11 (Fig. B.4). *(Streckeisen, 1973, p.26)*

FOID SYENITOID. Used as a general term in the provisional "field" classification (Fig. B.9) for rocks thought to contain essential foids and in which alkali feldspar is thought to be more abundant than plagioclase. *(Streckeisen, 1973, p.28)*

FOID-BEARING ALKALI FELDSPAR SYENITE. A collective term for alkali feldspar syenites containing small amounts of feldspathoids (less than 10% of the felsic minerals). Now defined modally in QAPF field 6' (Fig. B.4). *(Streckeisen, 1973, p.26)*

FOID-BEARING ALKALI FELDSPAR TRACHYTE. A collective term for alkali feldspar trachytes containing small amounts of feldspathoids (less than 10% of the felsic minerals). Now defined modally in QAPF field 6' (Fig. B.10). *(Streckeisen, 1978, p.4)*

FOID-BEARING DIORITE. A collective term for diorites containing small amounts of feldspathoids (less than 10% of the felsic minerals). Now defined modally in QAPF field 10' (Fig. B.4). *(Streckeisen, 1973, p.26)*

FOID-BEARING GABBRO. A collective term for gabbros containing small amounts of feldspathoids (less than 10% of the felsic minerals). Now defined modally in QAPF field 10' (Fig. B.4). *(Streckeisen, 1973, p.26)*

FOID-BEARING LATITE. A collective term for latites containing small amounts of feldspathoids (less than 10% of the felsic minerals). Now defined modally in QAPF field 8' (Fig. B.10). *(Streckeisen, 1978, p.4)*

FOID-BEARING MONZODIORITE. A collective term for monzodiorites containing small amounts of feldspathoids (less than 10% of the felsic minerals). Now defined modally in QAPF field 9' (Fig. B.4). *(Streckeisen, 1973, p.26)*

FOID-BEARING MONZOGABBRO. A collective term for monzogabbros containing small amounts of feldspathoids (less than 10% of the felsic minerals). Now defined modally in QAPF field 9' (Fig. B.4). *(Streckeisen, 1973, p.26)*

FOID-BEARING MONZONITE. A collective term for monzonites containing small amounts of feldspathoids (less than 10% of the felsic minerals). Now defined modally in QAPF field 8' (Fig. B.4). *(Streckeisen, 1973, p.26)*

FOID-BEARING SYENITE. A collective term for syenites containing small amounts of feldspathoids (less than 10% of the felsic minerals). Now defined modally in QAPF field 7' (Fig. B.4). *(Streckeisen, 1973, p.26)*

FOID-BEARING TRACHYTE. A collective term for trachytes containing small amounts of feldspathoids (less than 10% of the felsic minerals). Now defined modally in QAPF field 7' (Fig. B.10). *(Streckeisen, 1978, p.4)*

FOIDITE. A general term for volcanic rocks defined in QAPF field 15 (Fig. B.10), i.e. rocks containing more than 60% foids in total light-coloured constituents. If modes are not available, chemically defined in TAS field F (Fig. B.13). If possible the most abundant foid should be used in the name e.g. nephelinite, leucitite, etc. *(Streckeisen, 1965, p.485;*

Tomk. p.205)

FOIDITOID. Used as a general term in the provisional "field" classification (Fig. B.17) for rocks tentatively identified as foidite. *(Streckeisen, 1978, p.3)*

FOIDOLITE. A general term for plutonic rocks defined in QAPF field 15 (Fig. B.4), i.e. rocks containing more than 60% foids in total light-coloured constituents. If possible the most abundant foid should be used in the name e.g. nephelinolite, leucitolite, etc. *(Streckeisen, 1973, p.26)*

FORELLENSTEIN. A local German name applied to a spotted troctolite. *(Rath, 1855, p.551; from the German forelle = trout; Trög. 849; Joh. v.3, p.225; Tomk. p.206)*

FORTUNITE. A local name for a variety of trachyte characterised by the presence of phlogopite and bronzite. *(Adan de Yarza, 1893, p.350; Fortuna, Murcia, Spain; Trög. 233; Joh. v.3, p.20; Tomk. p.207)*

FOURCHITE. A melanocratic analcime lamprophyre containing abundant augite but devoid of olivine and feldspar. *(Williams, 1891, p.107; Fourche Mts., Arkansas, USA; Trög. 376; Joh. v.4, p.391; Tomk. p.207)*

FOYAITE. A hypersolvus nepheline syenite sometimes used as a group name for nepheline syenites. Now used as a term for nepheline syenites having a foyaitic (=trachytic) texture caused by the platy alkali feldspar crystals. *(Blum, 1861, p.426; Mt. Foia, Monchique, Portugal; Trög. 414; Joh. v.4, p.100; Tomk. p.208)*

FRAIDRONITE (FRAIDONITE). A local French name for a variety of biotite lamprophyre which was later described as a variety of kersantite consisting of biotite, chlorite, plagioclase and minor quartz. *(Dumas, 1846, p.572; Tomk. p.208)*

GABBREID. An obsolete field term for a coarse-grained igneous rock which contains more than 50% pyroxene. Feldspar is subordinate in amount. *(Johannsen, 1911, p.320; Tomk. p.214)*

GABBRIDE. A revised spelling recommended to replace the field term gabbreid. Now obsolete. *(Johannsen, 1926, p.182; Joh. v.1, p.57; Tomk. p.214)*

GABBRITE. An obsolete name for an aplitic

gabbro with only minor ferromagnesian minerals. *(Polenov, 1899, p.464; Trög(38). 849½; Joh. v.3, p.308; Tomk. p.214)*

GABBRO. A coarse-grained plutonic rock composed essentially of calcic plagioclase, pyroxene and iron oxides. If olivine is an essential constituent it is olivine gabbro – if quartz, quartz gabbro. Now defined modally in QAPF field 10 (Fig. B.4). *(Targioni-Tozzetti, 1768, p.432; from an old Tuscan name; Trög. 348; Joh. v.3, p.205; Tomk. p.214)*

GABBRO-NELSONITE. A local name for a melanocratic variety of dolerite which occurs as dykes and consists of labradorite, pyroxene, ilmenite and apatite. *(Watson & Taber, 1910, p.206; Nelson County, Virginia, USA; Trög. 770; Tomk. p.215)*

GABBRODIORITE. A term originally proposed for a gabbro in which the pyroxene is altered to uralitic hornblende. Now used for rocks which are mineralogically on the boundary of gabbro and diorite i.e. their plagioclase is An_{50}. *(Törnebohm, 1877b, p.391; Trög. 331; Joh. v.3, p.238; Tomk. p.214)*

GABBROGRANITE. An obsolete term for rocks intermediate between gabbro and granite and synonymous with farsundite when used in the sense of a comprehensive name for granitic rocks of the Farsund area. *(Törnebohm, 1881, p.21; Trög. 850)*

GABBROID. A term originally used as a group name for rocks of the gabbro-norite groups, although used by Shand (1927) for rocks of theralitic composition. Now used as a general term in the provisional "field" classification (Fig. B.9) for rocks tentatively identified as gabbro or monzogabbro. *(Gümbel, 1888, p.87; Tomk. p.214)*

GABBRONORITE. A collective name for a plutonic rock consisting of calcic plagioclase and roughly equal amounts of clinopyroxene and orthopyroxene. Now defined modally in the gabbroic rock classification (Fig. B.6). *(Streckeisen, 1973, p.27)*

GABBROPHYRE. A lamprophyric dyke-rock of basaltic composition consisting of labradorite and rare augite phenocrysts in a felted groundmass of hornblende needles with occasional quartz. Cf. malchite. Odinite was suggested as an alternative term. *(Chelius, 1892, p.3; Trög. 319; Joh. v.3, p.326; Tomk. p.215)*

GABBROPHYRITE. An obsolete name for a porphyritic gabbro-aplite. *(Polenov, 1899, p.464; Tomk. p.215)*

GABBROPORPHYRITE. An obsolete name for a porphyritic rock containing phenocrysts of plagioclase and some pyroxene in a fine-grained crystalline groundmass of calcic plagioclase. *(Rosenbusch, 1898, p.205; Trög. 371; Joh. v.3, p.309; Tomk. p.215)*

GABBROSYENITE. An obsolete name proposed for a rock mineralogically between monzonite and an orthoclase-bearing gabbro. *(Tarasenko, 1895, p.15; Trög. 852; Joh. v.3, p.212)*

GALLINACE (GALLINAZO). A local Peruvian name for a variety of obsidian, but later restricted by Cordier (1816) to basic glasses. *(Faujas de Saint-Fond, 1778, p.172; Tomk. p.215)*

GAMAICU. A local South American name for a variolite. *(Tomkeieff et al., 1983, p.215)*

GAREWAITE. An obsolete local name for an ultramafic lamprophyre composed of phenocrysts of corroded diopside in a groundmass of olivine, pyroxene, chromite, magnetite and kaolinised labradorite. *(Duparc & Pearce, 1904, p.154; Springs of Garewaïa River, Tilai Range, Urals, USSR; Trög. 741; Joh. v.3, p.336; Tomk. p.216)*

GARGANITE. A local name for a variety of lamprophyre containing olivine, hornblende, augite, orthoclase and plagioclase found in the centre of a kersantite dyke. *(Viola & Stefano, 1893, p.135; Punta delle Pietre Nere, Foggia, Italy; Trög. 853; Joh. v.3, p.40; Tomk. p.216)*

GARRONITE. An obsolete name for a variety of melanocratic diorite consisting mainly of biotite, augite and andesine with a little hornblende showing a reaction relationship with augite. Minor orthoclase and quartz occur. *(Reynolds, 1937, p.476; Slievegarron, County Down, Ireland; Tomk. p.216)*

GAUSSBERGITE. A variety of lamproite containing phenocrysts of leucite, augite and olivine in a glass matrix. *(Lacroix, 1926, p.599; Gaussberg, Kaiser Wilhelm II Land,* Antarctica; Trög. 504; Joh. v.4, p.262; Tomk. p.217)

GAUTEITE. A porphyritic variety of syenite consisting of phenocrysts of hornblende, pyroxene, plagioclase and occasionally biotite, set in a groundmass composed essentially of alkali feldspar. Analcime may occur. *(Hibsch, 1898, p.84; Gaute, S. of Děčín, N. Bohemia, Czechoslovakia; Trög. 245; Joh. v.4, p.39; Tomk. p.218)*

GEBURITE-DACITE. An obsolete name for a variety of dacite containing hypersthene. *(Gregory, 1902, p.203; from the aboriginal gebur = Mt. Macedon, Victoria, Australia; Trög. 854; Tomk. p.218)*

GESTELLSTEIN. An old German term for greisen. *(Brückmanns, 1778, p.214; Joh. v.2, p.18; Tomk. p.221)*

GHIANDONE. A local Italian name for a porphyritic granite or augen gneiss. *(Tomkeieff et al., 1983, p.221)*

GHIZITE. An obsolete name for a volcanic rock consisting of phenocrysts of Ti-augite, biotite and olivine in a groundmass of colourless glass with analcime and microlites of augite and olivine. Cf. scanoite. *(Washington, 1914b, p.753; Ghizu, Mt. Ferru, Sardinia; Trög. 589; Joh. v.4, p.216; Tomk. p.221)*

GIBELITE. A local name for a variety of trachyte consisting essentially of Na-microcline with small amounts of augite, hornblende (cossyrite) and quartz. *(Washington, 1913, p.691; Mt. Gibelé, Pantelleria Island, Italy; Trög. 211; Joh. v.3, p.18; Tomk. p.222)*

GIUMARRITE. A local name for a variety of monchiquite (a lamprophyre) containing hornblende. *(Viola, 1901, p.309; Giumarra, Sicily, Italy; Trög. 377; Tomk. p.222)*

GLADKAITE. An obsolete name for a diorite aplite composed of plagioclase, quartz, hornblende, and biotite with minor epidote. With less quartz it passes into plagiaplite. *(Duparc & Pearce, 1905, p.1614; Gladkaïa-Sopka ridge, N. Urals, USSR; Trög. 140; Joh. v.2, p.405; Tomk. p.223)*

GLEESITES. A collective name for subvolcanic haüyne rocks. *(Kalb & Bendig, 1938; Glees, Laacher See, near Koblenz, West Germany; Tomk. p.225)*

GLENMUIRITE. A local name for a variety of analcime gabbro or monzogabbro consisting of labradorite and augite with analcime, olivine and alkali feldspar. The rock was originally described as a teschenite. *(Johannsen, 1938, p.194; Glenmuir Water, near Lugar, Scotland; Trög(38). 566½; Tomk. p.225)*

GLIMMERITE. An ultramafic rock consisting almost entirely of biotite. The term was suggested as an alternative to biotitite to avoid the rather discordant sound of consecutive "tites". *(Larsen & Pardee, 1929, p.104; from the German glimmer = mica; Trög. 719; Joh. v.4, p.441; Tomk. p.226)*

GOODERITE. A coarse-grained variety of alkali feldspar syenite consisting of about 80% albite with a little nepheline and biotite. *(Johannsen, 1938, p.57; Gooderham, Glamorgan Township, Ontario, Canada; Trög(38). 179½; Tomk. p.228)*

GORDUNITE. An obsolete name for a variety of wehrlite composed of olivine, diopside and pyrope garnet. *(Grubenmann, 1908, p.129; Gordunotal, Bellinzona, Switzerland; Trög. 737; Joh. v.4, p.421; Tomk. p.228)*

GORYACHITE. A local name for a leucocratic plutonic rock composed of up to 70% nepheline, with basic to intermediate plagioclase, augite zoned to Ti-augite, with small amounts of alkali feldspar and olivine. Sporadic aegirine, alkali amphibole, and fluorite may also be present. *(Luchitskii, 1960, p.114; Mt. Goryachaya, Minusinsk depression, Siberia, USSR)*

GRAMMITE. An obsolete term for graphic granite. *(Langius, 1708, p.42; Tomk. p.230)*

GRANATINE. An obsolete general term for rocks consisting of three minerals, usually quartz and feldspar plus a ferromagnesian. *(Kirwan, 1794, p.342; Tomk. p.230)*

GRANATITE. A group name for ultramafic rocks containing abundant garnet. *(Tröger, 1935, p.289; Trög. 716 – 718)*

GRANEID. An obsolete field term for a coarse-grained igneous rock consisting of quartz and any kind of feldspar with biotite and/or amphibole and/or pyroxene. *(Johannsen, 1911, p.319; Tomk. p.230)*

GRANIDE. A revised spelling recommended to replace the field term graneid. Now obsolete.

(Johannsen, 1926, p.182; Joh. v.1, p.56)

GRANILITE. An obsolete term for a granite containing more than three essential minerals. *(Kirwan, 1794, p.346; Tomk. p.231)*

GRANITE. A plutonic rock consisting essentially of quartz, alkali feldspar and plagioclase in variable amounts usually with hornblende and/or biotite. Now defined modally in QAPF field 3 (Fig. B.4). See also A-type granite, I-type granite, M-type granite, S-type granite, two-mica granite, etc. *(a term of great antiquity usually attributed to Caesalpino, 1596, p.89 – see Johannsen for further discussion; Trög. 856; Joh. v.2, p.124; Tomk. p.231)*

GRANITEL. An obsolete term attributed to Saussure for a crystalline rock consisting of two of the minerals quartz, feldspar or a ferromagnesian mineral. *(Pinkerton, 1811a, p.203)*

GRANITELL. An obsolete term for a quartz tourmaline or quartz hornblende rock, attributed to D'Aubenton. *(Kirwan, 1784, p.343)*

GRANITELLE. An erroneous spelling of granitell. *(Tomkeieff et al., 1983, p.231)*

GRANITELLO. An obsolete Italian name for a fine-grained granite. *(Brückmanns, 1778, p.213; Joh. v.2, p.302; Tomk. p.232)*

GRANITIN. An obsolete term for a fine-grained granite. *(Pinkerton, 1811a, p.201; Tomk. p.232)*

GRANITITE. A confusing term redefined several times for various varieties of biotite and hornblende granite. Should not be used. *(Leonhard, 1823a, p.54; Trög. 857; Joh. v.2, p.226; Tomk. p.232)*

GRANITOID. A term originally used for rocks resembling granite but of different composition, it is now commonly used as a synonym for a granitic rock i.e. any plutonic rock consisting essentially of quartz, alkali feldspar and/or plagioclase. Now used as a general term in the provisional "field" classification (Fig. B.9) for rocks tentatively identified as granite, granodiorite or tonalite. *(Pinkerton, 1811a, p.209; Tomk. p.233)*

GRANITON. An obsolete term for a coarse-grained granite. *(Pinkerton, 1811a, p.202; Tomk. p.233)*

GRANITONE. An obsolete term originally proposed for rapakivi from Finland and mentioned by Ferber in his letters from Italy. Also used in Italy as an old name for gabbro. *(Kirwan, 1784, p.149; Tomk. p.233)*

GRANITOPHYRE. An obsolete group name for porphyritic granitic rocks. *(Gümbel, 1888, p.111; Tomk. p.233)*

GRANODIORITE. A plutonic rock consisting essentially of quartz, plagioclase and lesser amounts of alkali feldspar with minor amounts of hornblende and biotite. Name first used by Becker on maps of the Gold Belt of the Sierra Nevada. Now defined modally in QAPF field 4 (Fig. B.4). *(Lindgren, 1893, p.202; Trög. 105; Joh. v.2, p.318; Tomk. p.233)*

GRANODOLERITE. An obsolete term for a plutonic rock containing labradorite-bytownite and orthoclase. *(Shand, 1917, p.466; Trög. 119)*

GRANOFELSOPHYRE. An obsolete term for a porphyry with a groundmass texture between felsitic and granitic. *(Vogelsang, 1875, p.160; Tomk. p.234)*

GRANOGABBRO. An obsolete term for an orthoclase-bearing variety of quartz gabbro analogous to granodiorite. *(Johannsen, 1917, p.89; Trög. 110; Joh. v.2, p.367; Tomk. p.234)*

GRANOLIPARITE. A variety of liparite containing transparent feldspar. *(Lapparent, 1900, p.638; Tomk. p.234)*

GRANOLITE. An obsolete general term first suggested by L.V. Pirsson for all granular igneous rocks. *(Turner, 1899, p.141; Trög. 858; Tomk. p.234)*

GRANOMASANITE. An obsolete local term for a porphyritic granite with plagioclase phenocrysts. Effusive equivalent of eutectofelsite. *(Kotô, 1909, p.186; Ku-ryong, Masanpho, S Korea; Trög. 89)*

GRANOPHYRE. A term originally used for a granite porphyry with a microcrystalline groundmass. Rosenbusch (1877) later redefined it as it is used today, as a porphyritic rock of granite composition in which the groundmass alkali feldspar and quartz are in micrographic intergrowth. *(Vogelsang, 1872, p.534; Trög. 859; Joh. v.2, p.288; Tomk. p.234)*

GRANOPHYRITE. An obsolete term for nonporphyritic rocks with a microcrystalline texture. *(Vogelsang, 1872, p.534)*

GRANULITE. Apart from metamorphic usage it has also been used for fine-grained muscovite or two mica leuco-granites. *(Michel-Levy, 1874, p.177; Trög. 860; Tomk. p.235)*

GRANULOPHYRE. A quartz porphyry in which the groundmass has a microgranitic texture. *(Lapparent, 1885, p.1759; Morvan, France; Tomk. p.236)*

GRAPHIC GRANITE. A name originally suggested by Haüy for a variety of granite in which the quartz and feldspar are intergrown in such a way that it has the appearance of cuniform or runic writing. *(Brongniart, 1813, p.32; Joh. v.2, p.72; Tomk. p.236)*

GRAPHIPHYRE (GRAPHOPHYRE). Obsolete terms originally suggested to replace the textural term granophyre, graphophyre having a megacrystalline and graphiphyre a microcrystalline groundmass. Tröger (1938) uses graphopyre as a rock name, although he suggests it is better used as an adjective. Tomkeieff et al. (1983) suggests both are rock names. *(Cross et al., 1906, p.704; Trög. 861; Tomk. p.237)*

GRAZINITE. A variety of phonolitic nephelinite containing analcime but no olivine. *(Almeida, 1961, p.135; Mt. Grazinas, Trindade Island, Atlantic Ocean)*

GREENHALGHITE. A variety of quartz latite consisting of phenocrysts of oligoclase-andesine and biotite in a matrix of quartz and K-feldspar, corresponding to the yosemititic magma-type of Niggli. *(Niggli, 1923, p.113; Mt. Greenhalgh, Silverton, South Carolina, USA; Trög. 97; Tomk. p.238)*

GREENLANDITE. See grönlandite.

GREISEN. An old Saxon mining term originally used for a variety of granitic rock containing tin ore and virtually no feldspar. Later used for rocks which appear to be pneumatolytically altered granite and consisting essentially of quartz and mica, with or without topaz, tourmaline, fluorite and Sn, W, Be, Li, Nb, Ta, etc. mineralisation. *(Leonhard, 1823a, p.59; from the German greisstein = ash-coloured stone; Trög. 6; Joh. v.2, p.5; Tomk. p.239)*

GRENNAITE. A variety of agpaitic nepheline syenite with phenocrysts of catapleiite and/or eudialyte in a fine-grained groundmass rich in aegirine and showing gneissose textures. The rock is most probably a deformed and partly recrystallised lujavrite. *(Adamson, 1944, p.123; Grenna, now Grönna, Norra Kärr complex, Sweden; Tomk. p.239)*

GRIMMAITE. A variety of charnockite porphyry to quartz mangerite porphyry of QAPF fields 3b to 8* (Fig. B.4), respectively. It is suggested (Streckeisen, 1974, p.358) that this term should be abandoned. *(Ebert, 1968, p.1044; Grimma, Saxony, East Germany)*

GRIQUAITE. A variety of garnet clinopyroxenite occurring as inclusions in kimberlites, composed essentially of pyrope garnet, with diopside, and minor phlogopite, olivine, spinel, and ore. *(Beck, 1907, p.301; named after Griqualand, South Africa; Trög. 716; Tomk. p.240)*

GRÖBAITE. An obsolete name for a leucocratic variety of monzodiorite consisting of major amounts of sodic plagioclase, with lesser amounts of orthoclase, augite, hornblende and biotite. *(Grahmann, 1927, p.33; Gröba, Saxony, East Germany; Trög. 277; Tomk. p.240)*

GRÖNLANDITE (GREENLANDITE). An obsolete name for a variety of orthopyroxene hornblendite consisting essentially of hornblende and hypersthene. *(Machatschki, 1927, p.175; named after Greenland; from Island of Upernivik; Trög. 705; Joh. v.4, p.447; Tomk. p.238)*

GRORUDITE. A local name for a variety of peralkaline microgranite containing aegirine. *(Brögger, 1890, p.66; Grorud, Oslo district, Norway; Trög. 62; Joh. v.2, p.102; Tomk. p.240)*

GUARDIAITE. A variety of tephritic phonolite or nepheline-bearing latite containing plagioclase phenocrysts mantled with alkali feldspar in a groundmass of essential nepheline, augite, biotite and glass. *(Narici, 1932, p.234; Cape Guardia, Ponza Island, Italy; Trög. 531; Tomk. p.242)*

HAKUTOITE. A local name, attributed to F. Yamanari, for a variety of peralkaline trachyte that is characterised by sodic pyroxenes and amphiboles and small amounts of quartz. *(Lacroix, 1927a, p.1410; Hakuto San, Korea; Trög. 77; Tomk. p.244)*

HAMRONGITE. A local name for a variety of kersantite (a lamprophyre) containing phenocrysts of biotite in a groundmass of biotite, andesine and a little quartz. *(Eckermann, 1928b, p.13; Hamrångefjarden, Gävle, Sweden; Trög. 146; Joh. v.2, p.405; Tomk. p.245)*

HAPLITE. A name suggested as the etymologically correct spelling of aplite. *(Johannsen, 1932, p.91; Tomk. p.246)*

HAPLO-. A prefix suggested to denote a pure artifical mixture of various feldspars and diopside used in experimental work e.g. haplodiorite (diopside + sodic plagioclase), haplogabbro (diopside + calcic plagioclase), haplosyenite (diopside + albite). Cf. aplo-. *(Bowen, 1915, p.161; from the Greek haploos = simple)*

HAPLO-PITCHSTONE. A term suggested for an experimental composition approximating a natural pitchstone. *(Reynolds, 1958, p.388; Tomk. p.246)*

HAPLODIORITE, HAPLOGABBRO. See haplo-.

HAPLOPHYRE. An obsolete term for a granitic rock with a texture intermediate between porphyritic and equigranular. *(Stache & John, 1877, p.189; Tomk. p.246)*

HAPLOSYENITE. See haplo-.

HARRISITE. A variety of troctolite in which the large black lustrous olivines have a branching habit of growth and are orientated perpendicular to the layering. The matrix to the olivines is composed of highly calcic plagioclase and minor pyroxene. *(Harker, 1908, p.71; Glen Harris, Island of Rhum, Scotland; Trög. 401; Joh. v.3, p.349; Tomk. p.246)*

HARTUNGITE. A local name for a variety of melteigite consisting of nepheline, pyroxene and wollastonite with or without alkali feldspar. *(Eckermann, 1942, p.402; Hartung, Alnö Island, Westnorrland, Sweden; Tomk. p.247)*

HARZBURGITE. An ultramafic plutonic rock composed essentially of olivine and orthopyroxene. Now defined modally in the ultramafic rock classification (Fig. B.8). *(Rosenbusch, 1887, p.269; Harzburg, Harz Mts., Lower Saxony, West Germany; Trög. 732; Joh. v.4, p.438; Tomk. p.247)*

HATHERLITE. A local term for a variety of alkali feldspar syenite containing abundant anorthoclase, with biotite and hornblende. The rock was later called leeuwfonteinite. *(Henderson, 1898, p.46; Hatherly gun powder factory, Magaliesberg Range, South Africa; Trög. 182; Joh. v.3, p.10; Tomk. p.247)*

HAÜYNE BASALT. A general term originally used for a rock in which the place of feldspar is taken by haüyne. Later redefined as a rock of basaltic appearance consisting of haüyne, olivine and clinopyroxene, often with minor leucite and nepheline. The name should not be used as the term basalt is now restricted to a rock containing essential plagioclase. As the rock is a variety of foidite it should be given the appropriate name, e.g. olivine haüynite. *(Trimmer, 1841, p.172; Trög. 650; Joh. v.4, p.346; Tomk. p.247)*

HAÜYNE BASANITE. Now defined in QAPF field 14 (Fig. B.10) as a variety of basanite in which haüyne is the most abundant foid. *(Tröger, 1935, p.246; Trög. 598; Joh. v.4, p.238)*

HAÜYNE PHONOLITE. Now defined in QAPF field 11 (Fig. B.10) as a variety of phonolite in which haüyne is an important foid. *(Bořický, 1873, p.16; Trög. 473; Joh. v.4, p.131)*

HAÜYNFELS. An obsolete name based on the incorrect identification of haüyne in a rock that was later called ditroite. *(Haidinger, 1861, p.64; Ditro (now Ditrau), Transylvania, Romania; Joh. v.4, p.98; Tomk. p.247)*

HAÜYNITE. A term originally used for a basaltic rock with haüyne as the only foid. Now defined as a variety of foidite of field 15c of the QAPF diagram (Fig. B.10). *(Reinisch, 1917, p.68; Morgenberg, Blatt Wiesental, Saxony, East Germany; Trög. 650; Tomk. p.247)*

HAÜYNITITE. An obsolete term proposed for a volcanic rock consisting of haüyne and mafic minerals without olivine. *(Johannsen, 1938, p.336; Trög. 650; Tomk. p.248)*

HAÜYNOLITH. An obsolete term proposed for a monomineralic volcanic rock consisting of haüyne. *(Johannsen, 1938, p.337; Trög.633; Tomk. p.248)*

HAÜYNOPHYRE. A volcanic rock name equivalent to leucitophyre or nephelinite for rocks that are rich in haüyne and nosean. The matrix may contain glass, pyroxene, haüyne and other foids. *(Abich, 1839, p.337; Trög. 537; Joh. v.4, p.362; Tomk. p.248)*

HAWAIITE. A term originally defined as a variety of olivine-bearing basalt in which the normative plagioclase is oligoclase or andesine. Now defined chemically as the sodic variety of trachybasalt in TAS field S1 (Fig. B.13). *(Iddings, 1913, p.198; named after Hawaii; Trög. 327; Joh. v.3, p.171; Tomk. p.248)*

HEDRUMITE. A local name for a porphyritic fine-grained variety of alkali feldspar syenite with a trachytic or foyaitic texture consisting essentially of microcline-microperthite. The groundmass is rich in biotite and contains minor sodic amphibole, aegirine and nepheline. Cf. pulaskite. *(Brögger, 1898, p.183; Hedrum, Oslo district, Norway; Trög. 190; Joh. v.4, p.25; Tomk. p.248)*

HELSINKITE. A local name for an equigranular dyke-rock consisting essentially of albite and epidote, which was originally thought to be primary. *(Laitakari, 1918, p.3; named after Helsinki; Trög. 199; Joh. v.3, p.142; Tomk. p.249)*

HEPTORITE. A variety of lamprophyre consisting of phenocrysts of Ti-augite, barkevikite, olivine, and haüyne in a groundmass of glass and labradorite laths. *(Busz, 1904, p.86; from the Greek hepta = seven; Siebengebirge, near Bonn, West Germany; Trög. 561; Joh. v.4, p.244; Tomk. p.249)*

HERONITE. A variety of analcime monzosyenite which contains spheroidal aggregates of orthoclase in a matrix of analcime, labradorite and aegirine. *(Coleman, 1899, p.436; Heron Bay, Lake Superior, Ontario, Canada; Trög.516; Joh. v.4, p.287; Tomk. p.250)*

HEUMITE. A local name for a fine-grained variety of nepheline-bearing syenite consisting essentially of alkali feldspar, barkevikite and lepidomelane with minor amounts of plagioclase and nepheline. *(Brögger, 1898, p.90; Heum, Larvik, Norway; Trög.497; Joh. v.4, p.170; Tomk. p.251)*

HIGH-ALUMINA BASALT. A collective term for basalts from the tholeiitic, calc-alkaline and alkaline associations in which Al_2O_3 is generally greater than 16%. They usually, but not invariably, have phenocrysts of plagioclase. Tholeiitic basalts are commonly and calc-alkaline basalts are almost exclusively of this type. *(Kuno, 1960, p.122)*

HIGH-K. A chemical term originally applied to andesites with $K_2O > 2.5\%$ but later applied to volcanic rocks which plot above a line on the $SiO_2 - K_2O$ diagram i.e. for their SiO_2 values they are higher than usual in K_2O. The field is now defined as part of the TAS classification (Fig. B.15). *(Taylor, 1969, p.45)*

HIGH-MG ANDESITE. A term proposed as an alternative to boninite to avoid the equivalence between some ophiolite basalts and high-Mg andesites implied by some authors for the term boninite. *(Jenner, 1981, p.307)*

HIGHWOODITE. A variety of nepheline-bearing monzonite consisting essentially of anorthoclase, labradorite, augite and biotite. The light minerals only just exceed the dark minerals in abundance. *(Johannsen, 1938, p.40; Highwood Mts., Montana, USA; Trög(38). 237½; Tomk. p.252)*

HILAIRITE. A local name for a coarse-grained variety of sodalite nepheline syenite consisting of nepheline, alkali feldspar, sodalite and aegirine with a little eudialyte. *(Johannsen, 1938, p.289; Mt. St. Hilaire, Quebec, Canada; Trög(38). 415½; Tomk. p.252)*

HIRNANTITE. An obsolete name for an intrusive rock consisting essentially of albite with minor chlorite, iron ore and secondary quartz and calcite. Possibly an albite keratophyre or an albitised tholeiite. *(Travis, 1915, p.79; Craig-ddu, Hirnant, Berwyn Hills, Wales; Trög. 219; Joh. v.3, p.143; Tomk. p.253)*

HOBIQUANDORTHITE. An unwieldy name constructed by Johannsen to illustrate some of the possibilties of the mnemonic classification of Beliankin (1929) for a granitic rock consisting of *ho*rnblende, *bi*otite, *qu*artz, *an*desine and *orth*oclase. Cf. biquahororthandite and topatourbiolilepiquorthite. *(Johannsen, 1931, p.125)*

HOLLAITE. A local name for a coarse-grained variety of calcite melteigite consisting of pyroxene (55%), calcite and nepheline. *(Brögger, 1921, p.217; Holla Church, Fen complex, Telemark, Norway; Trög. 699;*

Tomk. p.253)

HOLMITE. An obsolete term suggested by Johannsen for a lamprophyre in which melilite was thought to occur. Apparently unknown to Johannsen, it had already been shown (Flett, 1935, p.185) that the mineral originally identified as melilite was, in fact, apatite. The rock is actually a monchiquite. *(Johannsen, 1938, p.378; Holm Island, Orkney Islands, Scotland; Trög(38). 746½; Tomk. p.253)*

HOLYOKEITE. A local name for an albitised dolerite with an ophitic texture. The principal minerals are albite, some zoisite, calcite, chlorite and opaques. *(Emerson, 1902, p.508; Holyoke, Massachusetts, USA; Trög. 197; Joh. v.3, p.139; Tomk. p.254)*

HONGITE. A name given to a hypothetical set of data used to illustrate the problems of the statistical interpretation of rock compositions. *(Aitchison, 1984, p.534)*

HOOIBERGITE. A local name for a melanocratic variety of what, according to the Subcommission classification, is a granite just inside QAPF field 3b (Fig. B.4) consisting essentially of green hornblende with minor amounts of labradorite, orthoclase and quartz. Traditionally the rock has been called a variety of gabbro. *(Westermann, 1932, p.46; Hooiberg, Aruba Island, Caribbean Sea; Trög. 281; Tomk. p.256)*

HÖRMANNSITE. An obsolete name for a variety of monzodiorite containing acid plagioclase and lesser amounts of orthoclase with biotite and calcite. Thought to have formed by the assimilation of marble. *(Ostadal, 1935, p.122; Hörmanns, NW Waldvirtel, Austria; Trög(38). 863½; Tomk. p.256)*

HORNBERG. An old German name for greisen. *(Brückmanns, 1778, p.214; Joh. v.2, p.18; Tomk. p.256)*

HORNBLENDE GABBRO. A variety of gabbro in which primary hornblende occurs. Now defined modally in the gabbroic rock classification (Fig. B.6). *(Streng & Kloos, 1877, p.113; Trög. 349; Joh. v.3, p.227; Tomk. p.256)*

HORNBLENDE PERIDOTITE. An ultramafic plutonic rock consisting essentially of olivine with up to 50% amphibole. Now defined modally in the ultramafic rock classification (Fig. B.8). *(Wyllie, 1967, p.2)*

HORNBLENDE PYROXENITE. A collective term for pyroxenites containing up to 50% of amphibole. Now defined modally in the ultramafic rock classification (Fig. B.8). *(Wyllie, 1967, p.2)*

HORNBLENDITE. An ultramafic plutonic rock composed almost entirely of hornblende. Now defined modally in the ultramafic rock classification (Fig. B.8). *(Phillips, 1848, p.40; Trög. 701; Joh. v.4, p.442; Tomk. p.257)*

HORTITE. An obsolete term for a melanocratic variety of monzosyenite consisting of alkali feldspar, plagioclase and minor calcite with abundant pyroxene and hornblende. Probably formed from gabbro by the assimilation of limestone. *(Vogt, 1915, p.5; Hortavaer, NW Leka, Trondheim, Norway; Trög. 261; Joh. v.3, p.66; Tomk. p.258)*

HORTONOLITITE. A variety of dunite composed essentially of hortonolite. *(Polkanov & Eliseev, 1941, p.28; Tomk. p.258)*

HOVLANDITE. A local name for a variety of monzogabbro composed of large crystals of bytownite, biotite, hypersthene and olivine. The rock is a hybrid type. *(Barth, 1944, p.68; Hovland Farm, Modum, Oslo district, Norway; Tomk. p.258)*

HRAFTINNA. An Icelandic name for obsidian. *(Johannsen, 1932, p.279; Tomk. p.258)*

HUDSONITE. A term given to a variety of pyroxene olivine hornblendite which was later replaced by the name cortlandtite as hudsonite was already in use as a mineral name. *(Cohen, 1885, p.242; Hudson River, New York, USA; Trög. 866; Joh. v.4, p.425; Tomk. p.259)*

HUNGARITE. A local term for a variety of andesite containing hornblende. *(Lang, 1877, p.196; named after Hungary; Trög. 867; Joh. v.3, p.169; Tomk. p.260)*

HURUMITE. A local name for a medium-grained variety of monzonite containing andesine and orthoclase, with biotite, clinopyroxene and quartz. *(Brögger, 1931, p.72; Hurum, Oslo district, Norway; Trög. 92; Tomk. p.260)*

HUSEBYITE. A local name for a medium-grained variety of nepheline syenite consisting of alkali feldspar, nepheline, aegirine-augite and alkali amphibole. The term is synonymous with essexite-foyaite. *(Brögger, 1933, p.35; Husebyås, Åker, Oslo district, Norway; Trög. 508; Joh. v.4, p.165; Tomk. p.261)*

HYALOCLASTITE. A consolidated pyroclastic rock composed of angular fragments of glass which may or may not be devitrifed. *(Rittmann, 1962, p.72)*

HYALOMELANE. An obsolete term for a basaltic glass insoluble in acids. *(Hausmann, 1847, p.545; from the Greek hyalos = glass and melas = dark; Trög. 869; Joh. v.3, p.290; Tomk. p.261)*

HYALOMICTE. An obsolete term for a greisen consisting of hyaline quartz and mica. *(Brongniart, 1813, p.34; Tomk. p.261)*

HYALOPSITE. An obsolete name for obsidian. *(Gümbel, 1888, p.41; Joh. v.2, p.279; Tomk. p.261)*

HYDROTACHYLYTE. A term originally used for a glass formed from a partially melted sandstone, but later used for tachylytes with high water contents (up to 13%). *(Petersen, 1869, p.32; from the Greek hydor = water, tachys = rapid, lytos = soluble; Trög. 870; Joh. v.3, p.290; Tomk. p.265)*

HYPABYSSAL. Pertaining to rocks whose type of emplacement is intermediate between plutonic and volcanic. Often applied to rocks from minor intrusions e.g. sills and dykes. *(Brögger, 1894, p.97; from the Greek hypo= under or from and abyssos = bottomless or unfathomable; Joh. v.1, p.176; Tomk. p.266)*

HYPERACIDITE. A term proposed for highly acid rocks. *(Loewinson-Lessing, 1896, p.175; Tomk. p.266)*

HYPERITE. An old name originally used for a variety of norite consisting of hypersthene, plagioclase and augite, but redefined with various other meanings e.g. diabase, gabbronorite and for noritic rocks with hypersthene coronas around olivine. *(Naumann, 1850, p.594; from an old Swedish name; Trög. 354; Joh. v.3, p.238; Tomk. p.266)*

HYPERITITE. An obsolete term for a bronzite-diabase. *(Törnebohm, 1877a, p.42; Tomk. p.266)*

HYPERSTHENITE. A variety of orthopyroxenite composed almost entirely of hypersthene. *(Naumann, 1850, p.595; Trög. 676; Joh. v.4,* *p.458; Tomk. p.267)*

HYPOBASITE. An obsolete term for an ultrabasic rock. *(Loewinson-Lessing, 1898, p.42; Tomk. p.267)*

HYSTEROBASE. A variety of diabase in which the augite is replaced by brown hornblende. *(Lossen, 1886, p.925; Trög. 321; Tomk. p.268)*

I-TYPE GRANITE. A general term for a range of metaluminous calc-alkali granitic rocks, mainly tonalites to granodiorites and granites, characterised by essential quartz, variable amounts of plagioclase and alkali feldspar, hornblende and biotite. Muscovite is absent. The prefix I implies that the source rocks have igneous compositions. *(Chappell & White, 1974, p.173)*

ICELANDITE. A variety of intermediate volcanic rock containing phenocrysts of andesine, clinopyroxene and/or orthopyroxene and/ or pigeonite and less commonly olivine in a groundmass of the same minerals. It differs chemically from a typical orogenic andesite in being poorer in alumina and richer in iron. *(Carmichael, 1964, p.442; named after Iceland)*

IGNEOUS ROCK. A rock that has solidified from a molten state either within or on the surface of the earth. *(Kirwan, 1794, p.455; from the Latin ignis = fire; Joh. v.1, p.177; Tomk. p.271)*

IGNIMBRITE. An indurated tuff consisting of crystal and rock fragments in a matrix of glass shards which are usually welded together. In some cases the welding is so extreme that the original texture shown by the glass shards is lost. The composition is usually acid to intermediate. *(Marshall, 1932, p.200; from the Latin ignis = fire and imber = shower; Trög(38). 871½; Tomk. p.271)*

IJOLITE. A plutonic rock consisting of pyroxene with 30% to 70% nepheline. Now defined modally as a variety of foidolite in QAPF field 15 (Figs. B.4 and B.7). *(Ramsay & Berghell, 1891, p.304; Iijoki, now Iivaara, Kuusamo, Finland; Trög. 607; Joh. v.4, p.313; Tomk. p.271)*

IJUSSITE. A local name for a variety of teschenite without olivine and with kaersutite and Ti-augite as the mafic phases. *(Rachkovsky, 1911, p.257; Ijuss River, Upper Yenisey River,* *Siberia, USSR; Trög. 567; Tomk. p.272)*

ILMEN-GRANITE. A biotite nepheline syenite later called miaskite. *(Tomkeieff et al., 1983, p.272; Ilmen Mts., Urals, USSR)*

ILMENITITE. A rock consisting almost entirely of ilmenite. *(Kolderup, 1896, p.14; Trög. 768; Joh. v.4, p.470; Tomk. p.272)*

ILVAITE. Alleged to be a variety of granite from the Island of Ilva, now Elba, Italy but the reference cited (Fournet, 1845) does not contain the name. However, the term should not be used rock name as it has been used since 1811 for a silicate of Fe and Ca. *(Tomkeieff et al., 1983, p.272)*

ILZITE. An obsolete local name for an aplite consisting of sodic plagioclase, quartz and biotite with minor alkali feldspar. *(Frentzel, 1911, p.176; Ilzgebirge, Passau, Bavaria, West Germany; Trög(38). 278½; Tomk. p.272)*

IMANDRITE. A local name for a variety of granite consisting of graphically intergrown quartz and albite with chloritised biotite. A hybrid rock formed by the interaction of nepheline syenite magma with greywacke. *(Ramsay & Hackman, 1894, p.46; Imandra, Kola Peninsula, USSR; Trög. 57; Tomk. p.272)*

INNINMORITE. A local name for an andesitic to dacitic rock composed of phenocrysts of calcic plagioclase and pigeonite in a fine-grained to glassy groundmass. *(Thomas & Bailey, 1915, p.209; Inninmore Bay, Morvern, Scotland; Trög. 126; Tomk. p.276)*

INTERMEDIATE. A commonly used chemical term now defined in the TAS classification (Fig. B.13) as a rock containing more than 52% and less than 63% SiO_2 i.e. lying between acid and basic rocks. *(Judd, 1886a, p.51; Joh. v.1, p.177; Tomk. p.277)*

INTRITE. An obsolete term for porphyritic rocks. *(Pinkerton, 1811a, p.75; Tomk. p.279)*

INTRODACITE. A term for a normal dacite that was later replaced by phanerodacite. *(Belyankin, 1923, p.99; Tomk. p.279)*

IOPHYRE. An obsolete term for blue to chocolate-brown porphyritic rocks containing amphiboles. *(Rozière, 1826, p.310; Tomk. p.281)*

ISENITE. A local name proposed for a variety of trachyandesite thought to contain hornblende, biotite and nosean. As the "nosean" was later shown to be apatite (Dannenberg, 1897), the name is not justified. *(Bertels, 1874, p.175; River Eis (Latin name Isena), Nassau, West Germany; Trög. 872; Tomk. p.281)*

ISOPHYRE. An obsolete name for obsidian. *(Tomkeieff et al., 1983, p.283)*

ISSITE. A local name for a fine-grained variety of hornblendite consisting essentially of hornblende, with cores of augite, and a little plagioclase. *(Duparc & Pamphil, 1910, p.1136; River Iss, Urals, USSR; Trög. 712; Joh. v.3, p.336; Tomk. p.283)*

ITALITE. A rock consisting almost entirely of leucite held together by small amounts of glass. Now defined modally as a leucocratic variety of leucititolite in QAPF field 15 (Figs. B.4 and B.7). *(Washington, 1920, p.33; Alban Hills, near Rome, Italy; Trög. 627; Joh. v.4, p.311; Tomk. p.284)*

ITSINDRITE. A local name for a fine-grained variety of nepheline syenite consisting essentially of microcline and nepheline in granophyric intergrowth, with aegirine, biotite and melanite. *(Lacroix, 1922, p.388; Itsindra Valley, Madagascar; Trög. 417; Joh. v.4, p.145; Tomk. p.284)*

IVERNITE. An obsolete name for a granite-like rock consisting of phenocrysts of orthoclase and plagioclase in a groundmass of euhedral feldspars with minor hornblende, mica and quartz. *(McHenry & Watt, 1895, p.93; Iverness, County Limerick, Ireland; Trög. 873; Tomk. p.284)*

IVOIRITE. A variety of clinopyroxene norite of QAPF field 10 (Fig. B.4) belonging to the charnockitic rock series. It is suggested (Streckeisen, 1974, p.358) that this term should be abandoned. *(Lacroix, 1910; Mt. Marny, Ivory Coast)*

IVREITE. A name suggested for a group of pyroxene quartz diorites. *(Marcet Riba, 1925, p.293; Ivrea, Piedmont, Italy; Tomk. p.284)*

JACUPIRANGITE. A variety of alkali pyroxenite consisting essentially of Ti-augite with minor amounts of Ti-magnetite, nepheline, apatite, perovskite, and melanite garnet. *(Derby, 1891, p.314; Jacupiranga, São Paulo, Brazil; Trög. 687; Joh. v.4, p.463; Tomk. p.285)*

JADEOLITE. A term for a green chrome-bearing variety of syenite which has the appearance of jade. *(Kunz, 1908, p.810; Tomk. p.285)*

JERSEYITE. A local name for a quartz-bearing variety of minette. *(Lacroix, 1933, p.187; Jersey, Channel Islands; Trög. 68; Tomk. p.286)*

JOSEFITE. An obsolete name for a poorly defined ultramafic rock composed of Ti-augite and olivine with secondary calcite and chlorite. *(Szádeczky, 1899, p.215; "Big Quarry", 2km. SE of Aswân, Egypt; Trög. 697; Joh. v.4, p.438; Tomk. p.286)*

JOTUN-NORITE. A member of the charnockitic rock series equivalent to hypersthene monzonorite in QAPF field 9 (Table B.4). The term is synonymous with Jotunite. *(Goldschmidt, 1916, p.34; Jotunheim, Norway)*

JOTUNITE. A member of the charnockitic rock series equivalent to hypersthene monzonorite in QAPF field 9 (Table B.4). The term is synonymous with Jotun-norite. *(Hødal, 1945, p.140; Jotunheim, Norway)*

JUMILLITE. A variety of olivine phonolitic leucitite composed essentially of leucite and diopside, and with subordinate olivine, alkali feldspar and phlogopite. *(Osann, 1906, p.306; Jumilla, Murcia, Spain; Trög. 505; Joh. v.4, p.263; Tomk. p.286)*

JUVITE. A local name for a coarse-grained variety of nepheline syenite in which K-feldspar is more abundant than Na-feldspar; sometimes used as a general term for potassic nepheline syenites. *(Brögger, 1921, p.93; Juvet, Fen complex, Telemark, Norway; Trög. 413; Joh. v.4, p.104; Tomk. p.286)*

KAGIARITE. See khagiarite.

KAHUSITE. A rock described as a silica-rich magnetite rhyolite consisting essentially of quartz, lesser amounts of magnetite and minor tourmaline, biotite and graphite. Possibly a re-melted iron-rich quartzite. *(Sorotchinsky, 1934, p.192; Kahusi volcano, SW of Lake Kivu, Zaire; Trög(38). 873½; Tomk. p.287)*

KAIWEKITE. A local name for a variety of trachyte that usually carries phenocrysts of anorthoclase and Ti-augite. *(Marshall, 1906, p.400; Kaiweke (=Long Beach), Kaikorai Valley, Otago, New Zealand; Trög. 216; Joh. v.4, p.38; Tomk. p.287)*

KAJANITE. An obsolete term for a melanocratic variety of leucitite largely composed of biotite and clinopyroxene and with subordinate leucite. *(Lacroix, 1926, p.600; Oele Kajan, E Borneo; Trög. 648; Joh. v.4, p.371; Tomk. p.287)*

KAKORTOKITE. A local name for a variety of agpaitic nepheline syenite displaying pronounced cumulate textures and igneous layering with a repetition of layers enriched in alkali feldspar, eudialyte and arfvedsonite. *(Ussing, 1912, p.43; Kakortok (now Qaqortoq), Ilimaussaq, Greenland; Trög. 878; Joh. v.4, p.118; Tomk. p.287)*

KALIOPLETE. An obsolete term for an igneous rock rich in potassium. *(Brögger, 1898, p.266; Tomk. p.288)*

KALMAFITE. A mnenonic name for a mixture of kalsilite plus mafic minerals. *(Hatch et al., 1961, p.358)*

KALSILITOLITE. A medium-grained holocrystalline rock mainly composed of kalsilite and clinopyroxene with small amounts of leucite, melanite, biotite and melilite occurring as ejected blocks in pyroclastic deposits. *(Federico, 1976, p.5; Colle Cimino, Alban Hills, Italy)*

KAMAFUGITE. A collective name proposed for the group of rocks KAtungite, MAFurite and UGandite. *(Sahama, 1974, p.96; Toro-Ankole, Uganda)*

KAMCHATITE. A local name for a porphyritic igneous rock consisting of orthoclase, bluish hornblende, pyroxene and epidote. Although similar in composition to monzonite, it contains no plagioclase. *(Morozov, 1938, p.19; Sredinnyi Range, Kamchatka, USSR; Tomk. p.288)*

KAMMGRANITE. An old term reused by Niggli for a dark coloured K-rich amphibole biotite granite. *(Groth, 1877, p.396; from Kamm = divide, granite of Vosges divide, France; Trög. 54; Tomk. p.289)*

KAMMSTEIN. An old German (Saxon) term for serpentinite. *(Tomkeieff et al., 1983, p.289)*

KAMPERITE. A local name for a fine- to medium-grained highly potassic dyke rock composed of almost equal amounts of orthoclase and biotite with minor oligoclase. The biotite is a late crystallisation product.

(Brögger, 1921, p.104; Kamperhoug, Fen complex, Telemark, Norway; Trög. 248; Joh. v.3, p.87; Tomk. p.289)

KANZIBITE. An local name for a K-rich variety of rhyolite in which the orthoclase phenocrysts are coloured black by graphite inclusions. *(Sorotchinsky, 1934, p.190; Kanzibi Lake, Kivu, Zaire; Trög(38). 878½; Tomk. p.289)*

KARITE. A term tentatively suggested for a quartz-rich variety of grorudite or peralkaline microgranite. *(Karpinskii, 1903, p.31; Kara River, Transbaikal Region, USSR; Trög. 63; Tomk. p.289)*

KARJALITE. A local name for a rock consisting essentially of albite with variable amounts of quartz, carbonate, amphibole, chlorite and iron ore. *(Väyrynen, 1938, p.74; from the Finnish word for Karelia; Tomk. p.289)*

KARLSTEINITE. A local name for a variety of peralkaline granite containing abundant microcline and some alkali amphiboles. *(Hackl & Waldmann, 1935, p.263; Karlstein, near Raabs, Lower Austria, Austria; Trög(38). 56½; Tomk. p.290)*

KÄRNÄITE. A glassy rock with the composition of dacite, containing phenocrysts of feldspar and abundant inclusions of tuff-like material. It has been identified as an impactite. *(Saksela, 1948, p.20; Kärnä Island, Lappajärvi, Finland; Tomk. p.290)*

KASANSKITE. See kazanskite. *(Osann, 1922, p.253)*

KÅSENITE (KOSENITE). A local name for a variety of calcite carbonatite consisting of sodic pyroxene, a little nepheline and 50% to 60% calcite. *(Brögger, 1921, p.222; Kåsene, Fen complex, Telemark, Norway; Trög. 756; Tomk. p.290)*

KASSAITE. A porphyritic dyke rock with phenocrysts of haüyne, labradorite with oligoclase rims, barkevikite and augite in a groundmass of hastingsite and andesine rimmed by oligoclase and orthoclase. *(Lacroix, 1918, p.542; Kassa Island, Los Archipelago, Guinea; Trög. 519; Joh. v.4, p.190; Tomk. p.290)*

KATABUGITE. A variety of jotunite or norite of QAPF fields 9-10 (Fig. B.4). It is a member of the bugite series with 50% to 58% SiO_2, 15% to 40% hypersthene, oligoclase-andesine antiperthite and quartz. It is suggested (Streckeisen, 1974, p.358) that this term should be abandoned. *(Bezborod'ko, 1931, p.145; Bug River, Podolia, Ukraine, USSR; Trög(38). 308½; Tomk. p.290)*

KATNOSITE. A local name for a variety of nordmarkite (a quartz-bearing syenite) containing biotite or aegirine. *(Brögger, 1933, p.86; Lake Katnosa, Nordmarka, Oslo district, Norway; Tomk. p.291)*

KATUNGITE. A potassic melanocratic variety of olivine melilitite composed essentially of olivine and melilite and with subordinate leucite, kalsilite and nepheline in a glassy matrix. *(Holmes, 1937, p.201; Katunga, Uganda; Trög(38). 672½; Joh. v.4, p.362; Tomk. p.291)*

KATZENBUCKELITE. A porphyritic variety of haüyne nepheline microsyenite containing phenocrysts of nepheline, nosean or analcime, biotite and olivine in a matrix of nepheline, analcime (leucite), alkali feldspar and sodic pyroxene and amphibole. *(Osann, 1902, p.403; Katzenbuckel, Odenwald, West Germany; Trög. 446; Joh. v.4, p.273; Tomk. p.292)*

KAUAIITE. A variety of monzodiorite with Ti-augite, a little olivine and feldspar zoned from labradorite to oligoclase and mantled by lime anorthoclase. *(Iddings, 1913, p.173; Kauai, Hawaiian Islands; Trög. 284; Joh. v.3, p.118; Tomk. p.292)*

KAUKASITE. See caucasite.

KAULAITE. A variety of olivine basalt containing anemousite instead of plagioclase, corresponding to the kaulaitic magma-type of Niggli (1936, p.366) This is the same as the olivine pacificite of Barth (1930). *(Tröger, 1938, p.67; Kaula Gorge, Ookala, Mauna Kea, Hawaii; Trög(38). 385; Tomk. p.292)*

KAXTORPITE. A local name for a coarse-grained, sometimes schistose, variety of nepheline syenite consisting of alkali feldspar, nepheline, alkali amphibole (eckermannite) with or without pectolite and aegirine. *(Adamson, 1944, p.188; Kaxtorp, Norra Kärr complex, Sweden; Tomk. p.292)*

KAZANSKITE (KASANSKITE). A local term for a mafic to ultramafic dyke rock consisting essentially of olivine with magnetite and small

amounts of bytownite. It is a variety of plagioclase-bearing dunite or melanocratic troctolite. *(Duparc & Grosset, 1916, p.106; Kazansky, Central Urals, USSR; Trög. 402; Joh. v.3, p.336; Tomk. p.292)*

KEDABEKITE. A local term for a gabbroic rock composed of bytownite and approximately equal amounts of hedenbergite and andradite. Probably a hybrid or skarn. *(Fedorov, 1901, p.135; Kedabek, Azerbaijan, Caucasus, USSR; Trög. 365; Joh. v.3, p.241; Tomk. p.292)*

KEMAHLITE. A local name for a fine-grained variety of monzonite containing pseudoleucite. *(Lacroix, 1933, p.190; Kemahl, Turkey; Trög. 520; Tomk. p.293)*

KENNINGITE. A local name for a leucocratic variety of basalt consisting mainly of labradorite with minor augite and serpentinised olivine. Thought to represent the volcanic equivalent of anorthosite. *(Eckermann, 1938a, p.277; Känningen Island, Rödö Archipelago, Sweden; Tomk. p.293)*

KENTALLENITE. A local term for a melanocratic variety of monzonite composed of olivine, augite, zoned plagioclase, brown and green micas and interstitial alkali feldspar. *(Hill & Kynaston, 1900, p.532; Kentallen Quarry, Ballachulish, Scotland; Trög. 260; Joh. v.3, p.99; Tomk. p.293)*

KENYTE (KENYITE). A variety of phonolitic trachyte characterised by rhomb-shaped phenocrysts of anorthoclase with or without augite and olivine set in a glassy *ne*-normative groundmass. *(Gregory, 1900, p.211; Mt. Kenya, Kenya; Trög. 467; Joh. v.4, p.130; Tomk. p.294)*

KERAMIKITE. A group name for a series of cordierite-bearing pumices and obsidians of rhyolitic composition which grade into cordierite-bearing microtinites (an obsolete term for trachyte). *(Koto, 1916a, p.197; Sakurajima volcano, Ryukyu Island, Japan; Trög(38). 40½; Tomk. p.294)*

KERATOPHYRE. A term originally used for a quartz-bearing orthoclase-plagioclase rock with a dense groundmass, but later used for albitised felsic extrusive rocks consisting essentially of albite with minor mafic minerals, often altered to chlorite. Potassic keratophyres are also recognised in which the feldspar is orthoclase. Commonly associated with spilites. *(Gümbel, 1874, p.43; from the Greek keras= horn; Trög. 213; Joh. v.3, p.47; Tomk. p.294)*

KERATOPHYRITE. A keratophyre with the composition of adamellite. *(Loewinson-Lessing, 1928, p.142; Trög(38). 880½; Tomk. p.294)*

KERSANTITE. A variety of lamprophyre consisting of phenocrysts of Mg-biotite, with or without hornblende, olivine or pyroxene in a groundmass of the same minerals plus plagioclase and occasional alkali feldspar. Now defined in the lamprophyre classification (Table B.3). *(Delesse, 1851, p.164; named after its similarity to the rock called "kersanton" from Kersanton, Brittany, France; Trög. 317; Joh. v.3, p.187; Tomk. p.295)*

KERSANTON. An obsolete local name for a rock that was later named kersantite. *(Rivière, 1844, p.537; Kersanton, Brittany, France; Trög. 881; Joh. v.3, p.187; Tomk. p.295)*

KHAGIARITE (KAGIARITE). An obsolete local name for a glassy pantellerite with distinct flow texture. *(Washington, 1913, p.708; Khagiar, Pantelleria Island, Italy; Trög. 73; Tomk. p.295)*

KHEWRAITE. A K-rich effusive rock composed of K-feldspar and enstatite, with minor ilmenite, hematite and apatite. The enstatite is altered to antigorite, vermiculite and quartz. *(Mosebach, 1956, p.200; Khewra Gorge, Pakistan)*

KHIBINITE (CHIBINITE). A variety of eudialyte-bearing nepheline syenite with aegirine, alkali amphibole and many accessory minerals, particularly those containing Ti and Zr. Although originally spelt chibinite, the preferred spelling is now khibinite. *(Ramsay & Hackman, 1894, p.80; Khibina complex, Kola Peninsula, USSR; Trög. 418; Joh. v.4, p.107; Tomk. p.98)*

KIIRUNAVAARITE (KIRUNAVAARITE). A local name for an ultramafic rock consisting almost entirely of magnetite. Cf. shishimskite. *(Rinne, 1921, p.182; Kiirunavaara, Lapland, Sweden; Trög. 761; Joh. v.4, p.466; Tomk. p.296)*

KILAUEITE. An obsolete term for an aphanitic basaltic rock rich in magnetite. *(Silvestri, 1888, p.186; Kilauea, Hawaii; Trög(38). 881½; Tomk. p.296)*

KIMBERLITE. An ultramafic rock consisting of major amounts of serpentinised olivine with variable amounts of phlogopite, orthopyroxene, clinopyroxene, carbonate and chromite. Characteristic accessory minerals include pyrope garnet, monticellite, rutile and perovskite. Now classified as a variety of lamprophyric rock (section B.7). *(Lewis, 1888, p.130; Kimberley, South Africa; Trög. 742; Joh. v.4, p.413; Tomk. p.296)*

KIVITE. A local name for a variety of leucite basanite largely composed of clinopyroxene and plagioclase with subordinate leucite and minor olivine and biotite. *(Lacroix, 1923, p.265; Lake Kivu, Zaire; Trög. 584; Joh. v.4, p.241; Tomk. p.297)*

KJELSÅSITE. A local name given to a plagioclase-rich larvikite which is a variety of augite syenite or monzonite. *(Brögger, 1933, p.45; Kjelsås, Sørkedal, Oslo district, Norway; Trög. 275; Joh. v.3, p.115; Tomk. p.297)*

KLAUSENITE. A local term used as a group name for two different types of rocks: 1) for dioritic norites and quartz gabbros from the Tyrol, 2) for a series of lamprophyric rocks ranging from diorite to tonalite in composition but with a graphic groundmass. *(Cathrein, 1898, p.275; Klausen, near Bressanone, Alto Adige, Italy; Trög. 338; Tomk. p.298)*

KLEPTOLITH. A term suggested for basic lamprophyric dyke rocks which intrude many Scandinavian granites. *(Sederholm, 1934, p.17; Tomk. p.298)*

KLINGHARDTITE. A variety of nepheline phonolite containing phenocrysts of sanidine. *(Kaiser-Gießen, 1913, p.597; Klinghardt Mts., Namibia; Trög(38). 881¾; Tomk. p.298)*

KLINGSTEIN (CLINKSTONE). An old term, used before the mineral composition of rocks was known, for rocks which ring when hit with a hammer. Later replaced by the name phonolite. *(Werner, 1787, p.11; from the German klingen = sound; Joh. v.4, p.121)*

KLOTDIORITE. A Swedish term for an orbicular diorite. *(Holst & Eichstädt, 1884, p.137; Tomk. p.298)*

KLOTGRANITE. A Swedish term for an orbicular granite. *(Holst & Eichstädt, 1884, p.137; Tomk. p.298)*

KODURITE. A local name suggested by L.L. Fermor for an intrusive rock consisting of K-feldspar, manganese garnet and apatite. *(Holland, 1907, p.22; Kodur Mines, Vizagapatam, now Vishakhapatnam, Andhra Pradesh, India; Trög. 224; Tomk. p.299)*

KOHALAITE. A general term proposed for intermediate volcanic rocks in which the normative feldspar is oligoclase; they sometimes contain normative or modal olivine. *(Iddings, 1913, p.193; Kohala, Waimea, Hawaii; Trög. 289; Joh. v.3, p.169; Tomk. p.299)*

KOHLIPHYRE. An obsolete term proposed for igneous rocks intruding coal formations. *(Ebray, 1875, p.291)*

KOKKITE (COCCITE). An obsolete group name for crystalline non-schistose igneous and sedimentary rocks. *(Gümbel, 1888, p.85; Tomk. p.108)*

KOLDERUPITE. A name suggested for a group of pyroxene tonalites. *(Marcet Riba, 1925, p.293; named after C.F. Kolderup)*

KOMATIITE. A variety of ultramafic lavas which crystallise from high temperature magmas with 18% to 32% MgO. They often form pillows and have chilled flow-tops and usually display well-developed spinifex textures with intergrown skeletal and bladed olivine and pyroxene crystals set in abundant glass. The more highly magnesian varieties are often termed peridotitic komatiite. Now defined chemically in the TAS classification (Fig. B.12). *(Viljoen & Viljoen, 1969, p.83; Komati river, Barberton, Transvaal, South Africa)*

KOMATIITIC BASALT. See basaltic komatiite.

KONGITE. A name given to a hypothetical set of data used to illustrate the problems of the statistical interpretation of rock compositions. *(Aitchison, 1984, p.534)*

KOSENITE. See kåsenite.

KOSWITE. A local term for a variety of olivine clinopyroxenite composed of clinopyroxene, olivine and magnetite. *(Duparc & Pearce, 1901, p.892; Koswinski Mts., Urals, USSR; Trög. 683; Joh. v.4, p.440; Tomk. p.300)*

KOTUITE. A local name for a dyke rock consisting of augite, nepheline, biotite and ore minerals with minor aegirine, apatite, perovskite and phlogopite. *(Butakova, 1956, p.213; Kotui river, S. of Khatanga, N. Siberia, USSR)*

KOVDITE. A local name for a variety of pyroxene hornblendite composed of green amphibole and orthopyroxene, with small amounts of biotite and plagioclase and occasional garnet. It may be a metamorphic rock. The description given in Tomkeieff et al. (1983) appears to be incorrect. *(Fedorov, 1903, p.215; Kovda, White Sea, USSR; Tomk. p.300)*

KOVDORITE. A local term for a variety of turjaite (a melilitolite) containing olivine. *(Zlatkind, 1945, p.92; Lake Kovdor, Kola Peninsula, USSR; Tomk. p.300)*

KRABLITE. An obsolete term for a rhyolitic crystal tuff. *(Preyer & Zirkel, 1862, p.317; Krabla, Iceland; Trög. 882; Tomk. p.300)*

KRAGERITE. An alternative spelling of krageröite. *(Watson, 1912, p.509; Trög. 313; Tomk. p.300)*

KRAGERÖITE. A local name for an aplite consisting essentially of sodic plagioclase and a considerable amount of rutile with minor quartz and orthoclase. *(Brögger, 1904, p.30; Kragerö, Arendal, Norway; Joh. v.3, p.124; Tomk. p.301)*

KRISTIANITE (CHRISTIANITE). A local name for a biotite granite of the Oslo province. *(Brögger, 1921, p.371; Kristiania, now Oslo, Norway; Trög. 883; Tomk. p.100)*

KUGDITE. A variety of olivine melilitolite consisting of melilite, olivine, pyroxene, nepheline, and Ti-magnetite. Now used as a special term in the melilitic rocks classification (section B.8.1) synonomous with olivine melilitolite. *(Egorov, 1969, p.29)*

KUGELDIORITE. An orbicular variety of gabbro composed essentially of calcic plagioclase with hornblende, minor hypersthene and quartz. This rock has also been called corsite, miagite and napoleonite. *(Leonhard, 1823a, p.107; Trög. 361; Joh. v.3, p.320)*

KULAITE. An obsolete term originally defined as a subgroup of basalts in which hornblende is more abundant than augite. Redefined (Washington, 1900) as a basic alkaline volcanic rock in which orthoclase and plagioclase are present in about equal amounts together with up to 25% nepheline. *(Wash-*

ington, 1894, p.115; Kula Devit, E by N of Izmir, Turkey; Trög. 578; Joh. v.4, p.198; Tomk. p.301)

KULLAITE. A local name for porphyritic variety of monzodiorite composed of phenocrysts of andesine and Na-orthoclase in an ophitic groundmass of oligoclase-andesine, chlorite and epidote after augite, quartz and magnetite. *(Hennig, 1899, p.19; Kullagården, Kullen, Sweden; Trög. 288; Joh. v.3, p.121; Tomk. p.303)*

KUSELITE. See cuselite.

KUSKITE. A discredited name for a light-coloured dyke rock originally thought to consist essentially of phenocrysts of quartz and scapolite in a groundmass of quartz, orthoclase and muscovite. The scapolite was later shown to be quartz and the rock a granite porphyry. *(Spurr, 1900b, p.315; Kuskokwim River, Holiknuk, Alaska, USA; Trög. 885; Joh. v.2, p.300; Tomk. p.302)*

KUZEEVITE (KUSEEVITE). A poorly-defined term used for a banded variety of the charnockitic rock series containing quartz, plagioclase, hypersthene, biotite, hornblende and garnet. Tomkeieff transliterates the name as Kuseevite. *(Ainberg, 1955, p.111; Kuzeeva River, tributary of Yenisey River, Siberia, USSR; Tomk. p.302)*

KVELLITE. A local name for an ultramafic dyke rock containing abundant phenocrysts of lepidomelane, olivine and barkevikite in a sparse groundmass of anorthoclase laths and some nepheline. *(Brögger, 1906, p.128; Kvelle, Larvik, Norway; Trög. 715; Joh. v.4, p.158; Tomk. p.302)*

KYLITE. A local term for a variety of nepheline-bearing gabbro containing abundant olivine predominating over Ti-augite, labradorite and minor nepheline. *(Tyrrell, 1912, p.121; Kyle district, near Dalmellington, Scotland; Trög. 554; Joh. v.4, p.226; Tomk. p.303)*

KYSCHTYMITE. A local name for a corundum-rich variety of anorthosite composed of 45% to 50% corundum, calcic plagioclase and minor poikilitic biotite, spinel, zircon and apatite. *(Morozewicz, 1899, p.202; Kyschtym, source of River Borsowska, Urals, USSR; Trög. 301; Joh. v.3, p.327; Tomk. p.303)*

LAACHITE. A local term for a variety of sani-

dinite containing anorthoclase and biotite occurring as ejecta in volcanic tuff. *(Kalb, 1936, p.190; Laacher See, near Koblenz, West Germany; Trög(38). 226½; Tomk. p.304)*

LABRADITE. An obsolete term proposed for a coarse-grained rock consisting essentially of labradorite. *(Turner, 1900, p.110; Trög. 296; Joh. v.3, p.198; Tomk. p.304)*

LABRADOPHYRE. An obsolete general term for rocks consisting of phenocrysts of labradorite in a groundmass of labradorite and pyroxene. *(Coquand, 1857, p.87; Trög. 887; Joh. v.3, p.307; Tomk. p.304)*

LABRADORITE. An old name once used as a group name for gabbros, basalts etc. Also used by the Soviets and French for labradorite-phyric basalts. *(Senft, 1857, Table II; Trög. 890; Tomk. p.304)*

LABRADORITITE. A name suggested for a variety of anorthosite composed almost entirely of labradorite. Johannsen (p.198) states that the name comes from an earlier 1920 publication, but this is in error. *(Johannsen, 1937, p.196; Trög. 296; Tomk. p.304)*

LADOGALITE. A melanocratic variety of alkali feldspar syenite, composed of clinopyroxene, alkali feldspar, hornblende, biotite, and apatite, with minor titanite and orthite. *(Khazov, 1983, p.1200)*

LADOGITE. A melanocratic apatite-rich variety of alkali feldspar syenite, composed of clinopyroxene, alkali feldspar, and apatite, with minor biotite, hornblende, titanite, and orthite. *(Khazov, 1983, p.1200; Lake Ladoga, near Lenigrad, USSR)*

LAHNPORPHYRY. An obsolete term for a rock variously described as a keratophyre and as an anchimetamorphic aegirine trachyte. The reference cited does not contain the name, although porphyries from the area are described. *(Koch, 1858, p.97; Lahntal, Hessen, West Germany; Trög. 891; Tomk. p.306)*

LAKARPITE. A local name for a coarse-grained variety of nepheline syenite consisting of alkali feldspar, altered nepheline, abundant arfvedsonite, aegirine and a little pectolite. *(Törnebohm, 1906, p.18; Lakarp, Norra Kärr complex, Sweden; Trög. 422; Joh. v.4, p.161; Tomk. p.306)*

LAMPROITE. A comprehensive term originally used for lamprophyric extrusive rocks rich in potassium and magnesium, corresponding to the lamproitic magma-type of Niggli. It is characterised by exceptionally high K, at intermediate to very low SiO_2, usually with K>Al and contains unusual phases such as K-Ti-richterite, priderite, wadeite, jeppite and Fe-orthoclase. Leucite may be present. Now classified as a variety of lamprophyric rock (section B.7). *(Niggli, 1923, p.184; Trög. 892; Tomk. p.307)*

LAMPROPHYRE. A name for a distinctive group of rocks which are strongly porphyritic in mafic minerals, typically biotite, amphiboles and pyroxenes, with any feldspars being confined to the groundmass. They commonly occur as dykes or small intrusions and often show signs of hydrothermal alteration. For further details see section 8 of the classification and Table B.3. *(Gümbel, 1874, p.36; from the Greek lampros = glistening; Trög. 893; Joh. v.3, p.32; Tomk. p.307)*

LAMPROSYENITE. A comprehensive term used for mesocratic biotite syenites corresponding to the lamprosyenitic magma-type of Niggli. *(Niggli, 1923, p.182; Trög. 894; Tomk. p.307)*

LANGITE. An obsolete name suggested for a group of pyroxene diorites. *(Marcet Riba, 1925, p.293; named after H.A. Lang)*

LAPIDITE. A variety of ignimbrite consisting of angular fragments of cognate rock in a fine welded matrix of glass shards. *(Marshall, 1935, p.358; Trög(38). 871½; Tomk. p.308)*

LAPILLI. Now defined in the pyroclastic classification (Table B.1) as a pyroclast of any shape with a mean diameter between 2 mm. and 64 mm. *(Lyell, 1835, p.391; from the Latin lapillus = little stone; Joh. v.1, p.179; Tomk. p.308)*

LAPILLI TUFF. Now defined in the pyroclastic classification (Table B.1) as a pyroclastic rock in which the average pyroclast size is between 2 mm. and 64 mm. *(Original reference uncertain)*

LARDALITE (LAURDALITE). A local name for a coarse-grained variety of nepheline syenite characterised by rhomb-shaped alkali or ternary feldspar crystals and large crystals of nepheline. *(Brögger, 1890, p.32; Lardal, Oslo district, Norway; Trög. 419; Joh. v.4, p.102; Tomk. p.308)*

LARVIKITE (LAURVIKITE). A variety of augite syenite or monzonite consisting of rhomb-shaped ternary feldspars (with a distinctive schiller), barkevikite, Ti-augite and lepidomelane. Minor nepheline, iron-rich olivine or quartz may be present. Common ornamental stone. *(Brögger, 1890, p.30; Larvik, Oslo district, Norway; Trög. 183; Joh. v.4, p.9; Tomk. p.308)*

LASSENITE. A local name for a fresh glass of trachyte composition. *(*Wadsworth, 1893, p.97; Lassen Peak, California, USA; Trög. 895; Joh. v.3, p.77; Tomk. p.309)*

LATHUS PORPHYRY. An obsolete local name for fine-grained acid rocks with flow structure and K_2O much greater than Na_2O. *(Holtedahl, 1943, p.32; Lathusås, Bærum, Norway)*

LATIANDESITE. A synonym for latite andesite. *(Rittmann, 1973, p.133)*

LATIBASALT. A synonym for latite basalt. *(Rittmann, 1973, p.133)*

LATITE. A term originally proposed for a rock which chemically lies between trachyte and andesite, but later used as a volcanic rock composed of approximately equal amounts of alkali feldspar and sodic plagioclase, i.e. the volcanic equivalent of monzonite. Now defined modally in QAPF field 8 (Fig. B.10) and, if modes are not available, chemically as the potassic variety of trachyandesite in TAS field S3 (Fig. B.13). *(Ransome, 1898, p.355; from Latium an Italian region; Trög. 270; Joh. v.3, p.100; Tomk. p.311)*

LATITE ANDESITE. A name orignally proposed as a comprehensive term for volcanic rocks in QAPF field 9, intermediate between latite and andesite and having a colour index less than 40, to cover such rock types as mugearite, shoshonite, benmoreite and hawaiite. The term was not widely accepted by English speaking petrologists and was dropped from later versions of the volcanic QAPF classification. The term is synonymous with andelatite. *(Streckeisen, 1967, p.161)*

LATITE BASALT. A name orignally proposed as a comprehensive term for volcanic rocks in QAPF field 9, intermediate between latite and basalt and having a colour index higher than 40. The term was not widely accepted by English speaking petrologists and was dropped from later versions of the volcanic QAPF classification. The term is synonymous with basalatite. *(Streckeisen, 1967, p.161)*

LAUGENITE. An obsolete name for a leucocratic rock originally described as a variety of diorite in which the plagioclase is oligoclase rather than the more common andesine. However, as the rock was said to be like an akerite and also contains alkali feldspar it is better described as a variety of monzodiorite of QAPF field 9 (Fig. B.4). *(Iddings, 1913, p.164; Tuft, Laugendal, Lardal, Oslo district, Norway; Trög. 896; Joh. v.3, p.118; Tomk. p.311)*

LAURDALITE. See lardalite.

LAURVIKITE. See larvikite.

LECKSTONE. A name given to teschenite used for lining the bottoms of ovens in Scotland. *(Tomkeieff et al., 1983, p.315)*

LEDMORITE. A local name for a coarse-grained variety of nepheline syenite consisting of alkali feldspar, altered nepheline, melanite, pyroxene and biotite. *(Shand, 1910, p.384; Ledmore River, Borralan complex, Assynt, Scotland; Trög. 486; Joh. v.4, p.117; Tomk. p.315)*

LEEUWFONTEINITE. A local term for a variety of alkali feldspar syenite containing abundant anorthoclase, with biotite and hornblende. The rock had previously been called hatherlite. *(Brouwer, 1917, p.775; Leeuwfontein, Bushveld, South Africa; Trög. 238; Joh. v.3, p.11; Tomk. p.315)*

LEHMANITE. An obsolete term for a feldspar quartz rock. *(Pinkerton, 1811a, p.206; named after J.G. Lehman; Joh. v.2, p.46; Tomk. p.315)*

LEIDLEITE. A local name for an andesitic rock with microlites of plagioclase, augite and iron ore in a fine-grained to glassy groundmass. *(Thomas & Bailey, 1915, p.207; Glen Leidle, Island of Mull, Scotland; Trög. 124; Tomk. p.315)*

LENGAITE (LENGAIITE). A carbonatite lava extruded in 1960 composed of Na-Ca-K carbonate minerals including nyerereite and gregoryite. The earlier name of natrocarbonatite is preferred. *(Dawson & Gale, 1970, p.222; Oldoinyo Lengai, Tanzania)*

LENNEPORPHYRY. An obsolete term for a quartz-rich keratophyre. Although the rock was originally described by Dechen (1845) he did not name the rock. *(Mügge, 1893, p.535; Trög. 11; Tomk. p.316)*

LENTICULITE. A variety of ignimbrite which contains elongated lenticles of glass in a welded matrix. *(Marshall, 1935, p.358; Trög(38). 871½; Tomk. p.316)*

LEOPARD ROCK. A local name for spotted rocks of several types e.g. a syenite from Ontario, a gabbro from Quebec, Canada. The reference states the name had been in use for some time. *(Gordon, 1896, p.99; Joh. v.3, p.86; Tomk. p.316)*

LEOPARDITE. A name suggested for a spotted rock. *(Hunter, 1853, p.377; Mecklenburg County, near Charlotte, New York, USA; Trög. 897; Tomk. p.316)*

LEPAIGITE. A colourless to pale brown or blue volcanic glass of rhyolite composition containing small euhedral phenocrysts of cristobalite, cordierite and sillimanite. It occurs as lapilli of globular habit thought to have been ejected during a fissure eruption of ignimbrites. *(Mueller, 1964, p.374; named after Padre G. le Paige; rock from San Pedro de Atacama, Chile)*

LESTIWARITE. A local name for a variety of microsyenite composed almost entirely of microperthite. *(Rosenbusch, 1896, p.464; Lestiware, Kola Peninsula, USSR; Trög. 170; Joh. v.3, p.25; Tomk. p.317)*

LEUCILITE. An obsolete name for leucitophyre. *(Naumann, 1850, p.655; Joh. v.4, p.351; Tomk. p.317)*

LEUCITE BASALT. A term used for a volcanic rock consisting essentially of leucite, olivine, and clinopyroxene. The name should not be used as the term basalt is now restricted to a rock containing essential plagioclase. As the rock is a variety of foidite it should be given the appropriate name, e.g. olivine leucitite. *(Zirkel, 1870, p.108; Trög. 900; Joh. v.4, p.351; Tomk. p.317)*

LEUCITE BASANITE. Now defined in QAPF field 14 (Fig. B.10) as a variety of basanite in which leucite is the most abundant foid. *(Rosenbusch, 1887, p.760; Trög. 595; Tomk. p.318)*

LEUCITE PHONOLITE. A term originally used for a rock consisting essentially of sanidine and leucite without nepheline. Zirkel (1894a) redefined it as a phonolite (sanidine + nepheline) in which leucite is an important foid. Now defined in QAPF field 11 (Fig. B.10) in the original sense. *(Rosenbusch, 1877, p.234; Trög. 471; Joh. v.4, p.132)*

LEUCITE TEPHRITE. Now defined in QAPF field 14 (Fig. B.10) as a variety of tephrite in which leucite is the most abundant foid. *(Rosenbusch, 1877, p.492; Trög. 581; Joh. v.4, p.235)*

LEUCITE TRACHYTE. A term originally used for a volcanic rock consisting of alkali feldspar, leucite and minor mafic minerals. As such a rock falls in the QAPF field 11 it would now be called leucite phonolite. *(Rath, 1868, p.297; Trög. 477; Joh. v.4, p.133; Tomk. p.318)*

LEUCITE-BASITE. A term originally used for a leucite basalt as originally defined i.e. a rock composed of leucite, olivine and clinopyroxene. Later used for all leucite-bearing ultrabasic rocks. Obsolete. *(Vogelsang, 1872, p.542; Tomk. p.318)*

LEUCITITE. A volcanic rock essentially composed of leucite, clinopyroxene and variable amounts of olivine. Now defined modally as a variety of foidite of QAPF field 15c (Fig. B.10) and, if modes are not available, chemically in TAS field F-U1 (Figs. B.13). *(Senft, 1857, p.287; Trög. 643; Joh. v.4, p.351; Tomk. p.318)*

LEUCITITE BASANITE. A variety of leucite basanite in which the leucite exceeds the plagioclase. *(Johannsen, 1938, p.301; Tomk. p.318)*

LEUCITITE TEPHRITE. A variety of leucite tephrite in which the leucite exceeds the plagioclase. *(Johannsen, 1938, p.301)*

LEUCITOID-BASALT. An obsolete name for a basalt which does not contain phenocrysts of leucite, but may contain leucite in the groundmass. *(Bořický, 1874, p.41; between Turtsch and Duppau (now Doupov), Bohemia, Czechoslovakia; Tomk. p.319)*

LEUCITOLITH. An obsolete term proposed for a monomineralic volcanic rock consisting of

leucite. *(Johannsen, 1938, p.336; Tomk. p.319)*

LEUCITONEPHELINITE. A general term for feldspar-free volcanic rocks that contain leucite and nepheline in almost equal amounts. *(Jung & Brousse, 1959, p.88)*

LEUCITOPHYRE. A variety of leucitite characterised by leucite phenocrysts and essentially composed of leucite, nepheline and clinopyroxene. *(Humboldt, 1837, p.257; Eifel district, near Koblenz, West Germany; Trög. 641; Joh. v.4, p.359; Tomk. p.319)*

LEUCO-. Originally used as a prefix for rocks with more than 95% felsic minerals, but now used in the modal QAPF classification as indicating that the rock has considerably more felsic minerals than would be regarded as normal for that rock type (Fig. B.7). *(Johannsen, 1920a, p.48; Joh. v.1, p.153; Tomk. p.319)*

LEUCO-APHANEID. An obsolete alternative term for felseid. *(Johannsen, 1911, p.321)*

LEUCOCRATE. A general term proposed for rocks rich in light-coloured minerals. *(Brögger, 1898, p.264)*

LEUCOCRATIC. Relating to rocks rich in light-coloured minerals. *(Brögger, 1898, p.264; from the Greek leukos = white, krateo = to predominate; Joh. v.1, p.179; Tomk. p.319)*

LEUCOLITE. An obsolete term for a leucocratic igneous rock. *(Loewinson-Lessing, 1901, p.118; Tomk. p.319)*

LEUCOPHYRE. An obsolete term for light-coloured altered diabases and also a variety of serpentinised peridotite. *(Gümbel, 1874, p.33; from the Greek leukos = white; Trög. 899; Joh. v.3, p.317; Tomk. p.320)*

LEUCOPHYREID. An obsolete alternative term for felseid porphyry. *(Johannsen, 1911, p.321)*

LEUCOPHYRIDE. A revised spelling recommended to replace the field term leucophyreid. Now obsolete. *(Johannsen, 1926, p.182; from the Greek leukos = white; Joh. v.1, p.58; Tomk. p.320)*

LEUCOPTOCHE. An obsolete adjectival term for igneous rocks poor in light coloured minerals. *(Loewinson-Lessing, 1901, p.118; Tomk. p.320)*

LEUCOSTITE. An obsolete group name for

porphyritic trachtyes, phonolites etc. *(Cordier, 1868, p.92; Tomk. p.320)*

LEUMAFITE. A mnemonic name for rocks consisting of *leu*cite and *ma*fic minerals. *(Hatch et al., 1961, p.385)*

LHERCOULITE. Alledged to be another name for lherzolite, however, the reference cited (Cordier, 1842) does not contain the name. *(Tomkeieff et al., 1983, p.320; Lhercoul, near Lhers, Ariège, Pyrénées, France)*

LHERZITE. A variety of hornblendite which occurs as dykes and consists essentially of hornblende with minor biotite; similar in composition to theralite. *(Lacroix, 1917c, p.385; Lac de Lherz, now Lhers, Ariège, Pyrénées, France; Trög. 704; Joh. v.4, p.444; Tomk. p.320)*

LHERZOLINE. An obsolete name for a fine-grained lherzolite. *(Cordier, 1868, p.128; Lac de Lherz, now Lhers, Ariège, Pyrénées, France; Tomk. p.320)*

LHERZOLITE. An ultramafic plutonic rock composed of olivine with subordinate orthopyroxene and clinopyroxene. Now defined modally in the ultramafic rock classification (Fig. B.8). *(Delamétherie, 1795, p.454; Lac de Lherz, now Lhers, Ariège, Pyrénées, France; Trög. 735; Joh. v.4, p.422; Tomk. p.320)*

LIMBURGITE. A basic volcanic rock containing phenocrysts of pyroxene, olivine and opaques in a glassy groundmass containing the same minerals. No feldspars are present. Now used as an approved synonym for a glassy basanite in TAS field F-U1 (Fig. B.13). *(Rosenbusch, 1872, p.53; Limburg, Kaiserstuhl, Baden, West Germany; Trög. 593; Tomk. p.321)*

LINDINOSITE. A local name for a variety of peralkaline granite containing nearly 60% riebeckite. *(Lacroix, 1922, p.580; Lindinosa, Corsica; Trög. 59; Tomk. p.322)*

LINDÖITE. A local name for a dyke-rock that is a leucocratic variety of trachyte or rhyolite containing minor amounts of arfvedsonite. *(Brögger, 1894, p.131; Lindö, Oslo district, Norway; Trög. 33; Joh. v.2, p.100; Tomk. p.322)*

LINOPHYRE. A porphyritic rock in which the phenocrysts occur in lines and streaks. *(Id-*

dings, 1909, p.224; Tomk. p.324)*

LINOSAITE. An obsolete local name for a variety of alkali basalt containing small amounts of nepheline. *(Johannsen, 1938, p.68; Linosa Island, Italy; Trög(38). 381½; Tomk. p.324)*

LIPARITE. A name given independently to the same rocks that were called rhyolites by Richthofen in the previous year, 1860. Although the term was also used in a broader sense later it has not been widely used. Now defined as a synonym for rhyolite. *(Roth, 1861, p.xxxiv; Lipari Island, Italy; Trög. 40; Joh. v.2, p.265; Tomk. p.324)*

LITCHFIELDITE. A coarse-grained, somewhat foliated variety of nepheline syenite consisting of K-feldspar, albite, nepheline, cancrinite, sodalite, and lepidomelane. *(Bayley, 1892, p.243; Litchfield, Maine, USA; Trög. 415; Joh. v.4, p.181; Tomk. p.325)*

LITHIC TUFF. Now defined in the pyroclastic classification (Fig. B.1) as a tuff in which lithic fragments are more abundant than either crystal or vitric fragments. *(Pirsson, 1915, p.193; Tomk. p.326)*

LITHOIDITE. A non-porphyritic rhyolite. *(Richthofen, 1860, p.183; Tomk. p.327)*

LLANITE. A local term for a granite porphyry containing abundant K-feldspar and quartz phenocrysts with groundmass albite. Although the name is attributed by Johannsen to Iddings, the cited reference only contains a description of the rock from Llano and not the name. *(Iddings, 1904; Llano, Texas, USA; Trög. 60; Joh. v.2, p.117; Tomk. p.329)*

LOSERO. See lozero.

LOSSENITE. A name suggested for a group of quartz gabbros "sensu stricto". *(Marcet Riba, 1925, p.293; named after K.A. Lossen)*

LOW-K. A chemical term originally applied to andesites with $K_2O < 0.7\%$ but later applied to volcanic rocks which plot below a line on the $SiO_2 - K_2O$ diagram i.e. for their SiO_2 values they are lower than usual in K_2O. The field is now defined as part of the TAS classification (Fig. B.15). *(Taylor, 1969, p.45)*

LOZERO (LOSERO). A local Mexican name for a bedded tuff used for tiles. *(Humboldt, 1823, p.217; from the Spanish lozas = thin plates of rock; Tomk. p.332)*

LUCIITE. An obsolete term for a coarse-grained

variety of the lamprophyric rock malchite consisting essentially of intermediate plagioclase and hornblende with minor quartz, orthoclase and biotite. Cf. orbite. *(Chelius, 1892, p.3; Luciberg, Melibocus, Odenwald, West Germany; Trög. 335; Tomk. p.332)*

LUGARITE. A local name for a variety of teschenite containing prominent phenocrysts of Ti-augite and kaersutite; labradorite is minor and analcime abundant. *(Tyrrell, 1912, p.121; Lugar, Scotland; Trög. 564; Joh. v.4, p.304; Tomk. p.332)*

LUHITE. A local name for a rock consisting of phenocrysts of olivine and pyroxene in a groundmass of pyroxene, melilite, haüyne, perovskite and biotite cemented by nepheline and calcite. Considered to be a nepheline-rich alnöite. *(Scheumann, 1922, p.523; Luhov, S. of Mimoň, N. Bohemia, Czechoslovakia; Trög. 666; Joh. v.4, p.387)*

LUJAVRITE (LUIJAURITE, LUJAUVRITE). A term, originally spelt luijaurite, for a melanocratic agpaitic variety of nepheline syenite rich in eudialyte, arfvedsonite and aegirine with perthitic alkali feldspar or separate microcline and albite. A pronounced igneous lamination is characteristic, as is the abundance in minerals rich in incompatible elements such as the REE, U, Th, Li, etc. *(Brögger, 1890, p.204; Luijaur, now Lujavr-Urt, Lovozero complex, Kola Peninsula, USSR; Trög. 421; Joh. v.4, p.106; Tomk. p.332)*

LUJAVRITITE. A variety of ijolite composed mainly of aegirine-augite and nepheline with much titanite and some apatite and characterised by the presence of small amounts of K-feldspar. *(Antonov, 1934, p.25; Khibina complex, Kola Peninsula, USSR; Trög(38). 607⅔; Tomk. p.332)*

LUNDYITE. A local name for a rock which is close to the boundary of alkali feldspar syenite and alkali feldspar granite and contains small amounts of katophorite. *(Hall, 1915, p.53; Lundy Island, Bristol Channel, England; Trög. 75)*

LUPATITE. An incompletely described rock named as a nepheline feldspar porphyry. *(Mennell, 1929, p.536; Lupata Gorge, Zambezi River, Mozambique; Trög(38). 906½; Tomk. p.333)*

LUSCLADITE. A variety of theralite poor in nepheline with the mode dominated by Ti-augite and labradorite. *(Lacroix, 1920, p.21; Ravin de Lusclade, Mont Dore, Auvergne, France; Trög. 544; Joh. v.4, p.197; Tomk. p.333)*

LUSITANITE. A melanocratic variety of alkali feldspar syenite containing riebeckite and aegirine. Also used as family name for rocks consisting essentially of alkali feldspar. *(Lacroix, 1916c, p.283; from Lusitania, the Roman name for Portugal; Trög. 222; Joh. v.3, p.46; Tomk. p.333)*

LUTALITE. An obsolete name for a variety of leucitite composed essentially of clinopyroxene and leucite and with subordinate olivine, nepheline and plagioclase. The value of Na_2O/K_2O is higher than in ordinary leucitites. *(Holmes & Harwood, 1937, p.10; Lutale, Birunga volcanic field, Uganda; Trög(38). 642½; Tomk. p.333)*

LUXULIANITE (LUXULLIANITE, LUXULYANITE). A local name for a variety of porphyritic granite containing abundant tourmaline replacing various minerals. It was spelt luxuliane in the original reference. *(Pisani, 1864, p.913; Luxulion, now Luxulyan, near Lostwithiel, Cornwall, England; Trög. 907; Joh. v.2, p.58; Tomk. p.333)*

M-CHARNOCKITE, M-ENDERBITE, ETC. The prefix m- was suggested for use with members of the charnockitic rock series which contain mesoperthite. *(Tobi, 1971, p.201)*

M-TYPE GRANITE. A general term for granitic rocks occurring in some continental margins and having the chemical and isotopic compositions of island arc volcanic rocks. The prefix M implies a mantle origin as they are assumed to have formed by partial melting of the subducted oceanic crust. *(White, 1979, p.539)*

MACEDONITE. A local name for a fine-grained variety of basaltic trachyandesite with trachytic texture consisting of plagioclase and alkali feldspar with minor biotite, hornblende, olivine and pyroxene. *(Skeats, 1910, p.205; Mt. Macedon, Victoria, Australia; Trög. 217; Joh. v.3, p.120; Tomk. p.335)*

MACUSANITE. An acid volcanic glass, with

phenocrysts of andalusite and sillimanite and more rarely staurolite and cordierite, chemically characterised by high alumina (16%-20%) and fluorine. Originally thought to be a tectite. *(Martin & de Sitter-Koomans, 1955, p.152; Macusani, S Peru)*

MADEIRITE. An obsolete name for a melanocratic picrite consisting of phenocrysts of Ti-augite and partially serpentinised olivine in a fine-grained groundmass of plagioclase, augite, and iron ore with some secondary calcite. *(Gagel, 1913, p.382; Ribeira de Massapez, Madeira, Atlantic Ocean; Trög. 404; Joh. v.4, p.73; Tomk. p.336)*

MADUPITE. An obsolete name for a melanocratic variety of leucitite essentially composed of diopside and phlogopite in a glassy matrix, which has the composition of nepheline and leucite. Wyomingite is a more felsic variety. *(Cross, 1897, p.129; from the Shoshone Indian madúpa = Sweetwater, Wyoming, USA; Trög. 645; Joh. v.4, p.370; Tomk. p.336)*

MAENAITE. A local name for a variety of trachyte which occurs as dykes and consists essentially of albite and orthoclase.*(Brögger, 1898, p.206; Lake Maena, Gran district, Norway; Trög. 193; Joh. v.3, p.123; Tomk. p.336)*

MAFIC. A collective term for modal ferromagnesian minerals, such as olivine, pyroxene, etc., which was introduced to stop the normative term femic incorrectly being used for that purpose. See also felsic and salic. *(Cross et al., 1912,p.561;Joh.v.1,p.180;Tomk.p.336)*

MAFITITE. A term originally proposed for rocks with a colour index (M) of 90-100, but later withdrawn in favour of ultramafitite. *(Streckeisen, 1967, p.163)*

MAFRAITE. A variety of monzogabbro in which kaersutite exceeds Ti-augite. *(Lacroix, 1920, p.22; Tifão de Mafra and Rio Touro, Cintra, Portugal; Trög. 285; Joh. v.4, p.55; Tomk. p.337)*

MAFURITE. An ultrabasic rock consisting of phenocrysts of olivine and minor pyroxene in a groundmass of diopside and kalsilite with small amounts of perovskite, olivine and biotite. *(Holmes, 1942, p.199; Mafuru craters, Uganda; Tomk. p.337)*

MAGMABASALT. An obsolete name for a basaltic rock consisting of augite, magnetite and glass and without visible feldspar. Cf. limburgite. *(Bořický, 1874, p.40; Trög. 908; Tomk. p.337)*

MAGNESIOCARBONATITE. A chemically defined variety of carbonatite (Fig. B.2) in which wt% MgO > (FeO+Fe$_2$O$_3$+MnO) and CaO / (CaO+MgO+FeO+Fe$_2$O$_3$+MnO) < 0.8. *(Woolley, 1982, p.16)*

MAGNETITE HÖGBOMITITE. An obsolete name for a variety of magnetitite containing up to 15% of högbomite (a mineral of the corundum-hematite group). On account of the small amount of högbomite present högbomite magnetitite would have been a more logical name (Tröger, 1935, Johannsen, 1938). *(Gavelin, 1916, p.306; Routivare, Norrbotten, Lapland, Sweden; Trög. 764; Joh. v.4, p.469; Tomk. p.339)*

MAGNETITE SPINELLITE. A rock consisting of Ti-magnetite and ilmenite, with abundant spinel, and minor olivine and hypersthene. *(Sjögren, 1893,p.63; Routivare,Norrbotten, Lapland, Sweden; Joh. v.4, p.469; Tomk. p.540)*

MAGNETITITE. An ultramafic rock consisting essentially of magnetite. *(Johannsen, 1920c, p.225; Trög. 761; Joh. v.4, p.466; Tomk. p.339)*

MALCHITE. A lamprophyric dyke rock consisting of small rare phenocrysts of hornblende and labradorite and occasionally biotite, in a groundmass of hornblende, andesine and quartz. Cf. gabbrophyre or odinite. *(Osann, 1892, p.386; Malchen or Melibokus, Odenwald, West Germany; Trög. 334; Joh. v.2, p.402; Tomk. p.339)*

MALIGNITE. A mesocratic nepheline syenite containing abundant aegirine-augite and some orthoclase and nepheline in equal amounts. Many other mafic minerals, such as amphibole, garnet, biotite, may be present. Now defined modally as a mesocratic variety of foid syenite in QAPF field 11 (Figs. B.4 and B.7). *(Lawson, 1896, p.337; Maligne River, Ontario, Canada; Trög. 487; Joh. v.4, p.115; Tomk. p.340)*

MAMILITE. A variety of lamproite essentially composed of leucite and magnophorite (an alkali amphibole) with subordinate phlogopite in a glassy matrix. *(Wade & Prider, 1940, p.68; Mamilu Hill, Kimberley district, West Australia; Tomk. p.340)*

MANDCHURITE (MANCHURITE, MANDCH-OURITE, MANDSCHURITE). A glassy variety of basanite in which plagioclase is not present as a mineral phase. *(Lacroix, 1928a, p.47; named after Manchuria, China; Trög. 592; Tomk. p.340)*

MANDELSTEIN. An old German name for an amygdaloidal rock. *(Werner, 1787, p.13; from the German mandel = almond, stein = rock; Trög. 910; Tomk. p.340)*

MANDSCHURITE. See mandchurite.

MANGERITE. An intermediate member of the charnockitic rock series equivalent to hypersthene monzonite of QAPF field 8 and frequently containing mesoperthite. *(Kolderup, 1903, p.109; Kalsaas, near Manger, Rado, Bergen district, Norway; Trög. 278; Joh. v.3, p.63; Tomk. p.341)*

MAPPAMONTE. A local name for a loosely textured grey tuff-like material found near Naples. *(Lorenzo, 1904, p.309; Tomk. p.341)*

MARCHITE. An obsolete name for a variety of websterite composed of diopside and enstatite. *(Kretschmer, 1917, p.149; March River (now Morava), Moravia, Czechoslovakia; Trög. 686; Joh. v.4, p.461; Tomk. p.341)*

MAREKANITE. A term given to the more or less rounded glass fragments, often showing concave indentations, that are formed when perlite fractures. *(Judd, 1886b, p.241; Marekanka River, Ochotsk Sea, Siberia, USSR; Trög. 911; Joh. v.2, p.284; Tomk. p.341)*

MAREUGITE. A variety of foid-bearing gabbro dominated by Ti-augite and bytownite and also containing haüyne. *(Lacroix, 1917a, p.587; Mareuges, Mont Dore, Auvergne, France; Trög. 553; Joh. v.4, p.222; Tomk. p.341)*

MARIANITE. A variety of boninite consisting of clinoenstatite, bronzite and augite phenocrysts and microlites in a glassy matrix with shaped psuedomorphs after olivine. The new name was proposed on the grounds that the type boninite was richer in olivine. *(Sharaskin et al., 1980, p.473; Mariana Trench, NW Pacific Ocean)*

MARIENBERGITE. A variety of natrolite phonolite with phenocrysts of Na-sanidine, andesine, and augite in a matrix of Na-sanidine, natrolite, and sodalite. *(Johannsen, 1938, p.169; Mt. Marienberg, near Ústí nad Labem, N. Bohemia, Czechoslovakia; Trög(38). 526; Tomk. p.342)*

MARIUPOLITE. A leucocratic variety of nepheline syenite characterised by the absence of K-feldspar and the presence of albite and aegirine. *(Morozewicz, 1902, p.244; Mariupol, Sea of Azov, Ukraine, USSR; Trög. 416; Joh. v.4, p.211; Tomk. p.342)*

MARKFIELDITE. A coarse-grained dioritic rock with idiomorphic andesine, augite or hornblende and micrographic intergrowths of quartz and orthoclase in the groundmass. The term was later used for lamprophyric rocks of similar mineralogy. *(Hatch, 1909, p.219; Markfield, Charnwood Forest, Leicester, England; Trög(38). 116; Joh. v.2, p.295; Tomk. p.342)*

MARLOESITE. A local name for an altered variety of trachyte containing olivine, hornblende and augite. *(Thomas, 1911, p.198; Marloes Bay, Dyfed, Wales; Trög. 214; Joh. v.3, p.175; Tomk. p.343)*

MAROSITE. An obsolete local name for a melanocratic variety of monzogabbro rich in biotite and augite with equal but small amounts of sanidine and bytownite and minor nepheline. *(Iddings, 1913, p.246; Pic de Maros, Sulawesi, Indonesia; Trög. 262; Joh. v.4, p.41; Tomk. p.344)*

MARSCOITE. A local name for a hybrid rock formed by the mixing of ferrodiorite with rhyolite. Variable in texture and composition the rock contains resorbed xenocrysts of quartz and andesine. *(Harker, 1904, p.175; Marsco, Island of Skye, Scotland; Trög. 332; Tomk. p.344)*

MARTINITE. An obsolete name for a variety of phonolitic tephrite composed essentially of plagioclase, leucite and alkali feldspar and with accessory clinopyroxene. *(Johannsen, 1938, p.200; Croce di San Martino, Vico volcano, near Viterbo, Italy; Trög(38). 569½; Tomk. p.344)*

MARUNDITE. A local mnemonic name for a *mar*garite cor*und*um pegmatite. *(Hall, 1922,* *p.43; Trög. 776; Tomk. p.344)*

MASAFUERITE. A variety of picrite basalt composed of 50% olivine phenocrysts in a groundmass of augite, calcic plagioclase and ores. *(Johannsen, 1937, p.334; Masafuera, Juan Fernandez Islands, Pacific Ocean; Trög(38). 410; Tomk. p.344)*

MASANITE. A variety of granite porphyry containing phenocrysts of zoned oligoclase in a micropegmatitic groundmass of quartz, orthoclase and minor biotite. *(Kotô, 1909, p.190; Ku-ryong Copper Mine, Ma-san-pho, S Korea; Trög. 89; Tomk. p.345)*

MASANOPHYRE. An obsolete name for a variety of masanite containing quartz and oligoclase phenocrysts in a micropegmatitic groundmass. *(Kotô, 1909, p.192; Trög. 89; Tomk. p.345)*

MASEGNA. A local name for a trachyte from the Euganean Hills, near Padova, Italy. *(Rio, 1822, p.350; Tomk. p.345)*

MATRAITE. A term proposed by Prof. Szabó for a variety of augite andesite in which the feldspar is anorthite. Now obsolete. *(Judd, 1876, p.302; Matra district, Hungary; Tomk. p.345)*

MEDIUM-K. A chemical term, now defined as part of the TAS classification (Fig. B.15), for rocks that lie between the high-K and low-K fields in the $SiO_2 - K_2O$ diagram.

MEIMECHITE (MEYMECHITE). An ultramafic volcanic rock composed of olivine phenocrysts in a groundmass of olivine, clinopyroxene, magnetite and glass. *(Moor & Sheinman, 1946, p.141; River Meimecha, tributary of River Kheta, N. Siberia, USSR; Tomk. p.348)*

MELA-. Originally used as a prefix for rocks with between 5% and 50% felsic minerals, but now used in the modal QAPF classification as indicating that the rock has considerably more mafic minerals than would be regarded as normal for that rock type (Fig. B.7). *(Johannsen, 1920a, p.48; Joh. v.1, p.153; Tomk. p.348)*

MELANEPHELINITE. Originally defined as a volcanic rock consisting of abundant pyroxene and some nepheline but with no olivine. Now commonly used for undersaturated basic and ultrabasic rocks consisting of abundant augite phenocrysts in a groundmass of inde-

terminate mineralogy. *(Johannsen, 1937, p.363; Joh. v.4, p.363)*

MELANIDE. An obsolete term suggested to replace names, such as greenstone, in which the ferromagnesian minerals cannot be distinguished in hand specimen. *(Milch, 1927, p.63; Joh. v.1, p.57; Tomk. p.348)*

MELANO-APHANEID. An obsolete alternative term for anameseid.*(Johannsen, 1911,p.321)*

MELANOCRATE. A general term proposed for rocks rich in dark-coloured minerals. *(Brögger, 1898, p.263; from the Greek melas = dark; Joh. v.1, p.180)*

MELANOCRATIC. Relating to rocks rich in dark-coloured minerals. *(Brögger, 1898, p.263; from the Greek melas = dark, krateo = to predominate; Joh. v.1, p.180; Tomk. p.348)*

MELANOLITE. An obsolete term for a melanocratic igneous rock. *(Loewinson-Lessing, 1901, p.118)*

MELANOPHYREID. An obsolete alternative term for anameseid porphyry. *(Johannsen, 1911, p.321; Tomk. p.349)*

MELANOPHYRIDE. A revised spelling recommended to replace the field term melanophyreid. Now obsolete. *(Johannsen, 1926, p.182; from the Greek melas = dark; Joh. v.1, p.58; Tomk. p.349)*

MELANOPTOCHE. An obsolete adjectival term for igneous rocks poor in dark coloured minerals. *(Loewinson-Lessing, 1901, p.118; Tomk. p.349)*

MELAPHYRE. A very old name replacing 'trapporphyr' and originally used for amygdaloidal rocks composed of plagioclase and augite but later used for Upper Palaeozoic rocks, usually basaltic. *(Brongniart, 1813, p.40; from the Greek melas = dark; Trög. 388; Joh. v.3, p.296; Tomk. p.349)*

MELAPORPHYRE. An obsolete name for a dark-coloured labradorite porphyry. *(Senft, 1857, p.271; Tomk. p.349)*

MELFITE. An obsolete local name for a haüynophyre. *(Lacroix, 1933, p.199; Melfi, Vulture, Italy; Trög. 649; Tomk. p.349)*

MELILITE BASALT. A term used for a volcanic rock consisting of phenocrysts of augite and olivine in a groundmass essentially of melilite, augite, olivine, and occasional nepheline. The name should not be used as the term basalt is now restricted to a rock containing essential plagioclase. As the rock is a variety of foidite it should be given the appropriate name, e.g. olivine melilitite. *(Stelzner, 1882, p.229; Hochbohl, near Owen, Württemberg, West Germany; Trög. 914; Joh. v.4, p.347; Tomk. p.349)*

MELILITHITE. An obsolete term proposed for a monomineralic volcanic rock consisting of melilite. The term is synonymous with melilitholith. *(Loewinson-Lessing, 1901, p.114; Trög. 749; Joh. v.4, p.346; Tomk. p.349)*

MELILITHOLITH. A term proposed to replace melilithite, for a monomineralic volcanic rock consisting of melilite. Now obsolete. *(Johannsen, 1938, p.337)*

MELILITITE. An ultramafic volcanic rock consisting essentially of melilite and pyroxene. Perovskite is also commonly present. Now defined modally in the melilitic rocks classification (Fig. B.3). *(Lacroix, 1893, p.627; Joh. v.4, p.346; Tomk. p.349)*

MELILITOLITE. An ultramafic plutonic rock consisting essentially of melilite, pyroxene and olivine. Now defined modally in the melilitic rocks classification (Fig. B.3). *(Lacroix, 1933, p.197; Trög. 744; Joh. v.4, p.337; Tomk. p.349)*

MELMAFITE. A mnemonic name for a mixture of *mel*ilite and *maf*ic minerals. *(Hatch et al., 1961, p.358)*

MELTEIGITE. The melanocratic member of the ijolite series, containing 10 to 30% of nepheline. Now defined modally as a melanocratic variety of foidite in QAPF field 15 (Figs. B.4 and B.7). *(Brögger, 1921, p.18; Melteig, Fen complex, Telemark, Norway; Trög. 609; Joh. v.4, p.327; Tomk. p.350)*

MESITE. An obsolete name for a rock of intermediate composition. *(Loewinson-Lessing, 1898, p.39; Tomk. p.351)*

MESO-. Originally used as a prefix for rocks with between 95% and 50% felsic minerals. *(Johannsen, 1920a, p.48; Joh. v.1, p.153; Tomk. p.351)*

MESOBUGITE. A member of the bugite series with 53% to 66% SiO_2, 5% to 20% hypersthene, oligoclase antiperthite and quartz. As all other members of the bugite series have

been recommended as terms to be abandoned (Streckeisen, 1974) it is suggested that this term should also be abandoned. *(Bezborod'ko, 1931, p.142; Bug River, Podolia, Ukraine, USSR; Tomk. p.80)*

MESTIGMERITE. A variety of melanocratic nepheline syenite or malignite with abundant aegirine-augite, nepheline, and orthoclase with some titanite and apatite. *(Duparc, 1926, p.b120; Mestigmer, Morocco; Trög(38). 487½; Tomk. p.353)*

METABOLITE. An obsolete term for an altered trachyte glass. *(*Wadsworth, 1893, p.97; Joh. v.3, p.77; Tomk. p.353)*

METALUMINOUS. A term used for rocks which are essentially undersaturated with respect to alumina and characterised mineralogically by hornblende and/or augite and/or biotite and chemically by hypersthene being in excess of corundum, which must be present, in the norm. *(Shand, 1927, p.128)*

MEYMECHITE. See meimechite.

MIAGITE. An orbicular variety of gabbro composed essentially of calcic plagioclase with hornblende, minor hypersthene and quartz. This rock has also been called corsite, kugeldiorite and napoleonite. *(Pinkerton, 1811b, p.63; Glacier Miage, Mont Blanc, France; Trög(38). 916½; Joh. v.3, p.232; Tomk. p.359)*

MIAROLITE. A textural variety of fine-grained granite containing many irregular drusy cavities. *(Fournet, 1845, p.495; from the Italian name of the rock = miarolo; Joh. v.2, p.130; Tomk. p.360)*

MIASKITE (MIASCITE). A leucocratic variety of biotite nepheline monzosyenite with oligoclase and perthitic orthoclase. Now used as a special term for oligoclase nepheline monzosyenite of QAPF field 12 (Fig. B.4). *(Rose, 1839, p.375; Miask, Ilmen Mts, Urals, USSR; Trög. 509; Tomk. p.360)*

MIASKITIC (MIASCITIC). A general term for nepheline syenites in which the molecular ratio of $Na_2O + K_2O / Al_2O_3 < 1$. Cf. agpaitic. *(Fersman, 1929, p.63; Tomk. p.360)*

MICROCLINITE. A variety of alkali feldspar syenite composed almost entirely of microcline. *(Loewinson-Lessing, 1901, p.114; Trög. 165; Joh. v.3, p.5; Tomk. p.361)*

MICROTINITE. An obsolete term for plagioclase-bearing trachytes, but later used by Lacroix (1900) for recrystallised ejected blocks of dioritic composition that contain glassy plagioclase (microtine). *(Wolf, 1866, p.33; Trög. 307; Tomk. p.364)*

MID-OCEAN RIDGE BASALT. A variety of low-K tholeiitic basalt, low in Ti, erupted at mid-ocean ridges consisting of Mg-olivine, Ca-rich clinopyroxene, plagioclase, Ti-magnetite, and variable amounts of pale brown glass. Orthopyroxene and ilmenite are very rare. *(Sun, S-S. et al, 1979, p.119)*

MIENITE. An obsolete name for a variety of rhyolite glass containing crystals of labradorite and minor hypersthene. *(Scheumann, 1925, p.283; Trög. 96; Tomk. p.365)*

MIHARAITE. An obsolete name for a variety of basalt with phenocrysts of bytownite and hypersthene in a groundmass of labradorite, augite, opaques and glass. *(Tsuboi, 1918, p.47; Volcano Mihara, Oshima Island, Japan; Trög. 162; Joh. v.3, p.284; Tomk. p.366)*

MIJAKITE. An obsolete term for a Mn-rich variety of basalt consisting of phenocrysts of bytownite, augite and hypersthene, in a groundmass of andesine, magnetite and pyroxene, which is probably Mn-rich. *(Petersen, 1890a, p.47; Mijakeshima, Bonin Islands, now Ogaswara-gunto Islands, Japan; Trög. 161; Joh. v.3, p.281; Tomk. p.366)*

MIKENITE. An obsolete term for a variety of leucitite essentially composed of leucite, nepheline and clinopyroxene and in which the Na_2O/K_2O ratio is higher than usual for rocks of this type. *(Lacroix, 1933, p.196; Mikeno volcano, Birunga volcanic field, Rwanda; Trög. 642; Tomk. p.366)*

MIMESITE. This name does not exist in the reference given (Cordier, 1868) but matches the description given for mimosite. *(Tröger, 1935, p.325; Trög. 918)*

MIMOPHYRE. An obsolete general term for volcanic ash, tuff, agglomerate etc. *(Brongniart, 1813, p.46; from the Greek mimos = imitation; Tomk. p.366)*

MIMOSE. An obsolete name originally suggested by Haüy for dolerite. *(Brongniart, 1813, p.32; from the Greek mimos = imitation; Joh. v.3, p.303; Tomk. p.366)*

MIMOSITE. An obsolete term originally used for melanocratic basaltic rocks, but redefined by Macdonald (1949) for highly undersaturated picrobasalts with >15% normative foids but no recognizable nepheline. *(Cordier, 1842, Vol.8, p.227; from the Greek mimos = imitation; Trög. 919; Joh. v.3, p.303; Tomk. p.366)*

MINETTE. Originally an old miners term for oolitic ironstones. Later used for a variety of lamprophyre consisting of phenocrysts of phlogopite-biotite and occasionally amphiboles in a groundmass of the same minerals plus orthoclase and minor plagioclase. Mg-olivine and diopsidic pyroxene may also be present. Now defined in the lamprophyre classification (Table B.3). *(Elie de Beaumont, 1822, p.524; Minkette Valley, Vosges, France; Trög. 247; Joh. v.3, p.33; Tomk. p.368)*

MINETTEFELS. An obsolete term suggested to replace the term minette, when used as a lamprophyre dyke rock name, because oolitic ironstones of Lorraine were also called minette. *(Kretschmer, 1917, p.21; Trög. 920; Tomk. p.368)*

MINIMITE. A term suggested for experimental compositions that crystallise at the minimum temperature for a given pressure. *(Reynolds, 1958, p.390; Tomk. p.368)*

MINVERITE. A local name for an albite dolerite with spilitic affinities that carries a primary brown hornblende. *(Dewey, 1910, p.46; St. Minver, Cornwall, England; Trög. 235; Joh. v.3, p.141; Tomk. p.369)*

MISSOURITE. A melanocratic intrusive rock composed of clinopyroxene and subordinate leucite and olivine. Now defined modally as a melanocratic variety of foidite in QAPF field 15 (Figs. B.4 and B.7). *(Weed & Pirsson, 1896, p.323; Missouri River, Highwood Mts., Montana, USA; Trög. 631; Joh. v.4, p.334; Tomk. p.369)*

MODLIBOVITE. A local term for a variety of polzenite (a lamprophyre) containing olivine and biotite phenocrysts in a groundmass of melilite, lazurite, biotite, augite and nepheline. *(Scheumann, 1922, p.496; Modlibov, N. Bohemia, Czechoslovakia; Trög. 664; Joh. v.4, p.388; Tomk. p.370)*

MODUMITE. A local name for an anorthositic facies of the Oslo alkali gabbro. *(Brögger,*

1933, p.35; Modum, Oslo district, Norway; Trög. 298; Joh. v.4, p.66; Tomk. p.370)

MONCHIQUITE. A variety of lamprophyre similar to camptonite except that the groundmass is feldspar-free and composed of combinations of glass and feldspathoids, especially analcime. Now defined in the lamprophyre classification (Table B.3). *(Hunter & Rosenbusch, 1890, p.447; Caldas de Monchique, Algarve, Portugal; Trög. 374; Joh. v.4, p.375; Tomk. p.371)*

MONDHALDEITE. An obsolete local name for a "camptonite-like" dyke rock largely composed of equal proportions of plagioclase and alkali feldspar, with subordinate clinopyroxene, amphibole and leucite, in a glassy matrix. *(Gruss, 1900, p.89; Mondhalde, Kaiserstuhl, Baden, West Germany; Trög. 264; Joh. v.4, p.55; Tomk. p.371)*

MONMOUTHITE. A variety of urtite composed essentially of nepheline and some hastingsite with minor albite and calcite. *(Adams & Barlow, 1910, p.277; Monmouth Township, Ontario, Canada; Trög. 606; Joh. v.4, p.317; Tomk. p.372)*

MONNOIRITE. A poorly defined local name for a coarse-grained porphyritic rock transitional between essexite and pulaskite. *(Osborne & Wilson, 1934, p.181; Monnoir, now Mt. Johnson, Quebec, Canada; Tomk. p.372)*

MONTREALITE. A highly melanocratic variety of alkali gabbro, dominated by olivine, Ti-augite and kaersutite. Labradorite can be in small amounts and minor nepheline may occur. *(Adams, 1913, p.38; Mount Royal, Montreal, Quebec, Canada; Trög. 397; Joh. v.4, p.430; Tomk. p.374)*

MONZODIORITE. A term suggested to replace syenodiorite for a plutonic rock intermediate between monzonite and diorite. Now defined modally in QAPF field 9 (Fig. B.4). *(Johannsen, 1920b, p.174; Trög. 994; Tomk. p.374)*

MONZOGABBRO. A term suggested to replace syenogabbro for a plutonic rock of gabbroic aspect that contains minor but essential orthoclase as well as calcic plagioclase. Now defined modally in QAPF field 9 (Fig. B.4). *(Johannsen, 1920c, p.212; Trög. 995; Joh. v.3, p.126; Tomk. p.374)*

MONZOGRANITE. A variety of granite having roughly equal amounts of alkali feldspar and plagioclase. Now defined modally in QAPF field 3b (Fig. B.4). Lacroix (1933, p.188) used the similar term "granite monzonitique". *(Streckeisen, 1967, p.166)*

MONZONITE. There has been considerable divergence in the use of the term due to the variety of rock types found in the Monzoni district but it is now commonly used for a plutonic rock containing almost equal amounts of plagioclase and alkali feldspar with minor amphibole and/or pyroxene. Now defined modally in QAPF field 8 (Fig. B.4). *(Lapparent, 1864, p.260; Mt. Monzoni, Alto Adige, Italy; Trög. 259; Joh. v.3, p.94; Tomk. p.374)*

MONZONORITE. A plutonic rock of gabbroic aspect enriched in hypersthene that contains plagioclase (oligoclase to labradorite) with minor but essential orthoclase. *(Johannsen, 1920c, p.212; Tomk. p.374)*

MONZOSYENITE. A collective name originally given to acid rocks of variable composition from the Monzoni region that contained both alkali feldspar and plagioclase. The same rocks were later called monzonite. *(Buch, 1824, p.344; Joh. v.3, p.95; Tomk. p.374)*

MOYITE. A term proposed for a variety of biotite granite in which quartz exceeds orthoclase. *(Johannsen, 1920b, p.158; Moyie Sill, Purcell Range, British Columbia, Canada; Trög. 55; Joh. v.2, p.28; Tomk. p.376)*

MUGEARITE. A volcanic rock, often exhibiting flow texture, containing small phenocrysts of olivine, augite and magnetite in a matrix of oligoclase, augite and magnetite with interstitial alkali feldspar. Now defined chemically as the sodic variety of basaltic trachyandesite in TAS field S2 (Fig. B.13). *(Harker, 1904, p.264; Mugeary, Island of Skye, Scotland; Trög. 290; Joh. v.3, p.118; Tomk. p.377)*

MUGODZHARITE. A local term for a plutonic rock consisting essentially of quartz and alkali felspar with minor epidote. Cf. alaskite. *(Chumakov, 1946, p.295; Mudgodzhar Hills, between Ural Mts. and Aral Sea, USSR; Tomk. p.378)*

MULATOPORPHYRY (MULATTOPHYRE). An obsolete name for a variety of quartz porphyry.

(Klipstein, 1843, p.78; Mt. Mulato, Predazzo, Alto Adige, Italy; Tomk. p.378)

MULDAKAITE. An ultramafic rock consisting of uralite, augite and hornblende. *(Karpinskii, 1869, p.231; Muldakaeva, Urals, USSR; Tomk. p.378)*

MUNIONGITE. A variety of phonolite with phenocrysts of nepheline in a groundmass of sanidine, aegirine-augite and glass. *(David et al., 1901, p.377; from the aboriginal muniong = Kosciusko Plateau, New South Wales, Australia; Trög. 456; Joh. v.4, p.268; Tomk. p.378)*

MURAMBITE. A melanocratic variety of leucite basanite composed of plagioclase, clinopyroxene, olivine and with subordinate leucite. *(Holmes & Harwood, 1937, p.136; Murambe volcano, Bufumbira, Uganda; Trög(38). 595¾; Tomk. p.379)*

MURAMBITOID. A term used for a melanocratic variety of murumbite containing very little leucite. *(Holmes & Harwood, 1937, p.145; Murambe volcano, Bufumbira, Uganda; Tomk. p.379)*

MURITE. A variety of phonolite containing phenocrysts of fayalite and Ti-augite (rimmed by aegirine-augite) in a groundmass of nepheline, sanidine and aegirine-augite. *(Lacroix, 1927b, p.32; Cape Muri, Rarotonga, Cook Islands, Pacific Ocean; Trög. 501; Joh. v.4, p.262; Tomk. p.379)*

NADELDIORITE. An obsolete local name for a diorite porphyrite containing needle-shaped hornblende. *(Gümbel, 1868, p.348; Rohrbach, Regen, Bavaria, West Germany; Trög. 316)*

NAPOLEONITE. An orbicular variety of gabbro composed essentially of calcic plagioclase with hornblende, minor hypersthene and quartz. This rock has also been called corsite, kugeldiorite and miagite. *(Cotta, 1866, p.155; named after Napoleon I; Trög. 361; Joh. v.3, p.232; Tomk. p.381)*

NATRIOPLETE. An obsolete term for a leucocratic igneous rock rich in sodium. *(Brögger, 1898, p.266; Tomk. p.381)*

NATROCARBONATITE. A rare variety of carbonatite lava, currently only known from one locality, consisting essentially of the Na-Ca-K carbonate minerals, nyerereite and gregoryite. It has also been called lengaite.

(Du Bois et al., 1963, p.446; Oldoinyo Lengai, Tanzania; Tomk. p.381)

NAUJAITE. A local name for an agpaitic variety of nepheline sodalite syenite characterised by a poikilitic texture with small crystals of sodalite enclosed in large grains of alkali feldspar, arfvedsonite, aegirine and eudialyte. The content of sodalite may exceed 50% of the rock. *(Ussing, 1912, p.32; Naujakasik, Ilimaussaq, Greenland; Trög. 635; Joh. v.4, p.250; Tomk. p.382)*

NAVITE. An obsolete name for a basaltic rock composed of phenocrysts of andesine, augite, enstatite and iddingsite in a groundmass of these minerals, opaques and glass. *(Rosenbusch, 1887, p.512; Nave, now Nahe district, Rheinland-Pfalz, West Germany; Trög. 346; Joh. v.3, p.298; Tomk. p.382)*

NAXITE. A corundum, phlogopite, plagioclase rock containing patches of blue tourmaline. Not found *in situ* but possibly from the contact zone between a muscovite granite and a peridotite. *(Papastamatiou, 1939, p.2089; Island of Naxos, Greece; Tomk. p.382)*

NEAPITE. A mnemonic name from *ne*pheline and *apa*tite, used for alkaline intrusive rocks consisting mainly of nepheline and apatite with minor aegirine and biotite. *(Vlodavets, 1930, p.34; Khibina complex, Kola Peninsula, USSR; Trög(38). 608¾; Tomk. p.382)*

NECROLITE. A local name for a variety of vesicular biotite latite occurring in the Viterbo and Tolfa regions in Tuscany; used for making Etruscan sarcophagi. *(Brocchi, 1817, p.156; Tomk. p.382)*

NEIVITE. A local name for a melanocratic variety of alkali feldspar syenite consisting of major amounts of hornblende with lesser amounts of albite and minor magnetite, pyrite, apatite, titanite, and calcite. *(Sobolev, 1959, p.115; River Neiva, Urals, USSR; Tomk. p.383)*

NELSONITE. A granular dyke-rock consisting essentially of ilmenite and apatite with or without rutile. Several varieties are distinguished according to mineral prefixes. *(Watson, 1907, p.300 states that the name was proposed "elsewhere" but gives no reference; Nelson County, Virginia, USA; Trög. 770; Joh. v.4, p.471; Tomk. p.383)*

NEMAFITE. A mnemonic name for a mixture of nepheline and *mafic* minerals. *(Hatch et al., 1961, p.358)*

NEMITE. A local name for a melanocratic variety of leucitite. *(Lacroix, 1933, p.199; Lake Nemi, Alban Hills, near Rome, Italy; Trög. 646; Tomk. p.383)*

NEPHELINE ANDESITE. A term suggested for volcanic rocks with sodic plagioclase and some nepheline but without olivine. They differ from basanite and tephrite which usually contain calcic plagioclase. *(Johannsen, 1938, p.215; Trög. 574; Tomk. p.385)*

NEPHELINE BASALT. A term used for a volcanic rock composed essentially of pyroxene, nepheline, and olivine. The name should not be used as the term basalt is now restricted to a rock containing essential plagioclase. As the rock is a variety of foidite it should be given the appropriate name, e.g. olivine nephelinite. *(Naumann, 1850, p.650; Wickenstein, Silesia, Poland; Trög. 923; Joh. v.4, p.338; Tomk. p.385)*

NEPHELINE BASANITE. Now defined in QAPF field 14 (Fig. B.10) as a variety of basanite in which nepheline is the most abundant foid. As this type of basanite is the commonest variety it is often known simply as basanite. *(Rosenbusch, 1887, p.763; Böhmisches Mittelgebirge (now České Středohoří), N. Bohemia, Czechoslovakia; Trög. 591; Joh. v.4, p.231; Tomk. p.385)*

NEPHELINE BASITE. An obsolete term used in two senses – for basanite and for all nepheline-bearing ultrabasic rocks. *(Vogelsang, 1872, p.542; Tomk. p.385)*

NEPHELINE BENMOREITE. A variety of benmoreite in which nepheline is present in small amounts. *(Coombs & Wilkinson, 1969, p.493)*

NEPHELINE DIORITE. Now defined in QAPF field 14 (Fig. B.4) as a variety of foid diorite in which nepheline is the most abundant foid. *(Johannsen, 1920b, p.177; Trög. 547; Joh. v.4, p.214)*

NEPHELINE GABBRO. Now defined in QAPF field 14 (Fig. B.4) as a variety of foid gabbro in which nepheline is the most abundant foid. The special term theralite may be used as an alternative. *(Lacroix, 1902, p.191; Trög. 552; Tomk. p.385)*

NEPHELINE HAWAIITE. A variety of hawaiite in which nepheline is present in small amounts. *(Coombs & Wilkinson, 1969, p.493)*

NEPHELINE LATITE. A term proposed in a classification for a volcanic rock in which foids > 10% of the total rock and alkali feldspar is 40%-60% of the total feldspar. *(Nockolds, 1954, p.1008)*

NEPHELINE MONZODIORITE. Now defined in QAPF field 13 (Fig. B.4) as a variety of foid monzodiorite in which nepheline is the most abundant foid. The special term essexite may be used as an alternative. *(Johannsen, 1920b, p.177)*

NEPHELINE MONZOGABBRO. Now defined in QAPF field 13 (Fig. B.4) as a variety of foid monzogabbro in which nepheline is the most abundant foid. The special term essexite may be used as an alternative. *(Johannsen, 1920c, p.216; Trög. 542)*

NEPHELINE MONZONITE. An alkaline plutonic rock with essential nepheline and roughly equal amounts of alkali feldspar and plagioclase. *(Lacroix, 1902, p.33; Trög. 510)*

NEPHELINE MONZOSYENITE. Now defined in QAPF field 12 (Fig. B.4) as a variety of foid monzosyenite in which nepheline is the most abundant foid. The term is synonymous with nepheline plagisyenite.

NEPHELINE MUGEARITE. A variety of mugearite in which nepheline is present in small amounts. *(Coombs & Wilkinson, 1969, p.493)*

NEPHELINE PLAGISYENITE. Now defined in QAPF field 12 (Fig. B.4) as a variety of foid plagisyenite in which nepheline is the most abundant foid. The term is synonymous with nepheline monzosyenite.

NEPHELINE SYENITE. Now defined in QAPF field 11(Fig. B.4) as a variety of foid syenite in which nepheline is the most abundant foid. *(Rosenbusch, 1877, p.203; Trög. 412; Joh. v.4, p.77; Tomk. p.385)*

NEPHELINE TEPHRITE. Now defined in QAPF field 14 (Fig. B.10) as a variety of tephrite in which nepheline is the most abundant foid. As this is the common variety it is often simply called tephrite. *(Rosenbusch, 1877, p.492; Trög. 576; Joh. v.4, p.231)*

NEPHELINE TRACHYANDESITE. A variety of trachyandesite in which nepheline is present in small amounts. *(Coombs & Wilkinson, 1969, p.493; Trög. 571)*

NEPHELINE TRACHYBASALT. A variety of trachybasalt in which nepheline is present in small amounts. *(Coombs & Wilkinson, 1969, p.493)*

NEPHELINE TRISTANITE. A variety of tristanite in which nepheline is present in small amounts. *(Coombs & Wilkinson, 1969, p.493)*

NEPHELINITE. A term originally used for a nepheline-bearing basaltic rock but now used for rocks consisting essentially of nepheline and clinopyroxene. Now defined modally as a variety of foidite of QAPF field 15c (Fig. B.10) and, if modes are not available, chemically in TAS field F-U1 (Figs. B.13). *(Cordier, 1842, Vol.8, p.618; Trög. 615; Joh. v.4, p.338; Tomk. p.385)*

NEPHELINITE BASANITE. A variety of nepheline basanite in which nepheline exceeds plagioclase. *(Johannsen, 1938, p.301; Tomk. p.386)*

NEPHELINITE TEPHRITE. A variety of nepheline tephrite in which nepheline exceeds plagioclase. *(Johannsen, 1938, p.301)*

NEPHELINITOID. An obsolete term for a variety of nephelinite in which the interstitial nepheline cannot be determined optically, but may be inferred chemically. *(Bořický, 1874, p.41; Hasenberg, Bohemia, Czechoslovakia; Joh. v.4, p.344; Tomk. p.386)*

NEPHELINOLITE. A plutonic rock now defined as a variety of foidolite of QAPF field 15c (Fig. B.4) in which nepheline is the most abundant foid. The term is subdivided into urtite, ijolite and melteigite on a basis of the mafic mineral content (Fig. B.7). The rock is the plutonic equivalent of the volcanic rock nephelinite. *(Streckeisen, 1976, p.21; Tomk. p.386)*

NEPHELINOLITH. An obsolete term proposed for a monomineralic volcanic rock consisting of nepheline. *(Loewinson-Lessing, 1901, p.114; Trög. 603; Joh. v.4, p.336; Tomk. p.386)*

NEVADITE. A local name for a porphyritic variety of rhyolite containing abundant phenocrysts of quartz, sanidine and plagioclase with minor biotite and hornblende. *(Richthofen, 1868, p.16; named after Nevada, USA; Trög. 42; Joh. v.2, p.273; Tomk. p.387)*

NEVOITE. A melanocratic, apatite-rich ultrabasic rock, composed of biotite, hornblende, apatite, alkali feldspar, and augite with minor titanite. *(Khazov, 1983, p.1200; Lake Nevo, now Ladoga, near Lenigrad, USSR)*

NEWLANDITE. An obsolete name for a variety of garnet websterite composed of garnet, Cr-diopside and enstatite. Occurs as inclusions in kimberlites. *(Bonney, 1899, p.315; Newlands diamond pipe, South Africa; Trög. 717; Tomk. p.387)*

NGURUMANITE. A medium-grained variety of melteigite composed of pyroxene, altered nepheline in an iron-rich mesostasis with vugs of zeolites and calcite. *(Saggerson & Williams, 1963, p.479; Nguruman Escarpment, Kenya)*

NIKLESITE. An obsolete name for a variety of websterite composed of diallage, enstatite and diopside, with lamellae of enstatite and diallage. *(Kretschmer, 1917, p.164; Nikles (now Raškov), Moravia, Czechoslovakia; Trög. 928; Joh. v.4, p.461; Tomk. p.387)*

NILIGONGITE. A medium-grained variety of melilite-bearing leucite ijolite with alkali pyroxenes, and with equal proportions of leucite and nepheline. *(Lacroix, 1933, p.198; Niligongo volcano, Birunga, Kivu, Zaire; Trög. 630; Joh. v.1(2nd.Ed.), p.269; Tomk. p.387)*

NIOLITE. An obsolete term for a variety of felsite containing spherules of radiating feldspar. *(Pinkerton, 1811b, p.74; Chain of Niolo, Corsica; Tomk. p.388)*

NONESITE. An obsolete local name for a variety of basalt composed of phenocrysts of labradorite, augite and olivine in a groundmass of andesine, augite, orthoclase, opaques and glass. *(Lepsius, 1878, p.163; Mendola Pass, Nonsberg, Alto Adige, Italy; Trög. 345; Joh. v.3, p.285; Tomk. p.389)*

NORDMARKITE. A variety of quartz-bearing alkali feldspar syenite composed mainly of microperthite with minor biotite, alkali amphibole or pyroxene. *(Brögger, 1890, p.54; Nordmarka district, near Oslo, Norway; Trög. 185; Joh. v.3, p.6; Tomk. p.389)*

NORDSJÖITE. A local name for a coarse-grained

variety of calcite nepheline syenite in which nepheline exceeds alkali feldspar. Aegirine-augite is usually present, sometimes with melanite. *(Johannsen, 1938, p.247; Stöve, Nordsjö, Fen complex, Telemark, Norway; Trög(38). 421½; Tomk. p.389)*

NORITE. A plutonic rock composed essentially of bytownite, labradorite or andesine and orthopyroxene. Now defined modally in the gabbroic rock classification (Fig. B.6). *(Esmark, 1823, p.207; Trög. 355; Joh. v.3, p.233; Tomk. p.390)*

NORTHFIELDITE. A variety of quartzolite occurring as a border phase in a granite and resembling vein quartz or greisen composed essentially of quartz with minor muscovite and traces of tourmaline and tremolite or actinolite. Cf. esmeraldite. *(Emerson, 1915, p.212; Crag Mtn., Northfield, Massachusetts, USA; Trög. 4; Joh. v.2, p.18; Tomk. p.390)*

NOSEAN BASANITE. Now defined in QAPF field 14 (Fig. B.10) as a variety of basanite in which nosean is the most abundant foid. *(Tröger, 1935, p.246; Trög. 597; Joh. v.4, p.238)*

NOSEANITE. A variety of nephelinite containing considerable amounts of nosean and amphibole. Now defined as a variety of foidite of QAPF field 15c (Fig. B.10). *(Bořický, 1874, p.41; St. Georgenberg, Říp, Bohemia, Czechoslovakia; Trög. 651; Joh. v.4, p.345; Tomk. p.390)*

NOSEANOLITH. An obsolete term proposed for a monomineralic volcanic rock consisting of nosean. *(Johannsen, 1938, p.337; Trög. 634; Tomk. p.391)*

NOSELITITE. An obsolete term proposed for a variety of noseanite consisting mainly of nosean, pyroxene and amphibole. The term is synonymous with noseanite. *(Johannsen, 1938, p.345; Tomk. p.391)*

NOSYKOMBITE. A local name for a variety of nepheline monzosyenite consisting of sanidine, nepheline, and kaersutite with small amounts of plagioclase; corresponds to the nosykombitic magma-type of Niggli. *(Niggli, 1923, p.157; Nosy-Komba Island, Madagascar; Trög. 507; Tomk. p.391)*

NOTITE. An obsolete name for a variety of porphyry containing quartz, feldspar and mica.

(Jurine, 1806, p.376; Mont Blanc, France; Tomk. p.391)

OBSIDIAN. A common term for a volcanic glass, usually with a water content <1%, often dark in colour, massive and with a conchoidal fracture. *(a term of great antiquity usually attributed to Theophrastus, 320BC – see Johannsen for further discussion; Trög. 930; Joh. v.2, p.276; Tomk. p.392)*

OCEANIC PLAGIOGRANITE. A term proposed for a series of plutonic rocks consisting of plagioclase, ranging in composition from oligoclase to anorthite, quartz and minor amounts of hornblende and pyroxene. They frequently shows the effects of low-grade metamorphism developing epidote, chlorite, actinolite and albite. *(Coleman & Peterman, 1975, p.1099)*

OCEANITE. A variety of melanocratic picritic basalt consisting of abundant phenocrysts of olivine and lesser amounts of augite in a groundmass of augite, olivine and plagioclase. *(Lacroix, 1923, p.49; Piton de la Fournaise, Réunion, Indian Ocean; Trög. 409; Joh. v.3, p.306; Tomk. p.392)*

ODINITE. A lamprophyric dyke-rock of basaltic composition consisting of labradorite and rare augite phenocrysts in a felted groundmass of hornblende needles with occasional quartz. Cf. malchite. Gabbrophyre was suggested as an alternative term. *(Chelius, 1892, p.3; Frankenstein, Odenwald, Hessen, West Germany; Trög. 319; Joh. v.3, p.326; Tomk. p.393)*

OKAITE. A local name for a variety of haüyne melilitolite consisting mainly of melilite and haüyne with some biotite and perovskite. Mineralogically the rock is similar to turjaite with haüyne taking the place of nepheline. *(Stansfield, 1923a, p.440; Oka Hills, Montreal, Quebec, Canada; Trög. 661; Joh. v.4, p.324; Tomk. p.394)*

OKAWAITE. A local name for a glassy aegirine-augite rhyolite. *(Nemoto, 1934, p.300; Okawa River, Tokati, Hokkaido, Japan; Trög(38). 47½; Tomk. p.394)*

OLIGOCLASE ANDESITE. A term originally given to a variety of andesite in which the normative plagioclase is oligoclase, but later (Washington & Keyes, 1928) applied to a volcanic rock consisting essentially of oligoclase with minor pyroxene. *(Iddings, 1913, p.193; Trög. 324; Tomk. p.394)*

OLIGOCLASE BASALT. A misnomer for a rock in which andesine was probably incorrectly identified as oligoclase (Johannsen, 1937). *(Bořický, 1874, p.43; Joh. v.3, p.289)*

OLIGOCLASITE. An obsolete term originally and inappropriately used for a plutonic rock consisting of orthoclase and lesser oligoclase, later replaced by the name cavalorite. Also used for both a plutonic rock consisting entirely of oligoclase (Kolderup,1898) and a fine-grained rock of trachytic texture composed essentially of oligoclase (Washington, 1923). *(Bombicci, 1868, p.79; Mt. Cavaloro, near Bologna, Italy; Trög. 292; Joh. v.3, p.145; Tomk. p.394)*

OLIGOPHYRE. An obsolete general term for rocks consisting of phenocrysts of oligoclase in a groundmass of the same mineral. *(Coquand, 1857, p.96; Trög. 932; Joh. v.3, p.183; Tomk. p.395)*

OLIGOSITE. An obsolete term proposed for a coarse-grained rock consisting essentially of oligoclase. *(Turner, 1900, p.110; Trög. 934; Joh. v.3, p.145; Tomk. p.395)*

OLIVINE BASALT. A commonly used term for a basalt containing olivine as an essential constituent. The reference cited actually contains the term olivine-bearing basalt. *(Rosenbusch, 1896, p.1018; Trög. 379; Joh. v.3, p.281; Tomk. p.395)*

OLIVINE CLINOPYROXENITE. An ultramafic plutonic rock consisting essentially of clinopyroxene and up to 50% olivine. Now defined modally in the ultramafic rock classification (Fig. B.8). *(Streckeisen, 1973, p.26)*

OLIVINE GABBRO. A commonly used name for a gabbro containing essential olivine. Now defined modally in the gabbroic rock classification (Fig. B.6). *(Lasaulx, 1875, p.310; Trög. 351; Joh. v.3, p.224; Tomk. p.396)*

OLIVINE GABBRONORITE. A collective term for plutonic rocks consisting of 10% to 90% calcic plagioclase and accompanied by olivine and pyroxene in various amounts. Now defined modally in the gabbroic rock classification (Fig. B.6). *(Streckeisen, 1973,*

p.27)*

OLIVINE HORNBLENDE PYROXEN-ITE. An ultramafic plutonic rock consisting of more than 30% pyroxene accompanied by amphibole and olivine in various amounts. Now defined modally in the ultramafic rock classification (Fig. B.8). *(Streckeisen, 1973, p.26)*

OLIVINE HORNBLENDITE. An ultramafic plutonic rock consisting essentially of amphibole and up to 50% olivine. Now defined modally in the ultramafic rock classification (Fig. B.8). *(Streckeisen, 1973, p.26; Trög. 707)*

OLIVINE MELILITITE. A collective term for ultramafic volcanic rocks consisting of melilite, clinopyroxene and olivine in various amounts. Now defined modally in the melilitic rocks classification (Fig. B.3). *(Streckeisen, 1978, p.13)*

OLIVINE MELILITOLITE. An ultramafic plutonic rock, defined modally in the melilitic rocks classification (Fig. B.3), consisting essentially of melilite and olivine with minor clinopyroxene. *(Streckeisen, 1978, p.13)*

OLIVINE NORITE. An old term for a norite containing essential olivine. Now defined modally in the gabbroic rock classification (Fig. B.6). *(Original reference uncertain; Joh. v.3, p.237)*

OLIVINE ORTHOPYROXENITE. An ultramafic plutonic rock consisting essentially of orthopyroxene and up to 50% olivine. Now defined modally in the ultramafic rock classification (Fig. B.8). *(Streckeisen, 1973, p.26)*

OLIVINE PACIFICITE. A term for an anemousite olivine basalt with the chemistry of nepheline basanite. Cf. pacificite. Niggli (1936, p.366) used the chemistry of this rock for the kaulaitic magma-type and Tröger (1938, p. 67) renamed it kaulaite. *(Barth, 1930, p.65; Kaula Gorge, Ookala, Mauna Kea, Hawaii; Trög. 385)*

OLIVINE PYROXENE HORN-BLENDITE. An ultramafic plutonic rock consisting of more than 30% amphibole accompanied by pyroxene and olivine in various amounts. Now defined modally in the ultramafic rock classification (Fig. B.8). *(Streckeisen, 1973, p.26)*

OLIVINE PYROXENE MELILITOLITE. An ultramafic plutonic rock, defined modally in the melilitic rocks classification (Fig. B.3), consisting essentially of melilite, clinopyroxene and lesser amounts of olivine. *(Streckeisen, 1978, p.13)*

OLIVINE PYROXENITE. An ultramafic plutonic rock consisting essentially of pyroxene and up to 50% olivine. Now defined modally in the ultramafic rock classification (Fig. B.8). *(Streckeisen, 1973, p.26)*

OLIVINE THOLEIITE. Chemically defined as an olivine-hypersthene normative basalt. It is an exceedingly abundant rock and contains phenocrysts of olivine and/or plagioclase and pyroxenes in a groundmass of Ca-poor pyroxene, labradorite, opaques and sometimes glass. Now regarded as a variety of subalkali basalt. *(Rosenbusch, 1887, p.515; Trög. 344)*

OLIVINE UNCOMPAHGRITE. A special term in the melilitic rocks classification (Fig. B.3) synonomous with olivine pyroxene melilitolite. *(Streckeisen, 1978, p.14)*

OLIVINE WEBSTERITE. An ultramafic plutonic rock consisting of 10% to 40% olivine with various amounts of clinopyroxene and orthopyroxene. Now defined modally in the ultramafic rock classification (Fig. B.8). *(Streckeisen, 1973, p.26)*

OLIVINITE. An old term for ultramafic plutonic rocks composed of olivine, pyroxene and amphibole. In the USSR the term is used for olivine rocks with accessory magnetite to distinguish them from dunite, which contains accessory chromite. *(Sjögren, 1876, p.58; Trög. 937; Joh. v.4, p.402; Tomk. p.396)*

ONGONITE. A name proposed for quartz keratophyres containing topaz which are the subvolcanic equivalents of REE-, Li-, F-rich granites. See also topaz rhyolite. *(Kovalenko et al., 1971, p.430; Ongon-Khairkhan, Mongolia)*

ONKILONITE. A variety of nephelinite containing olivine, augite, nepheline, leucite and perovskite in two generations. *(Backlund, 1915, p.307; Onkilone tribe, Vilkitzky Island, New Siberia Islands, USSR; Trög. 618; Joh. v.4, p.368; Tomk. p.396)*

OPDALITE. A member of the charnockitic rock series equivalent to hypersthene granodiorite of field 4 in the QAFP, consisting of zoned plagioclase, microcline, quartz, biotite, hypersthene and diopside. The term is synonymous with charno-enberbite. *(Goldschmidt, 1916, p.70; Opdal, Trondheim, Norway; Trög. 108; Joh. v.2, p.347; Tomk. p.399)*

OPHIGRANITONE. A variety of gabbro containing scaly serpentine, saussurite and diallage. *(Mazzuoli & Issel, 1881, p.326; Tomk. p.399)*

OPHIOLITE. A term originally applied to rocks consisting mainly of serpentine, but later extended to the rock suite of Alpine-type peridotites, gabbros, dolerites, spilites and keratophyres. Now used to define the association of basic to ultrabasic rocks thought to represent oceanic crust. *(Brongniart, 1813, p.37; from the Greek ophis = snake, on account of the appearence; Tomk. p.399)*

OPHITE. A term originally used by Pliny for a greenish mottled ornamental marble, but later applied to doleritic rocks from the Pyrénées, many of which were uralitised. *(a term of great antiquity usually attributed to Pliny, 77AD – see Johannsen for further discussion; from the Greek ophites = like a serpent; Trög. 938; Joh. v.3, p.319; Tomk. p.399)*

OPHITONE. An obsolete term for a greenish diabase. *(Cordier, 1842, Vol.8, p.135; Tomk. p.399)*

ORBICULITE. A group name for plutonic rocks with orbicular structure. *(Sederholm, 1928, p.72; Trög. 939; Tomk. p.400)*

ORBITE. A coarse-grained lamprophyric dyke rock composed of phenocrysts of hornblende in a plagioclase-rich groundmass. Cf. luciite. *(Chelius, 1892, p.3; Orbishöhe, near Zwingenberg, Odenwald, West Germany; Trög. 336; Joh. v.3, p.309; Tomk. p.400)*

ORDANCHITE. A local name for a volcanic rock, variously described as a variety of trachyandesite and tephrite, consisting essentially of andesine and haüyne with resorbed hornblende, augite and minor olivine. *(Lacroix, 1917a, p.582; Banne d'Ordanche, Auvergne, France; Trög.572; Joh. v.4, p.216; Tomk. p.400)*

ORDOSITE. A melanocratic variety of alkali

feldspar syenite containing abundant aegirine, microcline and a little phlogopite. *(Lacroix, 1925, p.482; Ordos province, China; Trög. 223; Joh. v.4, p.17; Tomk. p.401)*

ORENDITE. A variety of leucite phonolite essentially composed of leucite and alkali feldspar with subordinate clinopyroxene, mica and amphibole. *(Cross, 1897, p.123; Orenda Butte, Leucite Hills, Wyoming, USA; Trög. 478; Joh. v.4, p.260; Tomk. p.401)*

ORNÖITE. A local name for a variety of diorite consisting essentially of sodic plagioclase with minor hornblende, microcline, biotite and quartz. *(Cederström, 1893, p.107; Ornö Island, Stockholm, Sweden; Trög. 305; Joh. v.3, p.115; Tomk. p.402)*

OROGENIC ANDESITE. A term for a variety of andesite defined chemically as having SiO_2 between 53% and 63%, hy in the norm, TiO_2 < 1.75% and K_2O < (0.145 x SiO_2 – 5.135). *(Gill, 1981, p.2)*

OROTVITE. A melanocratic to mesocratic variety of nepheline-bearing diorite rich in kaersutite and with mafic minerals more abundant than the andesine-oligoclase. Nepheline and alkali feldspar can be present in minor amounts. *(Streckeisen, 1938, p.159; Orotva Valley, Ditrau, Transylvania, Romania; Trög(38). 308⅔; Joh. v.4, p.215; Tomk. p.402)*

ORTHO-. A prefix which has been applied to rock names in several senses. For example, to indicate the presence of orthopyroxene (e.g. orthoandesite, orthobasalt) or the abundance of orthoclase and lack of plagioclase (e.g. orthogranite) and later for rocks that are neither oversaturated nor undersaturated with respect to silica (e.g. orthosyenite, orthogabbro).

ORTHOANDESITE. An obsolete group name for an orthopyroxene-bearing andesite. The term is attributed to Oebbeke who supposedly used it for rocks from the Phillipines, but no actual reference is given. The only known publication is Oebbeke (1881), but it does not contain the name. *(Koto, 1916b, p.116; Joh. v.3, p.173)*

ORTHOBASALT. An obsolete term proposed for an olivine-bearing bronzite basalt. *(Koto, 1916b, p.119)*

ORTHOBASE. According to Tomkeieff et al.

(1983) the rock is a diabase or porphyrite containing orthoclase. However, the term does not appear in the reference cited. The same incorrect reference is also cited by Loewinson-Lessing (1932). *(Belyankin, 1911, p.363; Tomk. p.403)*

ORTHOCLASITE. A general term for a rock consisting essentially of orthoclase. Now used for fenitic rocks composed of K-rich feldspars (Sutherland, 1965) for which the terms orthoclase-rock and feldspar-rock have been used. If fluorite is present the rock is called borengite. *(Senft, 1857, p.51; Trög. 941; Joh. v.3, p.5; Tomk. p.403)*

ORTHOFELSITE. An obsolete term synonymous with orthophyre. *(Teall, 1888, p.291; Trög. 256; Tomk. p.404)*

ORTHOFOYAITE. An obsolete term proposed for a variety of foyaite in which plagioclase is practically absent. *(Johannsen, 1920b, p.161; Tomk. p.404)*

ORTHOGABBRO. A variety of gabbro that contains no quartz or foids i.e. it closely agrees with the definition of the "ideal" type. *(Hatch et al., 1949, p.279)*

ORTHOGRANITE. An obsolete term proposed for a variety of granite in which plagioclase is practically absent. The term was later withdrawn (Johannsen, 1932) in favour of kaligranite. *(Johannsen, 1920a, p.53; Joh. v.2, p.51; Tomk. p.404)*

ORTHOLITE. An obsolete name originally used in France for a rock consisting of orthoclase and mica. *(Lapparent, 1906, p.641; Tomk. p.404)*

ORTHOPHONITE. An obsolete term for nepheline syenite, being a mnemonic name from *ortho*clase and *phon*olite. *(Lasaulx, 1875, p.318; Tomk. p.405)*

ORTHOPHYRE. An obsolete name used for older trachytes and for rocks consisting of phenocrysts of orthoclase in a groundmass of orthoclase. *(Coquand, 1857, p.65; Trög. 256; Joh. v.3, p.80; Tomk. p.405)*

ORTHOPYROXENE GABBRO. A basic plutonic rock consisting mainly of calcic plagioclase, clinopyroxene and minor orthopyroxene. Now defined modally in the gabbroic rock classification (Fig. B.6). *(Johannsen, 1937, p.238)*

ORTHOPYROXENITE. An ultramafic plutonic rock consisting almost entirely of orthopyroxene. Now defined modally in the ultramafic rock classification (Fig. B.8). *(Wyllie, 1967, p.2)*

ORTHORHYOLITE. An obsolete term proposed for a variety of rhyolite in which plagioclase is practically absent. The term was later withdrawn (Johannsen, 1932) in favour of kalirhyolite. *(Johannsen, 1920b, p.159; Tomk. p.405)*

ORTHOSHONKINITE. An obsolete term originally proposed for a variety of shonkinite containing no plagioclase, but later withdrawn (Johannsen, 1938). *(Johannsen, 1920a, p.51; Joh. v.4, p.15)*

ORTHOSITE. A term proposed for a coarsegrained variety of alkali feldspar syenite consisting essentially of orthoclase. *(Turner, 1900, p.110; Trög. 163; Joh. v.3, p.4; Tomk. p.406)*

ORTHOSYENITE. A term originally used for a syenite in which the amount of alkali feldspar is more than 95% of the total feldspar, but later withdrawn (Johannsen, 1937) in favour of kalisyenite. The name was later redefined (Hatch et al., 1949, p.232) as a variety of syenite that contains no quartz or foids i.e. is exactly saturated with respect to silica. *(Johannsen, 1920b, p.160; Joh. v.3, p.8; Tomk. p.406)*

ORTHOTARANTULITE. An obsolete term proposed for rocks consisting of major amounts of quartz with minor orthoclase. The term was later withdrawn (Johannsen, 1932) in favour of arizonite of which Johannsen had been unaware. *(Johannsen, 1920a, p.53; Joh. v.2, p.32; Tomk. p.406)*

ORTHOTRACHYTE. A term suggested to replace alkali trachyte if the alternative name quartzfree liparite is regarded as a contradiction in terms. The name was later redefined (Hatch et al., 1949, p.247) as a variety of trachyte that contains no quartz or foids i.e. is exactly saturated with respect to silica. *(Rosenbusch, 1908, p.887; Joh. v.3, p.16)*

ORTLERITE. An altered variety of hornblende andesite or trachyandesite consisting of abundant phenocrysts of hornblende and a few of augite and biotite in a felted groundmass of andesine, chlorite, pyrite, calcite and devitrified glass. *(Stache & John, 1879, p.325; Ortler Alps, Alto Adige, Italy; Trög. 287; Joh. v.3, p.123; Tomk. p.406)*

ORVIETITE. A volcanic rock, close to the boundary between phonolitic tephrite and tephritic phonolite, essentially composed of equal proportions of plagioclase and alkali feldspar, with subordinate leucite and clinopyroxene. It contains less leucite than vicoite. *(Niggli, 1923, p.175; Orvieto, Vulsini district, near Viterbo, Italy; Trög. 540; Tomk. p.406)*

OSLO-ESSEXITE. A term suggested to replace the name essexite, which had been incorrectly applied to rocks occurring in the volcanic bosses in the Oslo province by Brögger (1933), as they do not contain nepheline. The term includes such types as kauaite, bojite, olivine gabbro, pyroxenite, etc. *(Barth, 1944, p.31; Tomk. p.407)*

OSLOPORPHYRY. A local name for an oligoclase porphyry. *(Brögger, 1898, p.207; Trög. 944; Tomk. p.407)*

OSSIPYTE (OSSIPITE, OSSYPITE). An obsolete local name for a variety of plagioclase-rich olivine gabbro. Dana states the name was suggested by Prof. Hitchcock. *(Dana, 1872, p.49; named after Ossipee Indians, New Hampshire, USA; Trög. 352; Joh. v.3, p.225; Tomk. p.407)*

ØSTERN PORPHYRY. A local term for a variety of porphyritic kjelsåsite or plagioclase-rich monzonite. *(Holtedahl, 1943, p.32; Østervann, Oslo district, Norway; Tomk. p.407)*

OSTRAITE. An obsolete name for a variety of clinopyroxenite composed of partly uralitised augite and spinel. *(Duparc, 1913, p.18; Ostraïa Sopka, Urals, USSR; Trög. 685; Joh. v.4, p.464; Tomk. p.407)*

OTTAJANITE. A variety of leucite tephrite essentially composed of plagioclase, leucite and clinopyroxene, with subordinate olivine. *(Lacroix, 1917b, p.208; Ottaviano, Mt. Somma, Naples, Italy; Trög. 945; Joh. v.4, p.201; Tomk. p.407)*

OUACHITITE. An ultramafic lamprophyre containing combinations of olivine, phlogopite, amphibole, and/or clinopyroxene phenocrysts with groundmass feldspathoids as well as carbonates. *(Kemp, 1890, p.393; Ouachita River, Arkansas, USA; Trög. 405; Joh. v.4, p.391; Tomk. p.408)*

OUENITE. A fine-grained variety of basalt which occurs as dykes and consists of anorthite with chrome-diopside, fringed by a little bronzite, and olivine. *(Lacroix, 1911a, p.817; Ouen Island, New Caledonia; Trög. 368; Joh. v.3, p.349; Tomk. p.408)*

OVERSATURATED. A term applied to igneous rocks in which there is an excess of SiO_2 over the other oxides, which gives silica minerals in the mode or quartz in the norm. *(Shand, 1913, p.510; Tomk. p.408)*

OWHAROITE. A local name for a strongly welded rhyolitic or dacitic tuff, previously called wilsonite. *(Grange, 1934, p.58; Owharoa, Waihi district, Auckland, New Zealand; Trög(38). 40⅔; Tomk. p.409)*

OXYPHYRE. An obsolete general term for porphyritic acid rocks which were thought to be complementary to lamprophyres. *(Pirsson, 1895, p.118; Tomk. p.410)*

OXYPLETE. An obsolete term for a leucocratic igneous rock in which $SiO_2 > 6$ R_2O_3. *(Brögger, 1898, p.266; Tomk. p.410)*

PACIFICITE. A tephritic rock composed of phenocrysts of augite, some labradorite and rare olivine in a groundmass of anemousite (plagioclase containing carnegieite), augite and opaques. *(Barth, 1930, p.60; named after Pacific Ocean; Haleakala, Maui, Hawaiian Islands; Trög. 384; Tomk. p.411)*

PAISANITE. A local name for a leucocratic variety of alkali feldspar microgranite consisting essentially of anorthoclase and quartz with minor amounts of riebeckite. *(Osann, 1893, p.131; Paisano Pass, Texas, USA; Trög. 29; Joh. v.2, p.100; Tomk. p.412)*

PALAEOPHYRE. An obsolete term for altered porphyrites. *(Gümbel, 1874, p.42; Trög.946; Joh. v.3, p.183; Tomk. p.412)*

PALAEOPHYRITE. An obsolete term for older post-Cretaceous porphyries. *(Stache & John, 1879, p.352; Trög. 947; Joh. v.3, p.183; Tomk. p.412)*

PALAEOPICRITE. An obsolete term for picrites of Palaeozoic age which were more altered than those of Tertiary age. *(Gümbel, 1874, p.38; Trög. 948; Tomk. p.412)*

PALAGONITE TUFF. A term for a tuff containing palagonite which is a yellow or brown completely devitrified basaltic glass. *(Waltershausen, 1846, p.402; Palagonia, Sicily, Italy; Trög.594; Joh. v.3, p.302; Tomk. p.412)*

PALATINITE. An old term applied to various rocks composed of augite and plagioclase. Rosenbusch redefined it as a bronzite tholeiite. *(Laspeyres, 1869, p.516; Palatia, now Pfalz, West Germany; Trög. 343; Joh. v.3, p.299; Tomk. p.413)*

PALLIOESSEXITE, PALLIOGRANITE. Obsolete terms suggested for varieties of igneous rocks which occur on the margin of an intrusion "a little way within the contact". *(Jevons et al., 1912, p.452; from the Latin pallium = cloak; Joh. v.4, p.61; Tomk. p.414)*

PANTELLERITE. A variety of leucocratic peralkaline rhyolite containing phenocrysts of aegirine-augite, anorthoclase and cossyrite. It is now defined and distinguished from comendite as a variety of peralkaline rhyolite of TAS field R (Fig. B.10) in which $Al_2O_3 < 1.33$ x total iron as $FeO + 4.4$ (Fig. B.16). *(Foerstner, 1881, p.537; Pantelleria Island, Italy; Trög. 72; Joh. v.2, p.64; Tomk. p.414)*

PARAMELAPHYRE. An obsolete term for a variety of mica porphyrite. *(Schmid, 1880, p.67; Tomk. p.416)*

PARCHETTITE. An obsolete name for a variety of leucite tephrite composed essentially of clinopyroxene, leucite and plagioclase and with accessory alkali feldspar. *(Johannsen, 1938, p.291; Fosso della Parchetta, Vico volcano, near Viterbo, Italy; Trög(38). 539½; Tomk. p.417)*

PAROPHITE. An obsolete term for a variety of serpentine. *(*Hunt, 1852, p.95; Tomk. p.418)*

PATRINITE. An obsolete term for a phonolite. *(Pinkerton, 1811a, p.167; named after E.L.M. Patrin; Tomk. p.419)*

PAWDITE. An obsolete name for a dark granular dyke-rock composed of plagioclase zoned from bytownite to oligoclase, hornblende, quartz, biotite and epidote. *(Duparc & Grosset, 1916, p.110; Pawdinskaya Datcha, Nikolai Pawda, Urals, USSR; Trög. 333; Joh. v.3, p.319; Tomk. p.419)*

PEARLSTONE. An obsolete term for perlite. The name is attributed to Jameson, but Johannsen gives no reference. *(Johannsen, 1932, p.284; Tomk. p.420)*

PECHSTEIN. The German name for pitchstone. *(*Schulz & Poetsch, 1759, p.267; Trög. 102; Joh. v.2, p.280)*

PEDROSITE. A peralkaline variety of hornblendite composed essentially of alkali amphibole (osannite), with some magnetite, and occasionally albite and analcime. *(Osann, 1922, p.260; Alter Pedroso, Alemtejo, Portugal; Trög. 703; Joh. v.4, p.444; Tomk. p.421)*

PEGMATITE. A term originally suggested by Haüy as a synonym for graphic granite, but later used for the coarse-grained facies of an igneous rock. *(Brongniart, 1813, p.32; from the Greek pegma = bond, framework; Trög. 65; Joh. v.2, p.72; Tomk. p.421)*

PEGMATITOID. An obsolete term proposed for coarse-grained veins, some of which may be late differentiates, occurring in basaltic rocks. *(Lacroix, 1928b, p.322; Tomk. p.422)*

PEGMATOID. An term proposed for both feldspathoidal pegmatites and for pegmatites without graphic texture. *(Shand, 1910, p.377; Trög. 951; Tomk. p.422)*

PEGMATOPHYRE. An obsolete term, originally proposed to replace granophyre and later for varieties of granophyre without plagioclase. *(Lossen, 1892, p.270; Trög. 952; Joh. v.2, p.70; Tomk. p.422)*

PÉLÉ'S HAIR. A name given to the fine hair-like threads of basaltic glass often formed during lava fountaining. The spelling is as found in the 2nd edition. *(*Ellis, 1825, p.264; named after Pélé, the Hawaiian Goddess of Fire; Joh. v.3, p.291; Tomk. p.423)*

PÉLÉ'S TEARS. Tear-drop-shaped lava filaments blown through open channels of lava vents. *(Perret, 1913, p.615; named after Pélé, the Hawaiian Goddess of Fire; Tomk. p.423)*

PELÉEITE. A comprehensive term for andesites and basaltic andesites that correspond to the peléeitic magma-type of Niggli (1923, p.123). *(Tröger, 1935, p.75; Mt. Pelée, Martinique, West Indies; Trög. 157; Tomk. p.423)*

PELEITI. An Italian name for Pélé's hair. *(Issel, 1916, p.657; Tomk. p.423)*

PENIKKAVAARITE. A local name for a variety of gabbro with 60% to 70% mafic minerals. Amphiboles are more abundant than Ti-augite

and plagioclase. Alkali feldspar may be present in minor amounts. *(Johannsen, 1938, p.52; Penikkavaara, Kuusamo, Finland; Trög(38). 285½; Tomk. p.426)*

PEPERIN-BASALT. An obsolete name for a tuff which forms mud flows and contains large crystals of augite and hornblende. *(Bořický, 1874, p.42; Kostenblatt (now Kostomlaty), Bohemia, Czechoslovakia; Tomk. p.426)*

PEPERINO. A local Italian name for a light coloured unconsolidated tuff from the Alban Hills containing many dark crystal fragments, giving it a peppery appearance. *(Buch, 1809, p.70; from the Italian pepe = pepper, as it resembles grains of pepper; Trög. 953; Joh. v.4, p.363; Tomk. p.426)*

PÉPÉRITE. A local term for a tuff or breccia formed by the intrusion of magma into wet sediments. Usually consists of fragments of glassy igneous rock and some sedimentary rock. *(Cordier, 1816, p.366; Tomk. p.426)*

PERACIDITE. A chemically derived term for an igneous rock consisting almost entirely of quartz. Such rocks should now be called quartzolite. *(Rinne, 1921, p.165; Trög. 1; Joh. v.2, p.11; Tomk. p.426)*

PERALBORANITE. A leucocratic variety of alboranite with less than 12.5% of pyroxene. Obsolete. *(Burri & Parga-Pondal, 1937, p.258; Alboran Island, near Cabo de Gata, Spain; Trög(38). 304½; Tomk. p.426)*

PERALKALINE. A chemical term for rocks in which the molecular amounts of Na_2O plus K_2O exceeds Al_2O_3. This produces acmite (ac) and sometimes sodium metasilicate (ns) in the CIPW norm and usually alkali pyroxenes and/or alkali amphiboles in the mode. *(Winchell, 1913, p.210; Tomk. p.426)*

PERALUMINOUS. A term used for rocks which are essentially oversaturated with respect to alumina and characterised mineralogically by primary muscovite, biotite, corundum, tourmaline, topaz, almandine and spessartine and chemically by corundum but with no hypersthene in the norm. *(Shand, 1927, p.128)*

PERIDOTEID. An obsolete field term for a coarse-grained igneous rock consisting of olivine with or without pyroxene, amphibole or biotite. Feldspar is absent. *(Johannsen, 1911, p.321; from the French péridot = oli-*

vine; Tomk. p.427)

PERIDOTIDE. A revised spelling recommended to replace the field term peridoteid. Now obsolete. *(Johannsen, 1926, p.182; Joh. v.1, p.57)*

PERIDOTITE. A collective term for ultramafic rocks consisting essentially of olivine with pyroxene and/or amphibole. Now defined modally in the ultramafic rock classification (Fig. B.8). *(Cordier, 1842, Vol.9, p.619; from the French péridot = olivine; Trög. 723; Joh. v.4, p.401; Tomk. p.427)*

PERIDOTITIC KOMATIITE. See komatiite. *(Viljoen & Viljoen, 1969, p.60; Komati river, Barberton, Transvaal, South Africa)*

PERIDOTITOID. An obsolete term proposed for an eclogitic rock similar to peridotite. *(Holmquist, 1908, p.292; Tomk. p.427)*

PERIDOTOID. A group name proposed for rocks composed essentially of olivine, pyroxene and iron ore. *(Gümbel, 1888, p.88; Tomk. p.427)*

PERKNIDE. A name recommended to replace the field term pyriboleid. Now obsolete. *(Johannsen, 1926, p.182; Joh. v.1, p.57; Tomk. p.428)*

PERKNITE. A collective name for ultramafic olivine-free rocks composed essentially of amphibole and/or pyroxene or biotite. *(Turner, 1901, p.507; from the Greek perknos = dark especially of fruit or bird; Trög. 954; Joh. v.4, p.399; Tomk. p.429)*

PERLITE. A term used for volcanic glasses which exhibit numerous concentric cracks so that when fragmented the pieces vaguely resemble pearls. Some are high in water content and expand when heated. The term is synonymous with pearlstone and perlstein. *(Beudant, 1822, p.360; from the French perle = pearl; in German perlstein; Trög. 955; Joh. v.2, p.284; Tomk. p.429)*

PERLSTEIN. The original German term for certain volcanic glasses now called perlite in English. Although the term is attributed to the reference cited, only the term "perlartig" is used. *(Fichtel, 1791, p.365; from the German perl = pearl, stein = rock; Trög. 955; Joh. v.2, p.284)*

PERTHITOPHYRE. An obsolete name for a variety of anorthosite or leucocratic monzogab-bro containing microperthite as an interstitial filling. *(Chrustschoff, 1888, p.476; Horoski, Volhynien, USSR; Trög. 956; Joh. v.3, p.126; Tomk. p.430)*

PERTHOSITE. A variety of alkali feldspar syenite consisting almost entirely of perthite. *(Phemister, 1926, p.41; Trög. 164; Joh. v.3, p.43; Tomk. p.430)*

PETRISCO. A local term for a variety of leucite trachyte from the Viterbo and Lake Vico areas, Italy. *(Rath, 1868, p.297; Tomk. p.431)*

PHANEREID. An obsolete field term for rocks whose different constituents can be seen megascopically. *(Johannsen, 1911, p.318; from the Greek phaneros = distinct, visible, clear; Tomk. p.433)*

PHANERIDE. A revised spelling recommended to replace the field term phanereid. Now obsolete. *(Johannsen, 1926, p.182; Tomk. p.433)*

PHANERODACITE. A term for a dacite containing the excess silica as quartz and not in glass. Cf. cryptodacite. *(Belyankin, 1923, p.100; Tomk. p.433)*

PHENO-. A prefix suggested for giving a provisional name to a volcanic rock based on the minerals that can be identified, usually phenocrysts. For example a glassy rock with phenocrysts of quartz and plagioclase could be called a phenodacite. *(Niggli, 1931, p.357)*

PHONOBASANITE. A synonym for phonolitic basanite of QAPF field 13 (Fig. B.10). *(Streckeisen, 1978, p.6)*

PHONOFOIDITE. A synonym for phonolitic foidite of QAPF field 15a (Fig. B.10). *(Streckeisen, 1978, p.7)*

PHONOLEUCITITE. A synonym for phonolitic leucitite of QAPF field 15a (Fig. B.10). *(Rittmann, 1973, p.135)*

PHONOLITE. Now defined in QAPF field 11 (Fig. B.10) in the sense of Rosenbusch (1877, p.234) as a volcanic rock consisting essentially of alkali feldspar and any foids. If nepheline is the only foid then the term phonolite may be used by itself but if, for example, leucite is the most abundant foid then the term leucitephonolite should be used, etc. If modes are not available phonolite is defined chemically in TAS field Ph (Fig. B.13). *(Cordier, 1816, p.151; from the Greek phone = sound, lithos = stone; Tróg. 465; Joh. v.4, p.120; Tomk. p.435)*

PHONOLITIC BASANITE. A collective term for alkaline basaltic rocks that are the volcanic equivalent to foid monzodiorites or monzogabbros and consist of plagioclase, feldspathoid, olivine, augite and often minor sanidine. If the amount of olivine is less than 10% it is a phonolitic tephrite. Now defined modally in QAPF field 13 (Fig. B.10). *(Streckeisen, 1978, p.6)*

PHONOLITIC FOIDITE. A collective term for alkaline volcanic rocks consisting of foids with some alkali feldspar as defined modally in QAPF field 15a (Fig. B.10). If possible the most abundant foid should be used in the name e.g. phonolitic nephelinite, phonolitic leucitite, etc. *(Streckeisen, 1978, p.7)*

PHONOLITIC TEPHRITE. A collective term for alkaline volcanic rocks consisting of plagioclase, feldspathoid, augite and often minor olivine and sanidine. If the amount of olivine is greater than 10% it is a phonolitic basanite. Now defined modally in QAPF field 13 (Fig. B.10). *(Streckeisen, 1978, p.6)*

PHONOLITOID. A term originally used as a collective name for phonolites and leucitophyres. Later used for a rock with the composition of phonolite but without modal nepheline (Lacroix, 1923). Now used as a general term in the provisional "field" classification (Fig. B.17) for rocks thought to contain essential foids and in which alkali feldspar is thought to be more abundant than plagioclase. *(Gümbel, 1888, p.86; Tomk. p.435)*

PHONOLITOID TEPHRITE. A term used for a rock consisting of sandine, nepheline, plagioclase, hornblende and olivine i.e. between phonolite and tephrite. *(Rosenbusch, 1908, p.1375; Tomk. p.435)*

PHONONEPHELINITE. A synonym for phonolitic nephelinite of QAPF field 15a (Fig. B.10). *(Rittmann, 1973, p.135)*

PHONOTEPHRITE. A synonym for phonolitic tephrite of QAPF field 13 (Fig. B.10), and also defined chemically in TAS field U2 (Fig. B.13). *(Rittmann, 1973, p.134)*

PHOSCORITE. A magnetite, olivine, apatite rock usually associated with carbonatites. The name is a mnemonic name from *phos-*

phate rock around a *core* of carbonatite. *(Russell et al., 1955, p.199; Tomk. p.435)*

PICOTITITE. A rock consisting essentially of picotite (85%) with small amounts of serpentine. *(Tröger, 1935, p.308; Tróg. 774; Tomk. p.437)*

PICRITE. A term originally used for a variety of dolerite or basalt extremely rich in olivine and pyroxene. Also used as the volcanic equivalent of a feldspar-bearing or alkali peridotite and for olivine-rich varieties of gabbro and teschenite. Now defined chemically as a group name for rocks with SiO_2 < 47%, total alkalis < 2% and MgO > 18% (Fig. B.12) and as a specific type within the group when total alkalis > 1%. *(Tschermak, 1866, p.262; from the Greek pikros = bitter, referring to high MgO; Tróg. 743; Joh. v.4, p.432; Tomk. p.437)*

PICRITE BASALT. A melanocratic variety of olivine-rich basalt containing abundant phenocrysts of olivine in a sparse groundmass of augite, labradorite, opaques and interstitial glass. *(Quensel, 1912, p.265; Masafuera, Juan Fernandez Islands, Pacific Ocean; Tróg. 410; Joh. v.3, p.334; Tomk. p.438)*

PICROBASALT. A chemical term for volcanic rocks, which will include certain picritic and accumulative rocks, introduced for TAS field Pc (Fig. B.13). *(Le Maitre, 1984, p.245)*

PICROPHYRE. An obsolete term for a variety of minette (a lamprophyre) containing augite and small amounts of olivine. *(Bořický, 1878, p.494; Libšice, Bohemia, Czechoslovakia; Tróg. 958; Tomk. p.438)*

PIENAARITE. A mafic variety of nepheline syenite containing abundant titanite, aegirine-augite, and anorthoclase. *(Brouwer, 1909, p.563; Pienaar River, Transvaal, South Africa; Tróg. 488; Joh. v.4, p.118; Tomk. p.438)*

PIKEITE. An obsolete name for a variety of phlogopite peridotite consisting of olivine phenocrysts and poikilitic phlogopite with inclusions of olivine and augite. *(Johannsen, 1938, p.427; Pike County, Arkansas, USA; Tróg(38). 721; Tomk. p.439)*

PIKROPTOCHE (PYCROPTOCHE). An obsolete adjectival term for igneous rocks poor in lime and magnesia. *(Loewinson-Lessing, 1901,* p.118; Tomk. p.471)

PILANDITE. A local term for a porphyritic variety of alkali feldspar syenite called hatherlite. (Henderson, 1898, p.48; Pilansberg, Bushveld, South Africa; Trög. 202; Joh. v.3, p.31; Tomk. p.439)

PIPERNO. A local name for a trachytic rock exhibiting eutaxitic texture in the form of light and dark streaks resembling flames. (Lorenzo, 1904, p.301; Trög. 960; Tomk. p.440)

PISSITE. An obsolete term for a pitchstone with a high melting point. (Delamétherie, 1795, p.461; Tomk. p.441)

PITCHSTONE. A volcanic glass with a lustre resembling pitch and usually containing a few phenocrysts and a water content between 4% and 10%. This is unlike obsidian which usually has a water content <1%. (Babington, 1799, p.94; from the German pechstein; Joh. v.2, p.280; Tomk. p.441)

PLÄDORITE. An obsolete term for a biotite hornblende granite, later renamed plethorite. (Lang, 1877, p.156; Tomk. p.445)

PLAGIAPLITE. A leucocratic diorite aplite with abundant zoned sodic plagioclase and only minor quartz. With more quartz it passes into gladkaite. (Duparc & Jerchoff, 1902, p.307; Koswinski Mts., Urals, USSR; Trög. 303; Joh. v.3, p.184; Tomk. p.442)

PLAGIDACITE. A special term used as a replacement for quartz andesite and now for dacites of QAPF field 5 (Fig. B.10), having the composition of tonalites. (Rittmann, 1973, p.133)

PLAGIOCLASE GRANITE. An obsolete term for a plutonic rock consisting of oligoclase or andesine, quartz and less than 10% of biotite and hornblende. A synonym for trondhjemite or leuco-tonalite. (Högbom, 1905, p.221; Tortola & Virgin Gorda, Virgin Islands, West Indies; Trög. 130; Joh. v.2, p.382)

PLAGIOCLASE-BEARING HORN-BLENDE PYROXENITE. A term for ultramafic plutonic rocks composed mainly of pyroxene with up to 50% amphibole and minor amounts of plagioclase and often quartz. Now defined modally in the ultramafic rock classification (Fig. B.8). (Streckeisen, 1973, p.27)

PLAGIOCLASE-BEARING HORN-BLENDITE. An ultramafic plutonic rock consisting essentially of amphibole with minor amounts of plagioclase. Now defined modally in the ultramafic rock classification (Fig. B.8). (Streckeisen, 1973, p.27)

PLAGIOCLASE-BEARING PYROXENE HORNBLENDITE. A term for ultramafic plutonic rocks composed mainly of amphibole with up to 50% pyroxene and minor amounts of plagioclase and often quartz. Now defined modally in the ultramafic rock classification (Fig. B.8). (Streckeisen, 1973, p.27)

PLAGIOCLASE-BEARING PYROXEN-ITE. An ultramafic plutonic rock consisting essentially of pyroxene with minor amounts of plagioclase. Now defined modally in the ultramafic rock classification (Fig. B.8). (Streckeisen, 1973, p.27)

PLAGIOCLASITE. A term originally used as a group name for a plagioclase-enriched gabbro. Now defined modally in QAPF field 10 (Fig. B.4) as a synonym for anorthosite. (Viola, 1892, p.121; Valle del Sinni, Basilicata, Italy; Trög. 961; Tomk. p.442)

PLAGIOCLASOLITE. An obsolete group name for plagioclase rocks ranging from albitite to gabbro. (Lacroix, 1933, p.191; Trög. 962)

PLAGIOGRANITE. A contraction of plagioclase granite commonly used in the USSR for a plutonic rock consisting of oligoclase or andesine, quartz and less than 10% of biotite and hornblende. Now defined as a synonym for trondhjemite and leucocratic tonalite of QAPF field 5 (Fig. B.4). (Zavaritskii, 1955, p.272)

PLAGIOLIPARITE. A variety of liparite containing phenocrysts of sodic plagioclase. (Duparc & Pearce, 1900, p.57; Cape Marsa, Ménerville, Algeria; Trög. 43; Tomk. p.442)

PLAGIOPHYRE. An obsolete group name for fine-grained intrusive rocks consisting of zoned plagioclase with subordinate altered mafic minerals. (Tyrrell, 1911, p.77; Trög. 120; Tomk. p.442)

PLAGIOPHYRITE. A term used for a porphyritic andesite or microdiorite. (Szentpétery, 1935, p.26; from a contraction of plagioclase-porphyrite; Tomk. p.443)

PLANOPHYRE. A porphyritic rock in which the

phenocrysts occur in layers. *(Iddings, 1909, p.224; Tomk. p.443)*

PLATEAU BASALT. A general term for basaltic lavas occurring in continental regions as vast accumulations of sub-horizontal flows that have been erupted, probably from fissures, in rapid succession on a regional scale. Many of the lavas are tholeiitic basalts, but transitional basalts are also common. The term is synonymous with flood basalt. *(*Geikie, 1903, p.344)*

PLAUENITE. An alternative term that was suggested, but not adopted, for the type syenite from Plauenschen Grund. *(Brögger, 1895, p.59; Plauenscher Grund, Dresden, Saxony, East Germany; Trög. 964; Joh. v.3, p.54; Tomk. p.444)*

PLETHORITE. An obsolete name for a biotite hornblende granite, previously called plädorite. *(Zirkel, 1894a, p.34; Tomk. p.445)*

PLUMASITE. A plutonic rock consisting largely of oligoclase with lesser amounts of corundum. *(Lawson, 1903, p.219; Spanish Peak, Plumas County, California, USA; Trög. 311; Joh. v.3, p.185; Tomk. p.446)*

PLUTONIC. A loosely defined term pertaining to those igneous processes that occur at considerable depth below the surface of the earth. Plutonic rocks are usually coarse-grained, but not all coarse-grained rocks are plutonic. *(Kirwan, 1794, p.455; named after Pluto, Greek God of the infernal regions; Joh. v.1, p.188; Tomk. p.446)*

PLUTONITE. A term proposed for deep-seated rocks. *(Scheerer, 1862, p.138; Joh. v.1(2nd.Ed.), p.275; Tomk. p.447)*

PLUTOVOLCANITE. An obsolete term for rocks that have been called hypabyssal i.e. with characteristics between plutonic and volcanic. *(Scheerer, 1864, p.403; Tomk. p.447)*

POENEITE. A variety of spilite in which K_2O is much greater than Na_2O. *(Roever, 1940, p.263; River Noil Poene, Moetis region, Timor)*

POGONITE. An obsolete term for what is now called Pélé's hair. *(Haüy, 1822, p.580; from the Greek pogon = beard; Tomk. p.449)*

POIKILEID. An obsolete field term for a porphyry. *(Johannsen, 1911, p.319; from the Greek poikilos = many coloured, spotted; Tomk. p.449)*

POLLENITE. A local name for a variety of tephritic phonolite containing phenocrysts of sanidine, plagioclase, hornblende and biotite in a groundmass of nepheline, other foids, olivine and glass. Cf. tautirite. *(Lacroix, 1907, p.138; Pollena Valley, Mt. Somma, Naples, Italy; Trög.470; Joh. v.4, p.167; Tomk. p.450)*

POLZENITE. A group name for olivine- and melilite-bearing lamprophyres containing some nepheline but no augite. It includes two varieties: modlibovite and vesecite, characterised by the absence and presence of monticellite, respectively. Now defined in the lamprophyre classification (Table B.3). *(Scheumann, 1913, p.728; Polzen River (now Ploučnice), N. Bohemia, Czechoslovakia; Trög. 965; Joh. v.4, p.388; Tomk. p.451)*

PONZAITE. A local name for a variety of peralkaline trachyte containing sodic pyroxenes and amphiboles and occasionally nepheline. *(Reinisch, 1912, p.121; Ponza Island, Italy; Joh. v.3, p.77; Tomk. p.452)*

PONZITE. A name given to the nepheline-free varieties of trachyte from Ponza Island. *(Washington, 1913, p.691; Ponza Island, Italy; Trög. 174; Joh. v.3, p.77; Tomk. p.452)*

PORPHYRIN. An obsolete term for a porphyritic rock with feldspar phenocrysts that can only be seen with a hand lens. *(Pinkerton, 1811a, p.87)*

PORPHYRITE. Originally used as a term for a quartz-free porphyry and later as a general term for porphyritic rocks of diorite composition. *(Naumann, 1854, p.664; from the Greek porphyreos = purple; Trög. 325; Joh. v.3, p.82; Tomk. p.453)*

PORPHYRON. An obsolete term for a porphyritic rock with discrete feldspar phenocrysts over 1 inch in length. *(Pinkerton, 1811a, p.88; Tomk. p.454)*

PORPHYRY. A general term applied to any igneous rock that contains phenocrysts in a finer-grained groundmass. It is often applied to rocks that contain two generations of the same mineral. *(Werner, 1787, p.12; from the Latin rock name porphyrite; Trög. 967; Joh. v.3, p.81; Tomk. p.454)*

POTASSIC TRACHYBASALT. A term introduced for the potassic analogue of hawaiite in TAS field S1 (Fig. B.13) to distinguish it from trachybasalt which is the collective name of the field. *(Le Maitre, 1984, p.245)*

POZZOLANA (POSSOLANA, PUZZOLANA). Probably the commonest spelling of an ancient local name for a porous variety of tuff, often pumice-rich and sometimes containing leucite, used for making hydraulic cement. *(Original reference uncertain; Pozzuoli, near Naples, Italy; Joh. v.3, p.20; Tomk. p.455)*

PRASOPHYRE. An obsolete name for a greenish porphyry. *(Le Puillon de Boblaye & Virlet, 1833, p.111; Tomk. p.457)*

PRESELITE. A local term proposed by archaeologists for a variety of greenish dolerite containing widely spaced clusters of sodic plagioclase and used for the megaliths of the inner circle of Stonehenge. *(Keiller, 1936, p.221; Prescelly Hills, Dyfed, Wales; Tomk. p.457)*

PROPYLITE. A seldom used term for rocks which have suffered propylitisation or hydrothermal alteration. Originally applied to greenstone-like rocks which were considered to be the "precursors of all other volcanic rocks" in the region. *(Richthofen, 1868, p.20; from the Greek propolos = a servant who goes before one; Trög. 970; Joh. v.3, p.177; Tomk. p.461)*

PROTEROBASE. An old term for altered rocks of basaltic composition which contain primary hornblende. *(Gümbel, 1874, p.14; from the Greek proteros = earlier; Trög. 394; Joh. v.3, p.318; Tomk. p.461)*

PROTOGINE. An old term given to the weakly cataclastic granites which occur abundantly in the Alps in the belief that they were part of the original crust of the earth. *(Jurine, 1806, p.372; from the Greek protos = first and gignomai = to be born; Joh. v.2, p.242; Tomk. p.462)*

PROTOKATUNGITE. A term for a variety of katungite without olivine. *(Holmes, 1942, p.199)*

PROTOPYLITE. An obsolete name for a propylitised porphyritic diorite. *(Stache & John, 1879, p.352; Tomk. p.463)*

PROTOTECTITE. An igneous rock which crystallises directly from a primary magma or its differentiation products. Cf. anatectite and syntectite. *(Loewinson-Lessing, 1934, p.7; Tomk. p.463)*

PROWERSITE. A term originally used for a potassic variety of minette (a lamprophyre) consisting of abundant biotite and orthoclase with lesser amounts of augite and altered olivine. Now regarded by some as a lamproite. *(Rosenbusch, 1908, p.1487; Prowers County, Colorado, USA; Trög. 228; Joh. v.3, p.41; Tomk. p.463)*

PUGLIANITE. A local name for a melanocratic variety of leucite gabbro (or theralite) largely composed of clinopyroxene and subordinate plagioclase, with minor mica and leucite. *(Lacroix, 1917b, p.210; Pugliani, Mt. Somma, Naples, Italy; Trög. 546; Joh. v.4, p.208; Tomk. p.469)*

PULASKITE. A variety of nepheline-bearing alkali feldspar syenite containing alkali feldspar and varying amounts of sodic pyroxenes and amphiboles, fayalite, biotite and minor amounts of nepheline. *(Williams, 1891, p.56; Fourche Mt., Pulaski County, Arkansas, USA; Trög. 186; Joh. v.4, p.5; Tomk. p.469)*

PULVERULITE. A variety of ignimbrite containing dust-like shards of glass surrounding the crystal grains. *(Marshall, 1935, p.358; from the Latin pulvis = dust; Trög(38). 871½; Tomk. p.470)*

PUMICE. A textural term applied to extremely vesiculated lavas which resemble froth or foam. *(a term of great antiquity usually attributed to Theophrastus, 320BC – see Johannsen for further discussion; from the Latin pumex = pumice; Trög. 821; Joh. v.2, p.282; Tomk. p.470)*

PUMICITE. A consolidated pumice. *(Chesterman, 1956, p.5; Tomk. p.470)*

PUMITE. An obsolete name for a pumice consisting of feldspathic glass. *(Cordier, 1816, p.372; Tomk. p.470)*

PUYS-ANDESITE. A term from an obsolete chemical classification, based on feldspar composition rather than SiO_2 alone, for a class of rocks in which $CaO : Na_2O : K_2O \approx 2 : 2 : 1$ and $SiO_2 \approx 63\%$. *(Lang, 1891, p.229; Chaine des Puys, Auvergne, France; Tomk. p.471)*

PUZZOLANA. See pozzolana.

PYCROPTOCHE. See pikroptoche.

PYNOSITE. A mnemonic name suggested for a rock consisting essentially of *pyroxene* and *nos*ean. *(Belyankin, 1929, p.22)*

PYRIBOLEID. An obsolete field term for a coarse-grained igneous rock consisting almost entirely of pyroxene and/or amphibole and/or biotite. It includes the pyroxeneids and amphiboleids. *(Johannsen, 1911, p.320; Tomk. p.471)*

PYRITOSALITE. An obsolete term for an extremely quartz-rich rock containing pyrites. *(Brögger, 1931, p.126; Tofteholmen, Hurum, Oslo district, Norway; Trög. 3; Tomk. p.471)*

PYROCLASTIC. Adjective applied to fragmental rocks produced as the result of volcanic eruption. *(Teall, 1887, p.493; from the Greek pyr = fire and clastos = broken; Joh. v.1(2nd.Ed.), p.230; Tomk. p.472)*

PYROCLASTIC DEPOSIT. Now defined in the pyroclastic classification (section B.5.2) as a general term to include both consolidated and unconsolidated assemblages of pyroclasts, which must make up more than 75% of the rock. *(Schmid, 1981, p.42)*

PYROCLASTS. Now defined in the pyroclastic classification (section B.5.1) as a general term for crystal, glass or rock fragments generated by disruption as a direct result of volcanic eruption which have not undergone any secondary re-deposition processes. *(Original reference uncertain; Tomk. p.472)*

PYROLITE. A term originally proposed for rocks formed at elevated temperatures within the Earth. Ringwood (1962) later used the term for an assumed Upper Mantle composition of 1 part basalt and 4 parts dunite. *(Preobrazhenskii, 1956, p.322; Tomk. p.473)*

PYROMERIDE. Originally applied to an orbicular diorite from Corsica described from the collection of R.J. Haüy. Later applied to spherulitic devitrified rhyolites having a nodular appearance. *(Monteiro, 1814, p.359 cites Brongniart as the original author of the name but gives no reference; from the Greek pyr = fire and meros = part, referring to the fact that feldspar is easily melted under the blowpipe, but quartz is not; Trög. 45; Tomk. p.473)*

PYROXENE HORNBLENDE GABBRONORITE. A collective term for basic plutonic rocks consisting essentially of calcic plagioclase, pyroxene and amphibole in various amounts. Now defined modally in the gabbroic rock classification (Fig. B.6). *(Streckeisen, 1973, p.27)*

PYROXENE HORNBLENDE PERIDOTITE. An ultramafic plutonic rock consisting of 40% to 90% olivine and various amounts of pyroxene and amphibole. Now defined modally in the ultramafic rock classification (Fig. B.8). *(Streckeisen, 1973, p.26)*

PYROXENE HORNBLENDITE. A term for ultramafic plutonic rocks composed mainly of amphibole with up to 50% pyroxene. Now defined modally in the ultramafic rock classification (Fig. B.8). *(Streckeisen, 1973, p.26)*

PYROXENE MELILITOLITE. An ultramafic plutonic rock, defined modally in the melilitic rocks classification (Fig. B.3), consisting essentially of melilite and clinopyroxene with minor olivine. *(Streckeisen, 1978, p.13)*

PYROXENE OLIVINE MELILITOLITE. An ultramafic plutonic rock, defined modally in the melilitic rocks classification (Fig. B.3), consisting essentially of melilite, olivine and lesser amounts of clinopyroxene. *(Streckeisen, 1978, p.13)*

PYROXENE PERIDOTITE. A term for ultramafic plutonic rocks composed mainly of olivine with up to 50% pyroxene. Now defined modally in the ultramafic rock classification (Fig. B.8). *(Wyllie, 1967, p.2)*

PYROXENEID. An obsolete field term for a coarse-grained igneous rock consisting almost entirely of pyroxene. *(Johannsen, 1911, p.320; Tomk. p.474)*

PYROXENIDE. A revised spelling recommended to replace the field term pyroxeneid. Now obsolete. *(Johannsen, 1926, p.182; Joh. v.1, p.57)*

PYROXENITE. A collective term named simultaneously in 1857 by Senft and Coquand for ultramafic plutonic rocks composed almost entirely of one or more pyroxenes and occasionally biotite, hornblende and olivine. Now defined modally in the ultramafic rock classification (Fig. B.8). *(Senft, 1857, p.42 and Coquand, 1857, p. 114; Trög. 673; Joh. v.4, p.400; Tomk. p.474)*

PYROXENOLITE. A term synonymous with pyroxenite. *(Lacroix, 1895, p.752; Trög.693; Joh. v.4, p.400; Tomk. p.474)*

PYTERLITE. A local name for a variety of rapakivi granite in which the ovoids of orthoclase are not mantled by plagioclase. *(Wahl, 1925, p.60; Pyterlahti, Virolahti, Finland; Trög. 79; Tomk. p.474)*

QUARTZ ALKALI FELDSPAR SYENITE. A felsic plutonic rock composed mainly of alkali feldspar, quartz and mafic minerals. Now defined modally in QAPF field 6* (Fig. B.4). *(Streckeisen, 1973, p.26)*

QUARTZ ALKALI FELDSPAR TRACHYTE. A felsic volcanic rock composed mainly of alkali feldspar, quartz and mafic minerals. Now defined modally in QAPF field 6* (Fig. B.10). *(Streckeisen, 1978, p.4)*

QUARTZ ANORTHOSITE. A leucocratic plutonic rock consisting essentially of calcic plagioclase, quartz and small amounts of pyroxene. Now defined modally in QAPF field 10* (Fig. B.4). *(Loughlin, 1912, p.108; Preston, Connecticut, USA; Trög. 128)*

QUARTZ DIOREID. An obsolete field term for a variety of dioreid containing quartz. *(Johannsen, 1911, p.320)*

QUARTZ DIORIDE. An obsolete field term for a variety of dioride containing quartz. *(Johannsen, 1931, p.57)*

QUARTZ DIORITE. The term was originally used for plutonic rocks consisting essentially of plagioclase, quartz and mafic minerals falling in field 5 of the QAPF, which are now called tonalite. Now defined modally in QAPF field 10* (Fig. B.4). *(Zirkel, 1866b, p.4; Trög. 131; Joh. v.2, p.378)*

QUARTZ DOLEREID. An obsolete field term for a variety of dolereid containing quartz. *(Johannsen, 1911, p.320)*

QUARTZ DOLERITE. A variety of dolerite composed mainly of plagioclase and pyroxenes with interstitial quartz or micropegmatite. The rock has tholeiitic affinities and its py-

roxenes are usually subcalcic augite accompanied by pigeonite or orthopyroxene. *(Tyrrell, 1926, p.120; Trög. 151)*

QUARTZ GABBREID. An obsolete field term for a variety of gabbreid containing quartz. *(Johannsen, 1911, p.320)*

QUARTZ GABBRIDE. An obsolete field term for a variety of gabbride containing quartz. *(Johannsen, 1931, p.57)*

QUARTZ GABBRO. A plutonic rock composed mainly of calcic plagioclase, clinopyroxene and quartz. Now defined modally in QAPF field 10* (Fig. B.4). *(Johannsen, 1932, p.409; Trög. 133; Tomk. p.476)*

QUARTZ LATITE. A term originally used for a volcanic rock composed of phenocrysts of quartz, plagioclase, biotite and hornblende in a glassy matrix potentially of quartz and alkali feldspar. Now commonly used for volcanic rocks composed of alkali feldspar and plagioclase in roughly equal amounts, quartz and mafic minerals. Now defined modally in QAPF field 8* (Fig. B.10). *(Ransome, 1898, p.372; Trög. 100)*

QUARTZ LEUCOPHYRIDE. An obsolete field term for a variety of leucophyride containing quartz. *(Johannsen, 1931, p.58)*

QUARTZ MONZODIORITE. A plutonic rock consisting essentially of sodic plagioclase, alkali feldspar, quartz and mafic minerals. Now defined modally in QAPF field 9* (Fig. B.4). *(Streckeisen, 1973, p.26)*

QUARTZ MONZOGABBRO. A plutonic rock consisting essentially of calcic plagioclase, alkali feldspar, mafic minerals and quartz. Now defined modally in QAPF field 9* (Fig. B.4). *(Streckeisen, 1973, p.26)*

QUARTZ MONZONITE. A plutonic rock consisting of approximately equal amounts of alkali feldspar and plagioclase and with essential quartz (5%-20% of felsic minerals) but not enough to make the rock a granite. Now defined modally in QAPF field 8* (Fig. B.4). The term was formerly used for granites of QAPF field 3b. *(Brögger, 1895, p.61; Trög. 86; Tomk. p.476)*

QUARTZ NORITE. A plutonic rock composed essentially of calcic plagioclase, quartz and orthopyroxene, Now defined modally as a variety of gabbro in QAPF field 10* (Fig.

B.4). *(Johannsen, 1932, p.409; Trög. 134)*

QUARTZ PORPHYRITE. An aphanitic rock of dacite composition containing phenocrysts of quartz and plagioclase in a glassy groundmass. Originally used by European petrologists only if such rocks were pre-Tertiary. *(Brögger, 1895, p.60; Trög. 149; Joh. v.2, p.396)*

QUARTZ PORPHYRY. An aphanitic rock of rhyolite composition containing phenocrysts of quartz and orthoclase in a glassy groundmass. Originally used by European petrologists only if such rocks were pre-Tertiary. *(Durocher, 1845, p.1281; Messangé, Brittany, France; Trög.41; Joh. v.2, p.286; Tomk. p.477)*

QUARTZ SYENITE. A plutonic rock consisting essentially of alkali feldspar, quartz and mafic minerals. Now defined modally in QAPF field 7* (Fig. B.4). *(Streckeisen, 1973, p.26; Trög. 239)*

QUARTZ TRACHYTE. A volcanic rock consisting of phenocrysts of alkali feldspar and quartz in a cryptocrystalline or glassy matrix. It is the volcanic equivalent of quartz syenite. Now defined modally in QAPF field 7* (Fig. B.10). *(Hauer & Stache, 1863, p.70; Trög. 50; Joh. v.2, p.266; Tomk. p.477)*

QUARTZ-RICH GRANITOID. A collective term for granitic rocks having a quartz content greater than 60% of the felsic minerals. Now defined modally in QAPF field 1b (Fig. B.9). *(Streckeisen, 1973, p.26)*

QUARTZOLITE. A collective term for plutonic rocks in which the quartz content is more than 90% of the felsic minerals. Now defined modally in QAPF field 1a* (Fig. B.4). *(Streckeisen, 1973, p.26; Tomk. p.476)*

RAABSITE. A local name for a variety of minette (a lamprophyre) consisting of microcline, sodic amphibole, biotite and olivine. *(Hackl & Waldmann, 1935, p.272; Raabs, Lower Austria, Austria; Trög(38). 229½; Tomk. p.479)*

RADIOPHYRE. An obsolete term for a porphyry with a subspherulitic texture. *(Bořický, 1882, p.74; Tomk. p.480)*

RADIOPHYRITE. An obsolete term for a radiophyre in which Na > K. *(Bořický, 1882, p.119; Tomk. p.480)*

RAFAELITE. A variety of analcime-bearing microsyenite containing amphibole and labradorite, but no nepheline. *(Johannsen, 1938, p.177; San Rafael Swell, Emery, Utah, USA; Trög(38). 508½; Tomk. p.480)*

RAGLANITE. A leucocratic variety of corundum-bearing nepheline diorite with oligoclase and biotite. *(Adams & Barlow, 1908, p.62; Raglan, Renfrew County, Ontario, Canada; Trög. 549; Joh. v.4, p.213; Tomk. p.480)*

RAPAKIVI. A term, used as both an adjective and noun, for a variety of granite usually containing amphibole and biotite and characterised by the presence of large oval grains (ovoids) of orthoclase which are usually mantled by plagioclase. Two generations of quartz are often present. *(Hjärne, 1694, p.V, No. 11; from the Finnish word for "rotten stone"; Joh. v.2, p.243; Tomk. p.481)*

RAQQAITE. A local term for an extrusive rock composed of phenocrysts of diopside and hypersthene, olivine, magnetite and apatite in a light-coloured analcime-like mesostasis containing some needles of natrolite. May be related to meimechite. *(Eigenfeld, 1965, p.46; Raqqa, E. of Aleppo, Syria)*

RAUHAUGITE. A poorly defined local name for a carbonatite composed mainly of dolomite. Now called dolomite carbonatite. *(Brögger, 1921, p.252; Rauhaug, Fen complex, Telemark, Norway; Trög. 758; Tomk. p.481)*

REDWITZITE. A local group name for lamprophyric rocks of variable composition characterised by the presence of large biotite phenocrysts. *(Willmann, 1919, p.5; Redwitz, Fichtelgebirge, West Germany; Trög. 975; Tomk. p.483)*

REGADITE. An obsolete name for a variety of biotite clinopyroxenite which occurs as dykes and consists of diallage and biotite in equal amounts. *(Cotelo Neiva, 1947b, p.127; A Regada, Bragança district, Portugal; Tomk. p.484)*

REGENPORPHYRY. An obsolete local term for a granite porphyry containing pseudomorphs after cordierite. *(Gümbel, 1868, p.420; Regen, Bavaria, West Germany; Trög. 959)*

RESINITE. An obsolete name for a variety of obsidian. More usually used as a component of coal. *(Haüy, 1822, p.580; Tenerife, Ca-*

nary Islands, Atlantic Ocean; Tomk. p.487)

RETICULITE. An extremely vesiculated variety of pumice consisting of a broken network of glass threads remaining after the vesicle walls collapse. *(Wentworth & Williams, 1932, p.47; from the Latin reticulum = small net; Tomk. p.488)*

RETINITE. An early term for obsidian, preceding the use of the term for fossil resin. Now used for water-rich rhyolite glasses which may have expanding properties. *(Dolomieu, 1794, p.103; from the Greek retina = resin; Tomk. p.488)*

RHAZITE. An obsolete term for a fine-grained basalt containing a ferromagnesian mineral. *(Pinkerton, 1811b, p.45; named after Rhazes, AD900)*

RHENOAPLITE. An erroneous spelling of rhenopalite. *(Tomkeieff et al., 1983, p.489; Tomk. p.489)*

RHENOPALITE. An obsolete group name for various aplitic rocks cutting gabbros, perhaps as magmatic segregation products. *(Schuster, 1907, p.43; named after Rheinpfalz, West Germany; Trög. 976)*

RHOMBPORPHYRY. A term applied to rocks which vary in composition from porphyritic trachytes to trachyandesites and carry rhomb-shaped phenocrysts of ternary feldspar. *(Buch, 1810, p.106; Oslo Region, Norway; Trög. 203; Joh. v.4, p.26; Tomk. p.490)*

RHÖN BASALT. A term from an obsolete chemical classification, based on feldspar composition rather than SiO_2 alone, for basaltic rocks in which $CaO : Na_2O : K_2O \approx 10 : 2.3 : 1$ and $SiO_2 \approx 40\%$. *(Lang, 1891, p.237; Tomk. p.490)*

RHYOBASALT. A term suggested for the volcanic equivalent of a granogabbro in order to maintain similar constructions between volcanic and plutonic rock names. *(Johannsen, 1920c, p.211; Trög. 119; Joh. v.2, p.369; Tomk. p.490)*

RHYODACITE. A term used for volcanic rocks intermediate between rhyolite and dacite, usually consisting of phenocrysts of quartz, plagioclase and a few ferromagnesian minerals in a microcrystalline groundmass. *(Winchell, 1913, p.214; Trög. 118; Joh. v.2, p.356; Tomk. p.490)*

RHYOLITE. A collective term for silicic volcanic rocks consisting of phenocrysts of quartz and alkali feldspar, often with minor plagioclase and biotite, in a microcrystalline or glassy groundmass and having the chemical composition of granite. Now defined modally in QAPF field 3 (Fig. B.10) and, if modes are not available, chemically in TAS field R (Fig. B.13). *(Richthofen, 1860, p.156; from the Greek rheo = to flow; Trög. 40; Joh. v.2, p.265; Tomk. p.490)*

RHYOLITOID. A rock with the chemical composition of rhyolite but without modal quartz. Now used as a general term in the provisional "field" classification (Fig. B.17) for rocks tentatively identified as rhyolite. *(Lacroix, 1923, p.2; Tomk. p.491)*

RIACOLITE. A local name for a sanidine trachyte. *(Tomkeieff et al., 1983, p.491; Puy de Dôme, Auvergne, France)*

RICOLETTAITE. A local name for a slightly alkaline variety of gabbro with very calcic plagioclase, Ti-augite, a little biotite, olivine and opaques. *(Johannsen, 1920c, p.224; Ricoletta, Mt. Monzoni, Italy; Trög. 359; Joh. v.3, p.131; Tomk. p.491)*

RIEDENITE. A local name for a melanocratic variety of foidolite (noseanolite) consisting of large biotite tablets in a granular aggregate of aegirine-augite, nosean and biotite. Cf. boderite and rodderite. *(Brauns, 1922, p.76; Rieden, Eifel district, near Koblenz, West Germany; Trög. 637; Joh. v.4, p.332; Tomk. p.492)*

RIKOTITE. A local name for a melanocratic banded variety of diorite composed mainly of clinopyroxene and anorthoclase with minor andesine, hornblende, biotite and opaques. *(Belyankin & Petrov, 1945, p.180; Rikotsky Gorge, River Dzirula, Georgia, USSR; Tomk. p.492)*

RINGITE. A local name for a coarse-grained carbonatite containing aegirine and alkali feldspar, considered to be a mixture of carbonatite and syenitic fenite. *(Brögger, 1921, p.199; Ringsevja, Fen complex, Telemark, Norway; Trög. 754; Tomk. p.493)*

RISCHORRITE. A variety of biotite-bearing nepheline syenite in which the nepheline crystals are poikilitically enclosed in microcline perthite. Aegirine-augite, apatite and opaques are often abundant. *(Kupletskii, 1932, p.36; Rischorr Plateau, Khibina, Kola Peninsula, USSR; Trög(38). 413½; Tomk. p.493)*

RIZZONITE. An obsolete name for a limburgitic dyke rock containing phenocrysts of Ti-augite, olivine, and opaques in a plentiful glassy base. *(Doelter, 1902, p.977; Mt. Rizzoni, Mt. Monzoni, Italy; Trög. 375; Tomk. p.494)*

ROCKALLITE. A local name for a variety of peralkaline granite containing abundant aegirine-augite and albite. *(Judd, 1897, p.57; named after the island of Rockall, Atlantic Ocean; Trög.58; Joh. v.2, p.41; Tomk. p.494)*

RØDBERG. An altered carbonatite composed of hematite-stained granular calcite, dolomite and sometimes ankerite. *(Vogt, 1918, p.78; from Norwegian "red rock"; Fen complex, Telemark, S. Norway; Tomk. p.495)*

RODDERITE. A term proposed for a melanocratic variety of nosean phonolite or noseanolite, with more nosean than alkali feldspar, biotite and augite. Cf. riedenite and rodderite. Originally called a nosean sanidine biotite augitite. *(Taylor et al., 1967, p.409 and Frechen, 1971, p.42 for the description; Rodderhöfen, near Rieden, Eifel district, near Koblenz, West Germany)*

RODINGITE. A metasomatically altered dyke rock, probably originally a dolerite, and composed mainly of augite and hydrogrossular. Associated with serpentinised peridotites and other rocks of the ophiolite suite. *(Bell et al., 1911, p.31; Roding River, Dun Mountain, Nelson, New Zealand; Trög. 718; Tomk. p.495)*

ROMANITE. An obsolete name suggested as a regional term for ciminites and vulsinites from the Roman province. *(Preller, 1924, p.154; Tomk. p.495)*

RONGSTOCKITE. A coarse-grained variety of foid-bearing monzodiorite with abundant Ti-augite and some biotite or brown hornblende. The foid is usually nepheline or cancrinite. *(Tröger, 1935, p.124; Rongstock (now Roztoky), Ústí nad Labem, N. Bohemia, Czechoslovakia; Trög. 282; Joh. v.4, p.50; Tomk. p.495)*

ROTENBURGITE. A name suggested for a group of amphibole diorites. *(Marcet Riba, 1925, p.293; Rotenburg, Kyffhäuser, S of Kassel,*

West Germany; Tomk. p.496)

ROUGEMONTITE. A variety of gabbro in which the plagioclase is anorthite and Ti-augite is abundant. *(O'Neill, 1914, p.64; Rougemont Mt., Quebec, Canada; Trög. 360; Joh. v.3, p.339; Tomk. p.496)*

ROUTIVARITE. A local name for a variety of anorthosite composed of labradorite and minor almandine. *(Sjögren, 1893, p.62; Routivare, Norrbotten, Lapland, Sweden; Trög. 297; Joh. v.3, p.201; Tomk. p.497)*

ROUVILLITE. A leucocratic variety of theralite with up to 80% plagioclase zoned from labradorite to bytownite and nepheline. *(O'Neill, 1914, p.28; Rouville County, Quebec, Canada; Trög.551; Joh. v.4, p.219; Tomk. p.497)*

RUNITE. An obsolete name for graphic granite. *(Pinkerton, 1811b, p.85; named after the texture which resembles runic characters; Trög. 38; Joh. v.2, p.84; Tomk. p.498)*

RUSHAYITE. A variety of olivine melilitite rich in phenocrysts of corroded forsteritic olivine in a groundmass of abundant melilite with olivine, perovskite and ore minerals with minor nepheline and augite. *(Denaeyer, 1965, p.2119; Chaîne du Rushayo, Nyiragongo volcano, Zaire)*

RUTTERITE. A medium-grained variety of nepheline-bearing syenite, with albite and perthitic microcline and subordinate hornblende. *(Quirke, 1936, p.179; Rutter, Bigwood Township, Sudbury district, Ontario, Canada; Trög(38). 178⅔; Joh. v.4, p.46; Tomk. p.498)*

S-TYPE GRANITE. A general term for a range of granitic rocks, mainly peraluminous granodiorites and granites, characterised by the presence of muscovite, alumino-silicates, garnet and/or cordierite in addition to essential quartz, alkali feldspar and plagiolcase. Hornblende is rare. The prefix S implies that the source rocks have sedimentary (pelitic) compositions. *(Chappell & White, 1974, p.173)*

SABAROVITE. A leucocratic variety of charnoenderbite of QAPF field 4 (Fig. B.4). It is an extreme member of the bugite series with high SiO_2, low K_2O, oligoclase-andesine antiperthite, quartz and hypersthene. It is suggested (Streckeisen, 1974, p.358) that this term should be abandoned. *(Bezborod'ko, 1931, p.142; Sabarovo village, near Vinnitsa, Podolia, Ukraine, USSR; Trög(38). 130½; Tomk. p.499)*

SAGVANDITE. A local name for a variety of orthopyroxenite composed of bronzite and small amounts of magnesite.*(Pettersen, 1883, p.247; Lake Sagvandet, Balsfjord, Tromsø, Norway; Trög. 677; Joh. v.4, p.459; Tomk. p.499)*

SAIBARITE (SAJBARITE). A local name for a variety of nepheline syenite with trachytoidal texture and banded structure consisting of nepheline, aegirine, alkali feldspar and albite, with minor alkali amphibole. *(Edel'shtein, 1930, p.23; Mt. Saibar, Minusinsk depression, W. Siberia, USSR)*

SAKALAVITE. A local name for a quartz normative basaltic rock containing intermediate plagioclase, augite, opaques and plentiful glass. *(Lacroix, 1923, p.15; Sakalava, Madagascar; Trög. 159; Tomk. p.500)*

SALIC. A name used in the CIPW normative classification for one of two major groups of normative minerals which includes quartz, feldspars, and feldspathoids as well as zircon, corundum, and the sodium salts. The other group is named femic. Cf. felsic.*(Cross et al., 1902, p.573; Joh. v.1, p.191; Tomk. p.500)*

SALITRITE. A local name for a variety of alkali pyroxenite consisting essentially of aegirine-augite and titanite with minor microcline. *(Tröger, 1928, p.202; Salitre Mts., Minas Geraes, Brazil; Trög. 695; Joh. v.3, p.40; Tomk. p.500)*

SANCYITE. A local name for a variety of trachyte that contains many large phenocrysts of sanidine and often tridymite. *(Lacroix, 1923, p.10; Aiguilles du Sancy, Mont Dore, Auvergne, France; Trög. 52; Tomk. p.502)*

SANDYITE. A melanocratic variety of foid-bearing alkali feldspar syenite with hastingsite, aegirine-augite, titanite, apatite, calcite, garnet, and ilmenite. Later said not to be an igneous rock but a metasomatised calcareous xenolith. *(Zavaritskii & Krizhanovskii, 1937, p.9; Sandy-Elga River, Ilmen Mts., Urals, USSR; Tomk. p.503)*

SANIDINITE. The term has been used in igneous

nomenclature for a rock consisting almost entirely of sanidine. To avoid conflict with the metamorphic usage it should not be used for igneous rocks.*(*Nose, 1808, p.156; Trög. 978; Joh. v.3, p.5; Tomk. p.504)*

SANIDOPHYRE. A local name for a variety of liparite with large phenocrysts of sanidine. *(Dechen, 1861, p.106; Siebengebirge, near Bonn, West Germany; Trög. 980; Tomk. p.504)*

SANNAITE. A variety of lamprophyre composed of combinations of olivine, Ti-augite, kaersutite, and Ti-biotite phenocrysts with alkali feldspar dominating over plagioclase in the groundmass which also contains nepheline. Now defined in the lamprophyre classification (Table B.3). *(Brögger, 1921, p.180; Sannavand, Fen complex, Telemark, Norway; Trög.498; Joh.v.4,p.159; Tomk. p.504)*

SANTORINITE. An obsolete term originally defined as an andesitic rock with calcic plagioclase and SiO_2 65% to 69% but later as a variety of hypersthene andesite (Becke, 1899). *(Washington, 1897b, p.368; Santorini, Greece; Trög. 154; Joh. v.3, p.174; Tomk. p.504)*

SANUKITE. A name given to a rock which was originally described as a bronzite andesite consisting of needles of bronzite in a groundmass of clear glass and abundant magnetite grains. Now used as a synonym for boninite. *(Weinschenk, 1891, p.150; Sanuki, Shikoko Island, Japan; Trög. 123; Joh. v.3, p.172; Tomk. p.505)*

SANUKITOID. A term originally given to all textural modifications of the sanukite magma type. *(Koto, 1916b, p.101; Joh. v.3, p.173; Tomk. p.505)*

SÄRNAITE. A leucocratic variety of cancrinite nepheline syenite with tablets of perthitic orthoclase, sometimes trachytoidal, and containing prisms of aegirine-augite. *(Brögger, 1890, p.244; Särna, Dalarne, Sweden; Trög. 426; Joh. v.4, p.108; Tomk. p.506)*

SATURATED. A term applied to igneous rocks which are neither oversaturated nor undersaturated with respect to silica, i.e. they have no silica minerals or foids in the mode or norm. *(Shand, 1913, p.508; Tomk. p.408)*

SAXONITE. A variety of harzburgite composed

of olivine and enstatite. *(Wadsworth, 1884, p.85; named after Saxony, East Germany; Trög. 731; Joh. v.4, p.434; Tomk. p.507)*

SCANOITE. An obsolete name for a volcanic rock consisting of phenocrysts of Ti-augite and olivine in a groundmass of colourless glass with minor analcime and microlites of augite and olivine. Cf. ghizite. *(Lacroix, 1924, p.531; Scano flow, Mt. Ferru, Sardinia; Trög. 600; Joh. v.4, p.218; Tomk. p.508)*

SCHALSTEIN. An old term originally used for diabase tuffs and altered diabases but now restricted to bedded Palaeozoic diabase tuffs. *(Cotta, 1855, p.53; from the German schal = flat and stein = rock; Trög. 981; Tomk. p.508)*

SCHILLERFELS. An obsolete term for a pyroxene (bastite) peridotite. *(Raumer, 1819, p.40; Trög. 709; Tomk. p.508)*

SCHÖNFELSITE. An obsolete name for a variety of picritic basalt containing phenocrysts of serpentinised olivine and augite in a groundmass of augite, bronzite, biotite, bytownite and abundant glass. *(Uhlemann, 1909, p.434; Altschönfels, Zwickau, Saxony, East Germany; Trög. 411; Joh. v.3, p.306; Tomk. p.510)*

SCHORENBERGITE. A local name for a variety of leucitite containing phenocrysts of nosean and sometimes leucite in a groundmass of abundant leucite, nepheline and aegirine. Feldspar is entirely absent. *(Brauns, 1922, p.46; Schorenberg, Rieden, Laacher See, near Koblenz, West Germany; Trög. 638; Joh. v.4, p.380; Tomk. p.510)*

SCHRIESHEIMITE. An obsolete name for a variety of pyroxene hornblende peridotite composed of large crystals of hornblende poikilitically enclosing olivine, which is altered to serpentine and talc, with minor phlogopite and diopside. *(Rosenbusch, 1896, p.348; Schriesheim, Odenwald, West Germany; Trög. 709; Joh. v.4, p.428; Tomk. p.510)*

SCORIA. A highly vesiculated lava or tephra which resembles clinker. Usually of basic composition. *(Cotta, 1866, p.97; Tomk. p.511)*

SCYELITE. An obsolete name for a variety of olivine hornblendite consisting of hornblende, olivine and phlogopite. *(Judd, 1885, p.401; Loch Scye, Scotland; Trög. 706; Joh. v.4, p.424; Tomk. p.512)*

SEBASTIANITE. A local name for fragments of plutonic rocks ejected from Mt. Somma, Italy, composed mainly of anorthite and biotite with minor augite and apatite. Heteromorph of puglianite with biotite instead of leucite. *(Lacroix, 1917b, p.210; San Sebastiano, Mt. Somma, Naples, Italy; Trög. 363; Joh. v.4, p.75; Tomk. p.512)*

SELAGITE. An obsolete name used for a variety of trachyte containing biotite. *(Haüy, 1822, p.544; from the Greek selageo = to beam brightly; Trög. 232; Joh. v.3, p.79; Tomk. p.515)*

SELBERGITE. A fine-grained variety of nosean leucite syenite consisting of phenocrysts of leucite, nosean (some haüyne), sanidine and aegirine-augite in a groundmass of nepheline, alkali feldspar and aegirine. *(Brauns, 1922, p.47; Selberg, Laacher See, near Koblenz, West Germany; Trög. 453; Joh. v.4, p.275; Tomk. p.516)*

SEMEITAVITE (SEMEJTAVITE, SEMEITOVITE). A variety of quartz alkali feldspar syenite containing anorthoclase, quartz, ilmenite, augite and riebeckite. *(Gornostaev, 1933, p.180; Semeitau Mts., Kazakh Republic, USSR; Tomk. p.516)*

SERPENTINITE. A plutonic rock composed almost entirely of serpentine minerals. Relics of original olivine and/or pyroxene may be present and chromite or chrome spinels are commonly present. An altered peridotite. *(Wells, 1948, p.95; from the mineral serpentine named after the French serpent = snake, on account of the appearence; Tomk. p.518)*

SESSERALITE. A local name for a variety of gabbro consisting essentially of plagioclase and hornblende with some corundum. *(Millosevich, 1927, p.30; Sessera Valley, Piedmont, Italy; Trög. 350; Tomk. p.519)*

SHACKANITE. A porphyritic variety of analcime phonolite consisting of phenocrysts of rhomb-shaped anorthoclase, augite and altered olivine in a groundmass of analcime, anorthoclase, biotite and glass. *(Daly, 1912, p.411; Shackan, British Columbia, Canada; Trög. 482; Joh. v.4, p.136; Tomk. p.519)*

SHASTAITE. A local term for a glassy dacite containing normative andesine c.f ungaite. *(Iddings, 1913, p.106; Mt. Shasta, Califor-*

nia, USA; Trög. 156; Tomk. p.520)

SHASTALITE. A local name for fresh andesite glass – the altered variety has been called weiselbergite. *(Wadsworth, 1884, p.97; Mt. Shasta, California, USA; Trög. 983; Joh. v.3, p.170; Tomk. p.520)*

SHIHLUNITE. A local name for a variety of trachyte containing augite, olivine and biotite. Some varieties contain small amounts of leucite. *(Ogura et al., 1936, p.92; Shihlung lava flow, Lung-chiang province, Manchuria; Trög(38). 232½; Tomk. p.520)*

SHISHIMSKITE. A local name for an ultramafic rock consisting essentially of magnetite with some perovskite and spinel. Cf. kiirunavaarite. *(Shilin, 1940, p.350; Shishim Mts., Urals, USSR; Tomk. p.521)*

SHONKINITE. A coarse-grained rock with abundant augite, some olivine, biotite or hornblende, and essential alkali feldspar and foids, usually nepheline. Now defined modally as a melanocratic variety of foid syenite in QAPF field 11 (Figs. B.4 and B.7). *(Weed & Pirsson, 1895a, p.415; from Shonkin, the Indian name for the Highwood Mts., Montana, USA; Trög. 489; Joh. v.4, p.13; Tomk. p.521)*

SHOSHONITE. A collective term, related to absarokite and banakite, for a trachyandesitic rock originally loosely described as an orthoclase-bearing basalt. It has since been used in several ways for potassic basaltic and intermediate volcanic rocks e.g. Joplin, 1964. Now defined chemically as the potassic variety of basaltic trachyandesite in TAS field S2 (Fig. B.13). *(Iddings, 1895a, p.943; Shoshone River, Yellowstone National Park, Wyoming, USA; Trög. 269; Joh. v.4, p.44; Tomk. p.521)*

SIDEROMELANE. A term for black to olive-green obsidian-like basaltic glass which is pale-coloured in thin section. *(Waltershausen, 1853, p.202; from the Greek sideros = iron and melas = dark; Sicily, Italy; Joh. v.3, p.324; Tomk. p.522)*

SIEVITE. An obsolete name given to a series of glassy rocks originally described as andesitic. *(*Marzari-Pencati, 1819, p.118; Sieva, Euganean Hills, Italy; Trög. 984; Tomk. p.522)*

SILEXITE. A term for an igneous rock consisting almost entirely of quartz. It is recommended that it is replaced by the term quartzolite, which is etymologically more correct. *(Miller, 1919, p.30; from the Latin silex = flint; Trög. 2; Joh. v.2, p.11; Tomk. p.523)*

SILICOCARBONATITE. An igneous rock comprising major amounts of both silicates and carbonates, but with silicates in excess of carbonates. The silicates may be Na-pyroxenes and amphiboles, biotite, phlogopite, olivine or feldspars. *(Brögger, 1921, p.350; Fen complex, Telemark, Norway; Tomk. p.525)*

SILICOGRANITONE. A variety of gabbro impregnated with silica. *(Mazzuoli & Issel, 1881, p.326; Tomk. p.525)*

SILICOTELITE. An obsolete name for rocks containing over 50% of non-silicate minerals. *(Rinne, 1921, p.144; Tomk. p.525)*

SILLAR. A local Peruvian name for a variety of ignimbrite which has been indurated primarily by pneumatolytic processes. *(Fenner, 1948, p.883; Arequipa, S Peru; Tomk. p.525)*

SILLITE. An obsolete name for a local rock variously called mica diorite, mica syenite, diabase porphyry and gabbro. *(Gümbel, 1861, p.187; Sill-Berge, near Berchtesgaden, West Germany; Trög. 985; Tomk. p.526)*

SINAITE. A term suggested (but never adopted) for plutonic alkali feldspar rocks to replace syenite – the type rock from Syene having been shown to be a hornblende granite. *(Rozière, 1826, p.306; Mt. Sinai, Egypt; Trög(38). 985½; Tomk. p.527)*

SIZUNITE. A local term for a variety of minette characterised by high K_2O (>10%) and P_2O_5 (2.5%) and consisting of microcline, biotite, apatite and opaques. *(Cogné & Giot, 1961, p.2569; Cap Sizun, Brittany, France)*

SKEDOPHYRE. An obsolete term for a porphyritic rock with evenly spaced phenocrysts (skedophyric texture). *(Iddings, 1909, p.224; Tomk. p.528)*

SKOMERITE. A local term for a volcanic rock composed of phenocrysts of albite-oligoclase, augite and minor olivine in a groundmass of albite, chlorite and opaques. *(Thomas, 1911, p.196; Skomer Island, Dyfed, Wales; Trög. 215; Joh. v.3, p.175; Tomk. p.529)*

SMALTO. A local Italian name for a glassy rhyolite from Lipari Island. *(Tomkeieff et al., 1983, p.531)*

SNOBITE. A local name for a hybrid variety of hypersthene dacite with inclusions of acid porphyry. *(Hills, 1958, p.551; Snob's Creek, near Eildon, Victoria, Australia)*

SODALITE BASALT. A term used for a volcanic rock consisting essentially of sodalite with minor amounts of olivine and other mafics. The name should not be used as the term basalt is now restricted to a rock containing essential plagioclase. As the rock is a variety of foidite it should be given the appropriate name, e.g. olivine-bearing sodalitite. *(Johannsen, 1938, p.346)*

SODALITE DIORITE. Now defined in QAPF field 14 (Fig. B.4) as a variety of foid diorite in which sodalite is the most abundant foid.

SODALITE GABBRO. Now defined in QAPF field 14 (Fig. B.4) as a variety of foid gabbro in which sodalite is the most abundant foid.

SODALITE MONZODIORITE. Now defined in QAPF field 13 (Fig. B.4) as a variety of foid monzodiorite in which sodalite is the most abundant foid.

SODALITE MONZOGABBRO. Now defined in QAPF field 13 (Fig. B.4) as a variety of foid monzogabbro in which sodalite is the most abundant foid.

SODALITE MONZOSYENITE. Now defined in QAPF field 12 (Fig. B.4) as a variety of foid monzosyenite in which sodalite is the most abundant foid. The term is synonymous with sodalite plagisyenite.

SODALITE PLAGISYENITE. Now defined in QAPF field 12 (Fig. B.4) as a variety of foid plagisyenite in which sodalite is the most abundant foid. The term is synonymous with sodalite monzosyenite.

SODALITE SYENITE. Now defined in QAPF field 11 (Fig. B.4) as a variety of foid syenite in which sodalite is the most abundant foid. *(Steenstrup, 1881, p.34; Trög.438; Joh. v.4, p.113; Tomk. p.532)*

SODALITHOLITH (SODALITHITE). The spellings adopted by Tröger (1935) and Johannsen (1938), respectively, for the plutonic rock that Ussing (1912) had called sodalitite.

SODALITITE. A term originally proposed for a leucocratic plutonic rock almost wholly composed of sodalite crystals together with minor alkali feldspar, aegirine and eudialyte. Cf. naujaite. However, the term is now used for volcanic rocks and is defined as a variety of foidite in QAPF field 15c (Fig. B.10). *(Ussing, 1912, p.156; Ilimaussaq, Greenland; Tomk. p.532)*

SODALITOPHYRE. A porphyritic rock consisting essentially of phenocrysts of sodalite, augite and hornblende in a glassy matrix. *(Hibsch, 1902, p.526; Böhmisches Mittelgebirge (now České Středohoří), N. Bohemia, Czechoslovakia; Trög. 986; Tomk. p.532)*

SOEVITE. An alternative spelling for sövite.

SOGGENDALITE. A local name for a pyroxene-rich variety of dolerite. *(Kolderup, 1896, p.159; Soggendal, Norway; Trög. 988; Joh. v.3, p.332; Tomk. p.533)*

SÖLVSBERGITE. A variety of peralkaline microsyenite or peralkaline trachyte, often occurring as minor intrusions, consisting essentially of alkali feldspar with minor alkali pyroxene and/or alkali amphibole. *(Brögger, 1894, p.67; Sölvsberg, Gran, near Trondheim, Norway; Trög. 192; Joh. v.3, p.107; Tomk. p.535)*

SOMMAITE. A medium- to coarse- grained variety of leucite monzosyenite consisting of phenocrysts of Ti-augite and lesser olivine in a groundmass of sanidine, labradorite and leucite. It occurrs abundantly among the ejected blocks of Mt. Somma. *(Lacroix, 1905, p.1189; Mt. Somma, Naples, Italy; Trög.511; Joh. v.4, p.193; Tomk. p.536)*

SORDAWALITE (SORDAVALITE). An obsolete term originally used to described a mineral and later redefined as a vitreous selvage to an olivine-rich dyke rock i.e. a tachylyte. *(Nordenskjöld, 1820, p.86; Sordavala, Lake Ladoga, near Leningrad, USSR; Trög. 989; Joh. v.3, p.323; Tomk. p.536)*

SÖRKEDALITE. A local name for a variety of olivine monzodiorite consisting of abundant antiperthitic andesine, which may be mantled by anorthoclase, and by olivine, opaques and apatite with minor clinopyroxene and biotite. *(Brögger, 1933, p.35; Kjelsås, Sörkedal, Oslo* district, Norway; Trög. 286; Joh. v.4, p.54; Tomk. p.536)

SÖVITE (SOEVITE). A special term in the carbonatite classification for the coarse-grained variety of calcite carbonatite in which the most abundant carbonate is calcite. Biotite and apatite are also frequently present. *(Brögger, 1921, p.246; Söve, Fen complex, Telemark, Norway; Trög. 753; Tomk. p.537)*

SPERONE. A local name for a yellowish, porous leucitite from the Alban Hills, Italy. *(*Gmelin, 1814; Trög. 990; Joh. v.4, p.363; Tomk. p.538)*

SPESSARTITE. A variety of lamprophyre consisting of phenocrysts of hornblende with or without biotite, olivine or pyroxene in a groundmass of the same minerals plus plagioclase and subordinate alkali feldspar. Now defined in the lamprophyre classification (Table B.3). *(Rosenbusch, 1896, p.532; Spessart, Bavaria, West Germany; Trög. 318; Joh. v.3, p.191; Tomk. p.538)*

SPHENITITE. A variety of pyroxenite, similar to jacupirangite, with up to 50% titanite (formerly called sphene). *(Allen, 1914, p.155; Ice River, British Columbia, Canada; Trög. 696; Joh. v.4, p.464; Tomk. p.538)*

SPIEMONTITE. An obsolete term for a variety of fine-grained diorite. *(Steininger, 1841, p.30; Spiemont Mt., Nahe district, Saarland, West Germany)*

SPILITE. A term originally used for altered phenocryst-poor or aphyric basaltic rocks but later (Flett, 1907) used for altered (often albitised) basaltic lavas. *(Brongniart, 1827, p.98; Trög. 329; Joh. v.3, p.299; Tomk. p.539)*

SPINELLITE. A rock in which spinel is the predominant mineral. However, Johannsen, to whom the term is attributed, only used the term in conjunction with a mineral qualifier e.g. magnetite spinellite. *(Tomkeieff et al., 1983, p.540)*

SPODITE. An obsolete term for volcanic ashes. *(Cordier, 1816, p.381; Tomk. p.541)*

SPUMULITE. A little used name for pumice stone. The reference implies the term was used earlier by Zavaritskii. *(Lebedinsky & Chu Tszya-Syan, 1958, p.15; Tomk. p.542)*

SPURINE. An obsolete name for a variety of porphyry containing quartz, feldspar and talc.

(Jurine, 1806, p.375; Mont Blanc, France; Tomk. p.542)

STAVRITE. A local name for a dyke rock consisting of large amounts of amphibole needles with interstitial biotite and minor quartz, opaques and apatite. The rock may be metamorphic. *(Eckermann, 1928a, p.405; Stavreviken, Alnö Island, Westnorrland, Sweden; Trög. 713; Joh. v.4, p.445; Tomk. p.544)*

STIGMITE. An obsolete name for a porphyritic pitchstone or obsidian. *(Brongniart, 1813, p.44; Tomk. p.545)*

STUBACHITE. An obsolete term originally used for an igneous rock composed of olivine and antigorite with some chrome spinel and pyroxene, but later shown to be a partially altered olivine pyroxene rock. *(Weinschenk, 1895, p.701; Stubachtal, Tyrol, Austria; Tomk. p.550)*

SUBALKALI. A term used for rocks that are not alkaline in character. *(Iddings, 1895b, p.183)*

SUBALKALI BASALT. Now defined chemically as a group name for varieties of basalt in TAS field B (Fig. B.13) which do not contain normative nepheline. Cf. alkali basalt. *(Chayes, 1966, p.137)*

SUBPHONOLITE. A variety of phonolite with a texture which suggests that it crystallised at depths below which phonolite would normally crystallise but not deep enough to be a nepheline syenite. *(Adamson, 1944, p.241; Tomk. p.552)*

SUDBURITE. A local name for a melanocratic volcanic rock containing hypersthene and augite phenocrysts in an equigranular groundmass of these minerals, bytownite and biotite. The original rock is probably a hornfels. *(Coleman, 1914, p.215; Sudbury, Ontario, Canada; Trög. 393; Joh. v.3, p.305; Tomk. p.553)*

SUEVITE. A glassy material found in breccia or tuff around the Ries meteorite impact crater and containing the high pressure polymorphs of silica, coesite and stishovite. *(Sauer, 1919, p.15; named after the Suevi who lived around Nördlinger, Ries, Bavaria, West Germany in Roman times; Trög. 992; Tomk. p.553)*

SULDENITE. An obsolete name for a variety of

quartz microgabbro (monzogabbro) containing phenocrysts of labradorite, hornblende and quartz in a groundmass of these minerals with orthoclase, diopside and biotite. *(Stache & John, 1879, p.382; Suldenferner, Ortler Alps, Alto Adige, Italy; Trög. 117; Tomk. p.554)*

SUMACOITE. An obsolete name for a variety of porphyritic trachyandesite containing abundant phenocrysts of intermediate plagioclase and augite in a groundmass of augite, sodic plagioclase, orthoclase and minor nepheline, haüyne and altered sodalite. The rock was originally described as an andesitic tephrite. *(Johannsen, 1938, p.188; Sumaco volcano, Ecuador; Trög(38). 267; Tomk. p.554)*

SUSSEXITE. A leucocratic medium-grained variety of nepheline syenite with abundant large crystals of nepheline in a matrix of nepheline, aegirine-augite needles and alkali feldspar. Katophorite and biotite may also be present. *(Brögger, 1894, p.173; Beemerville, Sussex County, New Jersey, USA; Trög. 462; Joh. v.4, p.269; Tomk. p.556)*

SUZORITE. A local name for a coarse-grained rock consisting essentially of biotite, with small amounts of red orthoclase, augite and apatite. *(Faessler, 1939, p.47; Suzor Township, Laviolette County, Quebec, Canada; Tomk. p.556)*

SVARTVIKITE. A variety of microsyenite consisting essentially of albite, in which abundant diopsidic pyroxene grains occur, intergrown with muscovite and calcite. Originally described as an albitophyre with a chemical similarity to holyokeite. *(Eckermann, 1938b, p.425; Svartviken Cove, N. Ulfö Island, Sweden; Tomk. p.556)*

SVIATONOSSITE. A local term for syenites characterised by the presence of andradite garnet. Probably hybrid rocks. *(Eskola, 1921, p.41; Sviatoi Nos, Transbaikalia, Siberia, USSR; Trög. 188; Joh. v.3, p.111; Tomk. p.556)*

SYENEID. An obsolete field term for a coarse-grained igneous rock consisting of feldspar with biotite and/or amphibole and/or pyroxene. *(Johannsen, 1911, p.320; Tomk. p.557)*

SYENIDE. A revised spelling recommended to replace the field term syeneid. Now obsolete. *(Johannsen, 1926, p.182; Joh. v.1, p.57)*

SYENILITE. An obsolete name for an amphibole granite. *(Cordier, 1868, p.78; Syene, now Aswân, Egypt; Tomk. p.557)*

SYENITE. A plutonic rock consisting mainly of alkali feldspar with subordinate sodic plagioclase, biotite, pyroxene, amphibole and occasional fayalite. Minor quartz or nepheline may also be present. Now defined modally in QAPF field 6 (Fig. B.4). *(a term of great antiquity usually attributed to Pliny, 77AD – see Johannsen for further discussion; Syene, now Aswân, Egypt; Trög. 240; Joh. v.3, p.52; Tomk. p.557)*

SYENITELLE. An obsolete name for a variety of banded syenite. *(Rozière, 1826, p.250; Syene, now Aswân, Egypt; Tomk. p.557)*

SYENITITE. A term proposed (but not adopted) for both biotite-bearing syenitic aplites and plagioclase-bearing syenites. *(Polenov, 1899, p.464; Trög. 993; Joh. v.3, p.61; Tomk. p.558)*

SYENITOID. Used as a general term in the provisional "field" classification (Fig. B.9) for rocks tentatively identified as syenite or monzonite. *(Streckeisen, 1973, p.28)*

SYENODIORITE. A comprehensive term for plutonic rocks intermediate between syenite and diorite. *(Johannsen, 1917, p.89; Trög. 994; Joh. v.3, p.110; Tomk. p.558)*

SYENOGABBRO. A comprehensive term for plutonic rocks intermediate between syenite and gabbro. *(Johannsen, 1917, p.89; Trög. 995; Joh. v.3, p.126; Tomk. p.558)*

SYENOGRANITE. A variety of granite consisting of alkali feldspar with subordinate plagioclase. Now defined modally in QAPF field 3a (Fig. B.4). *(Streckeisen, 1967, p.166)*

SYENOID. An obsolete term proposed for all syenites containing feldspathoids. *(Shand, 1910, p.377; Trög. 996; Joh. v.4, p.78; Tomk. p.558)*

SYNNYRITE. A variety of pseudoleucite nepheline syenite, with a micropegmatitic texture, composed of K-feldspar and kalsilite with biotite, albite, minor aegirine-augite, titanite, apatite, fluorite, magnetite, garnet, and alkali amphibole. *(Zhidkov, 1962, p.33; Synnyr massive, Baikal rift, Siberia, USSR)*

SYNTECTITE. An igneous rock produced by the

contamination of primary magmas with crustal rocks or melts. Cf. anatectite and prototectite. *(Loewinson-Lessing, 1934, p.7; Tomk. p.560)*

TABONA. A local name used on the island of Tenerife for an obsidian without phenocrysts. *(Tomkeieff et al., 1983, p.562)*

TACHYLYTE. A black basaltic glass usually containing crystallites and generally occurring as the selvages of dykes and sills. *(Breithaupt, 1826, p.112; from the Greek tachys = rapid and lytos = soluble; Trög. 387; Joh. v.3, p.290; Tomk. p.562)*

TAHITITE. A volcanic rock consisting of phenocrysts of haüyne in a glassy groundmass with augite microlites, Ti-magnetite, haüyne, and occasional orthoclase and leucite. Although regarded as an extrusive equivalent of a nepheline monzonite (Johannsen, 1938), with which it is associated, it would be classified chemically in TAS as a haüyne tephriphonolite or phonotephrite. *(Lacroix, 1917a, p.583; Papenoo, Tahiti, Pacific Ocean; Trög. 537; Joh. v.4, p.189; Tomk. p.562)*

TAIMYRITE. A local name for a variety of nosean phonolite consisting essentially of anorthoclase and nosean. *(Chrustschoff, 1894, p.427; Taimyr River, N. Siberia, USSR; Trög. 434; Joh. v.4, p.134; Tomk. p.563)*

TAIWANITE. A local name for an intrusive magnesium-rich basaltic glass containing phenocrysts of labradorite and olivine. *(Juan et al., 1953, p.1; named after Taiwan; Tomk. p.563)*

TALZASTITE. A local name for a coarse-grained variety of ijolite containing Ti-augite. The nepheline is altered to hydronepheline and cancrinite. *(Termier et al., 1948, p.81; N'Talzast volcano, Azrou, Morocco; Tomk. p.563)*

TAMARAITE. A local name for a mafic lamprophyric rock consisting of augite, brown hornblende, biotite, and titanite with nepheline (or cancrinite and analcime) and small amounts of plagioclase. *(Lacroix, 1918, p.544; Cape Topsail, Tamara Island, Los Archipelago, Guinea; Trög. 523; Joh. v.4, p.297; Tomk. p.563)*

TANDILEOFITE. A dyke rock composed of phenocrysts of labradorite, microcline and amphibole in a groundmass of quartz, microcline, albite and labradorite. *(Pasotti, 1954, p.5; Cerro Tandileofú, Buenos Aires, Argentina)*

TANNBUSCHITE. A local name for a melanocratic variety of olivine nephelinite consisting largely of pyroxene with smaller amounts of nepheline and olivine. *(Johannsen, 1938, p.364; Mt. Tannbusch (now Jedlová), near Ústí nad Labem, N. Bohemia, Czechoslovakia; Trög(38). 623½; Tomk. p.563)*

TARANTULITE. A local name for a quartz-rich alaskite. *(Johannsen, 1920a, p.54; Missouri Mine, Tarantula Spring, Nevada, USA; Trög. 8; Joh. v.2, p.35; Tomk. p.564)*

TASMANITE. A zeolitised variety of ijolite with nepheline and Ti-augite, lesser amounts of opaques, melilite and olivine, and minor amounts of apatite and perovskite. *(Johannsen, 1938, p.318; Shannon Tier, near Bothwell, Tasmania, Australia; Trög(38). 607½; Tomk. p.564)*

TAURITE. An obsolete local name for a variety of spherulitic alkali feldspar rhyolite containing phenocrysts of anorthoclase in a groundmass of quartz, orthoclase, aegirine-augite and arfvedsonite. *(Lagorio, 1897, p.5; Taurida, Crimea, USSR; Trög. 74; Tomk. p.564)*

TAUTIRITE. A local name for a variety of tephritic phonolite consisting of small phenocrysts of hornblende, titanite, augite, and biotite in a groundmass of alkali feldspar and andesine with some nepheline and sodalite. Cf. pollenite. *(Iddings & Morley, 1918, p.117; Tautira Beach, Taiarapu, Tahiti; Trög. 532; Joh. v.4, p.166; Tomk. p.564)*

TAVOLATITE. A variety of leucite phonolite essentially composed of leucite with subordinate alkali feldspar, nepheline, haüyne, plagioclase and clinopyroxene. *(Washington, 1906, p.50; Osteria del Tavolato, Alban Hills, near Rome, Italy; Trög. 530; Joh. v.4, p.285; Tomk. p.565)*

TAWITE. A variety of sodalitite composed largely of crystals of sodalite together with aegirine. Some nepheline, alkali feldspar and eudialyte are usually present. Feldspathic varieties have been called beloeilite and feldspattavite. *(Ramsay & Hackman, 1894, p.93; Tawajok Valley, Lovozero complex,*

Kola Peninsula, USSR; Trög. 636; Joh. v.4, p.319; Tomk. p.565)

TEPHRA. A collective term used in the pyroclastic classification (section B.5.2) for pyroclastic deposits that are predominatly unconsolidated. *(Thorarinsson, 1944, p.6; from the Greek tephra = ashes; Tomk. p.567)*

TEPHRILEUCITITE. A synonym for tephritic leucitite of QAPF field 15b (Fig. B.10). *(Rittmann, 1973, p.135)*

TEPHRINE. An obsolete term for tephrite. *(Brongniart, 1813, p.40; Tomk. p.567)*

TEPHRINEPHELINITE. A synonym for tephritic nephelinite of QAPF field 15b (Fig. B.10). *(Rittmann, 1973, p.135)*

TEPHRIPHONOLITE. A synonym for tephritic phonolite of QAPF field 12 (Fig. B.10), and also defined chemically in TAS field U3 (Fig. B.13). *(Rittmann, 1973, p.134)*

TEPHRITE. An alkaline basaltic rock composed essentially of calcic plagioclase, clinopyroxene and feldspathoid. Now defined modally in QAPF field 14 (Fig. B.10) and, if modes are not available, chemically in TAS field U1 (Fig. B.13). *(a term of great antiquity usually attributed to Pliny, 77AD – see Johannsen for further discussion; from the Greek tephra = ash; Trög. 999; Joh. v.4, p.230; Tomk. p.567)*

TEPHRITIC FOIDITE. A collective term for alkaline volcanic rocks consisting of foids with some plagioclase as defined modally in QAPF field 15b (Fig. B.10). It is distinguished from basanitic foidite by having less than 10% modal olivine. If possible the most abundant foid should be used in the name e.g. tephritic nephelinite, tephritic leucitite, etc. *(Streckeisen, 1978, p.7)*

TEPHRITIC PHONOLITE. A collective term for alkaline volcanic rocks consisting of alkali feldspar, sodic plagioclase, feldspathoid and various mafic minerals. Now defined modally in QAPF field 12 (Fig. B.10). *(Streckeisen, 1978, p.6)*

TEPHRITOID. A term originally proposed for rocks intermediate between olivine-free basalt and tephrite. Later used as an adjective to describe tephrite-like rocks. Now used as a general term in the provisional "field" classification (Fig. B.17) for rocks thought to

contain essential foids and in which plagioclase is thought to be more abundant than alkali feldspar. *(Bücking, 1881, p.157; Joh. v.4, p.69; Tomk. p.568)*

TEREKTITE. A local name for an effusive equivalent of semeitavite, which is a variety of quartz alkali feldspar syenite. *(Gornostaev, 1933, p.191; Terekty Hill, Semeitau Mts., USSR; Tomk. p.568)*

TERZONTLI. An erroneous spelling of tezontli. *(Tomkeieff et al., 1983, p.569)*

TESCHENITE (TESCHINITE). A variety of analcime gabbro consisting of olivine, Ti-augite, labradorite and analcime. Originally spelt teschinite, Zirkel (1866b, p.318) changed the spelling to teschenite. Now defined as a synonym for analcime gabbro of QAPF field 14 (Fig. B.4). *(Hohenegger, 1861, p.43; Teschen, (now divided into Český Těšín, Czechoslovakia and Cieszyn, Poland); Trög. 565; Joh. v.4, p.226; Tomk. p.569)*

TETIN. A local name for a volcanic ash used in the Azores for cement making. *(Lea & Desch, 1935, p.244; Tomk. p.570)*

TEXONTLI. A Mexican name for a cellular amygdaloidal lava or pumice. *(Humboldt, 1823, p.358; Tomk. p.570)*

TEZONTLI. A local term, probable Aztec in origin, originally applied to scoriaceous basalt, but now applied to any other type of rock except granite or marble. *(Ives, 1956, p.122)*

THERALITE. A variety of nepheline gabbro consisting essentially of Ti-augite, labradorite and nepheline. Olivine is a variable but sometimes major constituent. Now defined as a synonym for nepheline gabbro of QAPF field 14 (Fig. B.4). *(Rosenbusch, 1887, p.247; from the Greek therao = eagerly looked for; Trög. 514; Joh. v.4, p.222; Tomk. p.571)*

THEROLITE. According to Zirkel the correct spelling of theralite. *(Zirkel, 1894b, p.800; Kola Peninsula, USSR; Tomk. p.571)*

THOLEIITE (THOLEITE, THOLEYITE). This term has caused considerable confusion. It was originally used for a "doleritic trapp" said to consist of albite and ilmenite. Later Rosenbusch (1887) redefined it as an olivine-poor or olivine-free plagioclase augite rock with intersertal texture. It then became commonly used as a variety of basalt composed of labradorite, augite, hypersthene or pigeonite, with olivine (often showing a reaction relationship) or quartz, and often with interstitial glass. Yoder & Tilley (1962) defined it chemically as a hypersthene normative basalt in which sense it is now most commonly used. However, Jung (1958) showed the type rock was not a tholeiite, as chemically defined above, but a leucocratic subvolcanic variety of monzodiorite for which he proposed the name tholeyite. *(Steininger, 1840, p.99; Tholey, Nahe district, Saarland, West Germany; Trög. 344; Joh. v.3, p.298; Tomk. p.572)*

THOLEIITIC BASALT. The term the Subcommission recommends should be used instead of tholeiite.

TholerITE. An obsolete early form of dolerite. *(Leonhard, 1823a, p.118; from the Greek toleros = dirty, gloomy; Tomk. p.572)*

Thuresite. A local term for a variety of alkali feldspar syenite composed of microcline, Na-amphibole and hornblende with augite cores. *(Hackl & Waldmann, 1935, p.260; Thures, near Raabs, Lower Austria, Austria; Trög(38). 178½; Tomk. p.572)*

TilaITE. A mafic variety of gabbro composed of green chrome-diopside, olivine and minor highly calcic plagioclase. *(Duparc & Pearce, 1905, p.1614; Tilai-Kamen, Koswa region, N. Urals, USSR; Trög. 399; Joh. v.3, p.245; Tomk. p.572)*

Timazite. An obsolete local term for an altered hornblende biotite andesite. *(Breithaupt, 1861, p.51; Timok valley, Serbia, Yugoslavia; Trög. 1000; Joh. v.3, p.170; Tomk. p.573)*

Tinguaite. A variety of phonolite consisting of alkali feldspar, nepheline with or without other foids, aegirine and sometimes biotite and characterised by "tinguaitic texture" in which needles of aegirine occur interstitially in a mozaic of alkali feldspar and foids. *(Rosenbusch, 1887, p.628; Serra de Tingua, Rio de Janeiro, Brazil; Trög. 445; Joh. v.4, p.145; Tomk. p.573)*

Tipyneite. A mnemonic name suggested for a rock consisting essentially of titaniferous pyroxene and nepheline. *(Belyankin, 1929, p.22)*

Tirilite. A local term for a rapakivi-diabase

hybrid of granodiorite composition consisting of andesine, microcline-perthite, hornblende, biotite and serpentinised pyroxene. *(Wahl, 1925, p.69; Tirilä, near Lappeenranta, Finland; Trög. 81; Tomk. p.573)*

Titanolite. A name for a variety of alkali pyroxenite rich in titanite with magnetite and calcite. Tröger (1938) suggests the rock may be a skarn. *(Kretschmer, 1917, p.189; Trög. 1001; Tomk. p.573)*

Tjosite. A local name for a lamprophyric rock or micromelteigite similar to jacupirangite with phenocrysts of augite, abundant magnetite, ilmenite and apatite with some biotite in a groundmass of anorthoclase and nepheline. Similar to cocite but with nepheline instead of leucite. *(Brögger, 1906, p.128; Tjose, near Larvik, Norway; Trög. 495; Joh. v.4, p.277; Tomk. p.574)*

Toadstone. An obsolete term for an amygdaloidal basalt named because it can look like the skin of a toad. *(Pinkerton, 1811a, p.93; Joh. v.3, p.281; Tomk. p.574)*

Toellite. See töllite.

Töienite. A name proposed for a rock which was later found to be identical to windsorite. The name was later withdrawn. *(Brögger, 1931, p.65; Trög. 1002)*

Tokéite. A local name for a melanocratic variety of basalt containing abundant phenocrysts of augite and some olivine and opaques in a fine-grained groundmass of these minerals, labradorite and biotite. *(Duparc & Molly, 1928, p.24; Toké-Grat, Gouder Valley, near Addis Ababa, Ethiopia; Trög. 407; Joh. v.3, p.306; Tomk. p.574)*

Töllite (Toellite). An obsolete local name for a garnet-bearing variety of quartz diorite porphyry. *(Pichler, 1875, p.926; Töll, near Merano, Alto Adige, Italy; Trög. 115; Joh. v.2, p.400; Tomk. p.574)*

TONALITE. A plutonic rock consisting essentially of quartz and sodic plagioclase, usually with biotite and amphibole. Now defined modally in QAPF field 5 (Fig. B.4). *(Rath, 1864, p.249; Tonale Pass, Adamello, Alto Adige, Italy; Trög. 132; Joh. v.2, p.378; Tomk. p.574)*

Tönsbergite. A local term for a variety of alkali feldspar syenite in which the alkali feldspars are rhomb-shaped. Occurs as a red altered variety of larvikite. *(Brögger, 1898, p.328; Tonsberg, Norway; Trög. 184; Tomk. p.574)*

TOPATOURBIOLILEPIQUORTHITE. An unwieldy name constructed by Johannsen to illustrate some of the possibilities of the mnemonic classification of Beliankin (1929) for a variety of granite consisting of *top*az, *tour*maline, *bio*tite, *oligo*clase, *lepi*dolite, *qu*artz and *orth*oclase. Cf. biquahororthandite and hobiquandorthite. *(Johannsen, 1931, p.125)*

TOPAZ RHYOLITE. A variety of rhyolite rich in fluorine, silica and the lithophile elements, characterised by the presence of topaz in gas cavities. Although not present in all cavities, the topaz can usually be found in 15 to 30 minutes diligent searching with a hand lens – not the most suitable characteristic for naming a rock! Stated to be synonymous with ongonite. *(Burt et al., 1982, p.1818)*

TOPAZITE. A term suggested for rocks consisting of topaz and quartz. *(Johannsen, 1920a, p.53; Trög. 6; Joh. v.2, p.21; Tomk. p.575)*

TOPSAILITE. A local name for a dyke rock containing phenocrysts of plagioclase, augite, apatite, and Ti-magnetite in a groundmass of andesine, biotite, barkevikite, augite, and titanite. *(Lacroix, 1911b, p.79; Cape Topsail, Tamara Island, Los Archipelago, Guinea; Trög. 558; Tomk. p.575)*

TORDRILLITE. An obsolete group name proposed for quartz alkali feldspar lavas and chemically equivalent to alaskite. *(Spurr, 1900a, p.189; Tordrillo Mts., Alaska, USA; Trög. 44; Joh. v.2, p.113; Tomk. p.576)*

TORRICELLITE. An obsolete term for a fine-grained basalt containing quartz. *(Pinkerton, 1811b, p.50; named after Torricelli, 1640; Tomk. p.576)*

TORYHILLITE. A variety of nepheline syenite composed mainly of nepheline with albite, aegirine-augite and minor calcite, but without K-feldspar. *(Johannsen, 1920b, p.163; Toryhill, Monmouth Township, Ontario, Canada; Trög. 423; Joh. v.4, p.298; Tomk. p.576)*

TOSCANITE. A name for a variety of rhyodacite (a rock intermediate between rhyolite and dacite) containing calcic plagioclase, ortho-

clase, hypersthene and biotite in a glassy matrix of rhyolite composition. Cf. dellenite. *(Washington, 1897a, p.37; Mt. Amiata, Tuscany, Italy; Trög. 101; Joh. v.2, p.310; Tomk. p.576)*

TOURMALITE. A term suggested for rocks consisting of tourmaline and quartz only. *(Johannsen, 1920a, p.53; Joh. v.2, p.22; Tomk. p.577)*

TRACHIVICOITE. An obsolete name for a variety of vicoite rich in sanidine. *(Washington, 1906, p.57; Tomk. p.577)*

TRACHORHEITE. An old field term to cover andesite, trachyte and rhyolite when they are difficult to distinguish in the field. *(Endlich, 1874, p.319; Tomk. p.577)*

TRACHYANDESITE. A term originally used for volcanic rocks intermediate in composition between trachyte and andesite and containing approximately equal amounts of alkali feldspar and plagioclase. Later used for volcanic rocks containing foids as well as alkali feldspar and plagioclase (Rosenbusch, 1908). Now defined chemically in TAS field S3 (Fig. B.13). *(Michel-Levy, 1894, p.8; Trög. 1003; Joh. v.3, p.118; Tomk. p.577)*

TRACHYBASALT. Although originally used for foidal rocks, the term has mainly been used for basaltic volcanic rocks containing labradorite and alkali feldspar. Now defined chemically in TAS field S1 (Fig. B.13). *(Bořický, 1874, p.44; Trög. 1004; Joh. v.3, p.128; Tomk. p.577)*

TRACHYDACITE. A term originally used for a variety of rhyolite containing bronzite and an alkali feldspar to oligoclase ratio of 2:1. Now defined chemically as rocks with more than 20% normative quartz in TAS field T (Fig. B.13). *(Millosevich, 1908, p.418; Riu Mannu, Sassari, Sardinia; Trög. 94; Tomk. p.578)*

TRACHYDOLERITE. A term mainly used for volcanic rocks containing both orthoclase and labradorite. Cf. trachybasalt. *(Abich, 1841, p.100; Trög. 1005; Joh. v.3, p.128; Tomk. p.578)*

TRACHYLABRADORITE. A term for a leuco-trachybasalt in which the plagioclase is more calcic than An_{50}. *(Jung & Brousse, 1959, p.112)*

TRACHYLIPARITE. A term originally used for a variety of rhyolite with an alkali feldspar to oligoclase ratio of 3:1. *(Derwies, 1905, p.71; Medovka, Pyatigorsk, N Caucasus, USSR; Trög. 93; Tomk. p.578)*

TRACHYPHONOLITE. An obsolete term for a phonolite with trachytic texture. *(Bořický, 1873, p.18; Trög. 1006)*

TRACHYTE. A volcanic rock consisting essentially of alkali feldspar. Now defined modally in QAPF field 7 (Fig. B.10) and, if modes are not available, chemically in TAS field T (Fig. B.13). *(Brongniart, 1813, p.43; from the Greek trachys = rough; Trög. 251; Joh. v.3, p.67; Tomk. p.578)*

TRACHYTE-ANDESITE. An obsolete term for the extrusive equivalents of monzonite, later superseded by the term latite. *(Brögger, 1895, p.60; Trög. 1007; Joh. v.3, p.100)*

TRACHYTOID. A term proposed for volcanic rocks consisting mainly of sanidine, plagioclase, hornblende or mica. Later used as a textural term (Johannsen, 1931). Now used as a general term in the provisional "field" classification (Fig. B.17) for rocks tentatively identified as trachyte. *(Gümbel, 1888, p.86; Tomk. p.579)*

TRANSITIONAL BASALT. A variety of basalt transitional between typical tholeiitic basalt and alkali basalt. It consists of olivine, Ca-rich augite, plagioclase, and Ti-magnetite plus variable, but small, amounts of alkali feldspar. Ca-poor pyroxenes are absent. It is often associated with peralkaline rhyolite and peralkaline trachyte. In the USSR the term mid-alkali basalt is used as a synonym. *(Kuno, 1960, p.121)*

TRAPP (TRAP). A term that came to be used imprecisely for volcanic and medium-grained rocks of basaltic composition. Now sometimes used for plateau basalts. *(Rinman, 1754, p.293; from the Swedish trappar = steps i.e. jointing; Trög. 1009; Joh. v.3, p.247; Tomk. p.580)*

TRAPPIDE. A name recommended to replace the field term anameseid. Now obsolete. *(Johannsen, 1926, p.181; Joh. v.1, p.58; Tomk. p.580)*

TRAPPITE. An obsolete term for trapp. *(Brongniart, 1813, p.40; Tomk. p.580)*

TRASS. A local Italian name for a non-stratified tuff consisting of small fragments of altered trachytic pumice. *(Leonhard, 1823b, p.692; Joh. v.3, p.20; Tomk. p.580)*

TRASSOITE. An obsolete term for a volcanic tuff consisting of fragments of feldspar and glass. *(Cordier, 1816, p.366; Tomk. p.580)*

TREMATODE. An obsolete name for a vesicular andesite. *(Haüy, 1822, p.578; Volvic, Auvergne, France; Tomk. p.581)*

TRIAPHYRE. An obsolete term proposed for igneous rocks intruding Triassic formations. *(Ebray, 1875, p.291; Tomk. p.581)*

TRISTANITE. Originally described as a member of a potassic volcanic rock series falling in Daly's compositional gap between trachyandesite and trachyte, with a differentiation index between 60 and 75. It consists essentially of phenocrysts of plagioclase, zoned from labradorite to oligoclase and often rimmed with alkali feldspar, Fe-olivine, and pyroxene in a groundmass of Fe-olivine and Ti-augite with andesine and interstitial alkali feldpar. *(Tilley & Muir, 1963, p.439; Tristan da Cunha, Atlantic Ocean)*

TROCTOLITE. A variety of gabbro composed essentially of highly calcic plagioclase and olivine with little or no pyroxene. Now defined modally in the gabbroic rock classification (Fig. B.6). *(Lasaulx, 1875, p.317; from the Greek troktes = trout; Trög. 353; Joh. v.3, p.225; Tomk. p.581)*

TRONDHJEMITE. A leucocratic variety of tonalite consisting essentially of sodic plagioclase and quartz with minor biotite. Orthoclase is characteristically absent and hornblende is rare. Now used as a synonym for plagiogranite and leucocratic tonalite of QAPF field 5 (Fig. B.4). *(Goldschmidt, 1916, p.75; Trondhjem, now Trondheim, Norway; Trög. 129; Joh. v.2, p.387; Tomk. p.582)*

TROWLESWORTHITE. An obsolete name for a coarse-grained pneumatolytic vein found in granite consisting of red orthoclase, tourmaline and fluorite with minor quartz. *(Worth, 1884, p.177; Trowlesworthy Tor, Cornwall, England; Joh. v.3, p.5; Tomk. p.582)*

TSINGTAUITE. A local name for a variety of granite porphyry containing phenocrysts of microperthite and sodic plagioclase. *(Rinne,*

1904, p.142; Tsingtau, Shantung, China; Trög. 88; Joh. v.2, p.300; Tomk. p.582)

TUFAITE. An obsolete term for a tuff consisting of fragments of pyroxene and other components. *(Cordier, 1816, p.366; Tomk. p.582)*

TUFF. Now defined in the pyroclastic classification (Table B.1) as a pyroclastic rock in which the average pyroclast size is 2 mm. to 1/16 mm. The term is synonymous with ash tuff. *(Phillips, 1815, p.172; Tomk. p.583)*

TUFFITE. A term used in the pyroclastic classification (Table B.2) for rocks consisting of mixtures of pyroclasts and epiclasts. *(Schmid, 1981, p.43)*

TURJAITE. A name for a variety of melilitolite mainly consisting of melilite, biotite and nepheline with minor perovskite, melanite garnet and apatite. *(Ramsay, 1921, p.489; Turja, now Cape Turi, Kola Peninsula, USSR; Trög. 659; Joh. v.4, p.323; Tomk. p.585)*

TURJITE. A melanocratic lamprophyre rich in biotite, analcime, calcite and melanite garnet. *(Belyankin & Kupletskii, 1924, p.35; Turja, now Cape Turi, Kola Peninsula, USSR; Trög. 640; Tomk. p.585)*

TUSCULITE. A local name for a leucocratic variety of melilite leucitite composed mainly of leucite, minor melilite, pyroxene, alkali feldspar and magnetite. *(Cordier, 1868, p.118; Tusculum, Frascati, Alban Hills, near Rome, Italy; Trög. 1011; Tomk. p.585)*

TUTVETITE. A local name for a variety of trachyte. *(Johannsen, 1938, p.49; Tutvet, Hedrum, Oslo district, Norway; Trög(38). 171½; Tomk. p.585)*

TUVINITE. A local name for a variety of urtite in which the predominant mineral is nepheline but in which calcite is also present. *(Yashina, 1957, p.35; Tuva region, S. Siberia, USSR; Tomk. p.585)*

TVEITÅSITE. A local name for a melanocratic variety of fenite composed mainly of aegirine-augite and sometimes perthite. Titanite, apatite, and sometimes nepheline may be present. *(Brögger, 1921, p.155; Tveitåsen, Fen complex, Telemark, Norway; Trög. 226; Joh. v.4, p.33; Tomk. p.585)*

UGANDITE. A melanocratic variety of leucitite largely composed of clinopyroxene, olivine and leucite with subordinate plagioclase in a glassy matrix. *(Holmes & Harwood, 1937, p.11; Bufumbira, Uganda; Trög(38). 642⅔; Joh. v.1(2nd.Ed.), p.285; Tomk. p.587)*

UKRAINITE. A local name for a quartz monzonite. *(Bezborod'ko, 1935, p.198; Mariupol, Sea of Azov, Ukraine, USSR; Trög(38). 86½; Tomk. p.587)*

ULRICHITE. A porphyritic variety of micro nepheline syenite or phonolite containing phenocrysts of alkali feldspar, nepheline and barkevikite in a groundmass of alkali feldspar, Na-amphibole and pyroxene. *(Marshall, 1906, p.397; named after Ulrich; Dunedin, New Zealand; Trög. 455; Joh. v.4, p.153; Tomk. p.587)*

ULTRABASIC. A commonly used chemical term now defined in the TAS classification (Fig. B.13) as a rock containing less than 45% SiO_2. See also basic, intermediate and acid. *(Judd, 1881, p.317; Joh. v.1, p.194; Tomk. p.588)*

ULTRABASITE. A collective term that had been incorrectly used as a synonym for ultramafic rocks. Streckeisen (1967) suggested replacing the term with mafitite, which was later withdrawn in favour of ultramafitite. *(Original reference uncertain)*

ULTRAMAFIC. A term for rocks consisting essentially of mafic minerals e.g. peridotite, dunite. *(Hess, 1937, p.263; Tomk. p.588)*

ULTRAMAFITE. A term suggested as a less correct alternative to ultramafitolite i.e. a collective name for ultramafic plutonic rocks. However, as a mafite is a mineral it is now recommended that it should not be used. *(Streckeisen, 1976, p.15)*

ULTRAMAFITITE. A collective term for ultramafic volcanic rocks of QAPF field 16 (Fig. B.10) i.e. composed of more than 90% of mafic minerals. *(Streckeisen, 1978, p.7)*

ULTRAMAFITOLITE. A collective term for ultramafic plutonic rocks. Now defined modally in QAPF field 16 (Fig. B.4). *(Streckeisen, 1976, p.15)*

UMPTEKITE. A variety of alkali feldspar syenite consisting of microperthite, arfvedsonite and aegirine. Nepheline is sometimes present. *(Ramsay & Hackman, 1894, p.81; Umptek, Khibina complex, Kola Peninsula, USSR;*

Trög. 181; Joh. v.4, p.8; Tomk. p.588)

UNAKYTE (UNAKITE). A local term for a variety of granite containing appreciable amounts of epidote. *(Bradley, 1874, p.519; Unaka Range, Great Smoky Mts., N. Carolina – Tennessee, USA; Trög. 70; Joh. v.2, p.59; Tomk. p.589)*

UNCOMPAHGRITE. A variety of pyroxene melilitolite consisting of more than 65% melilite with pyroxene. Now used as a special term in the melilitic rocks classification (section B.8.1) synonymous with pyroxene melilitolite. *(Larsen & Hunter, 1914, p.473; Uncompahgre, Colorado, USA; Trög. 745; Joh. v.4, p.320; Tomk. p.589)*

UNDERSATURATED. A term applied to igneous rocks which are undersaturated with respect to silica, i.e. they have foids or Mg-olivine in the mode or norm. *(Shand, 1913, p.510; Tomk. p.408)*

UNGAITE. A local term for a glassy dacite containing normative oligoclase. Cf. shastaite. *(Iddings, 1913, p.106; Unga Island, Kamchatka, USSR; Trög. 122; Tomk. p.589)*

URALITITE. An obsolete term for a diabase in which all the pyroxene has been altered to uralite. *(Kloos, 1885, p.87; Trög. 1012; Joh. v.3, p.319)*

URBAINITE. An obsolete term for a rock consisting essentially of ilmenite and rutile with less hematite and sapphirine occurring as dykes in an anorthosite. *(Warren, 1912, p.276; St. Urbain, Quebec, Canada; Trög. 769; Joh. v.4, p.470; Tomk. p.590)*

URTITE. A plutonic rock consisting of over 70% nepheline with some aegirine-augite, but no feldspar. Now defined modally as a leucocratic variety of foidolite in QAPF field 15 (Figs. B.4 and B.7). *(Ramsay, 1896, p.463; Lujavr-Urt, Lovozero complex, Kola Peninsula, USSR; Trög. 604; Joh. v.4, p.316; Tomk. p.590)*

USSURITE. A local name for a volcanic rock with an unusual poikilitic texture composed of large crystals of oligoclase (36%) with numerous inclusions of microlites of augite (40%) and olivine (16%). Small amounts of analcime are present. *(Gapeeva, 1959, p.157; Ussuri River, E. Siberia, USSR)*

VALAMITE. A local name for a variety of

hypersthene quartz dolerite with abundant iron ores. Iron-rich hypersthene is the only mafic silicate mineral. *(Wahl, 1907, p.69; Valamo Island, Lake Ladoga, near Lenigrad, USSR; Trög(38). 342½; Tomk. p.591)*

VALBELLITE. An obsolete name for a variety of peridotite consisting of olivine, bronzite, hornblende, magnetite and some pyrrhotite. Cf. weigelite. *(Schaefer, 1898, p.501; Val Bella, Ivrea zone, Piedmont, Italy; Trög. 736; Joh. v.4, p.430; Tomk. p.591)*

VALLEVARITE. An obsolete local name for a variety of anorthosite, consisting essentially of andesine and microcline antiperthite with minor diopside and biotite. *(Gavelin, 1915, p.19; Vallevara, Routivare, Norrbotten, Lapland, Sweden; Trög. 295; Tomk. p.591)*

VARIOLITE. An old term that came to be used for basaltic rocks with a pock-marked appearance, composed of radial aggregates of feldspar and pyroxene microlites in a microcrystalline or devitrified glassy base. The variolitic texture denotes rapid cooling and the quench growth of the crystals, such as is often found in submarine basalts, and should not be confused with spheroidal texture which indicates devitrification. *(Aldrovandi, 1648, p.882; from the Latin variola = smallpox; Trög. 1014; Joh. v.3, p.300; Tomk. p.591)*

VÄRNSINGITE. A local name for a coarse-grained albite dolerite pegmatite dyke rock with some augite, minor hornblende and chlorite. *(Sobral, 1913, p.169; Västra Värnsingen Island, Nordingrå, Sweden; Trög. 198; Joh. v.3, p.143; Tomk. p.592)*

VAUGNERITE. A dyke rock consisting of major amounts of biotite, hornblende and plagioclase and a little orthoclase. *(Fournet, 1861, p.606; Vaugneray, near Lyon, France; Trög. 109; Joh. v.2, p.405; Tomk. p.592)*

VENANZITE. A melanocratic variety of leucite olivine melilitite largely composed of melilite, leucite, kalsilite and olivine. Rosenbusch (1899) called this rock euktolite unaware that it had already been named venanzite. *(Sabatini, 1899, p.60; San Venanzo, Umbria, Italy; Trög. 672; Joh. v.4, p.361; Tomk. p.593)*

VENTRALLITE. A consistent misspelling of vetrallite in Johannsen on p.174 and in the index. *(Johannsen, 1938, p.174)*

VERITE. A black lamproite with phenocrysts of phlogopite and olivine in a glassy matrix containing microlites of diopside and phlogopite. *(Osann, 1889, p.311; Vera, Cabo de Gata, Spain; Trög. 234; Joh. v.3, p.21; Tomk. p.593)*

VESBITE. A variety of leucititolite largely composed of leucite with subordinate clinopyroxene and melilite. *(Washington, 1920, p.46; from Mons Vesbius, the Latin name for Vesuvius; Trög. 656; Joh. v.4, p.360; Tomk. p.594)*

VESECITE. A local name for a variety of polzenite (a lamprophyre) consisting of olivine, considerable amounts of monticellite, and melilite in a matrix of monticellite, phlogopite, and nepheline. *(Scheumann, 1922, p.496; Vesec, N. Bohemia, Czechoslovakia; Trög. 663; Joh. v.4, p.388; Tomk. p.594)*

VESUVITE. A variety of tephritic leucitite largely composed of leucite, clinopyroxene and subordinate plagioclase. *(Lacroix, 1917d, p.483; Mt. Vesuvius, Naples, Italy; Trög. 582; Tomk. p.594)*

VETRALLITE. A variety of tephritic phonolite containing essential labradorite. *(Johannsen, 1938, p.173; Vetralla, Vico volcano, near Viterbo, Italy; Trög(38). 525½; Tomk. p.594)*

VIBETOITE (VIPETOITE). A local name for a variety of pyroxenite containing abundant Ti-augite and hornblende with biotite, primary calcite, and occasional albite and nepheline. As pointed out by Sæther (1957) the name was actually misspelt by Brögger in the belief that the locality was named Vibeto instead of Vipeto. The strictly correct spelling should have been vipetoite but vibetoite is in common usage. *(Brögger, 1921, p.76; Vibeto, Fen complex, Telemark, Norway; Trög. 711; Tomk. p.594)*

VIBORGITE. See wiborgite.

VICOITE. A volcanic rock, close to the boundary between phonolitic tephrite and tephritic phonolite, largely composed of leucite, with lesser equal amounts of alkali feldspar and plagioclase and subordinate clinopyroxene. Cf. orvietite. *(Washington, 1906, p.57; Vico volcano, near Viterbo, Italy; Trög. 539; Joh. v.4, p.293; Tomk. p.594)*

VINTLITE. An obsolete local name for a variety

of microdiorite or tonalite containing phenocrysts of hornblende, quartz and oligoclase. *(Pichler, 1875, p.927; Vintl, near Bressanone, Alto Adige, Italy; Trög. 144; Joh. v.2, p.399; Tomk. p.595)*

VIPETOITE. See vibetoite.

VITERBITE. A variety of tephritic phonolite consisting of alkali feldspar, leucite, labradorite and augite. *(Washington, 1906, p.40; Viterbo, Italy; Trög. 528; Joh. v.4, p.175; Tomk. p.595)*

VITRIC TUFF. Now defined in the pyroclastic classification (Fig. B.1) as a tuff in which vitric fragments are more abundant than either crystal or lithic fragments. *(Pirsson, 1915, p.193; Tomk. p.595)*

VITROPHYRE. A term for variety of porphyry in which the groundmass is glassy. Also applied to the basal portions of many welded ignimbrites. *(Vogelsang, 1872, p.534; Trög. 1015; Joh. v.2, p.275; Tomk. p.597)*

VITROPHYREID. An obsolete field term for a porphyritic glass. *(Johannsen, 1911, p.322; Tomk. p.597)*

VITROPHYRIDE. A revised spelling recommended to replace the field term vitrophyreid. Now obsolete. *(Johannsen, 1926, p.182; Joh. v.1, p.58)*

VITROPHYRITE. An obsolete term for nonporphyritic rocks with a glassy texture. Cf. pitchstone. *(Vogelsang, 1872, p.534; Trög. 1016; Tomk. p.597)*

VOGESITE. A variety of lamprophyre in which amphibole is more abundant than biotite and alkali feldspar is more abundant than plagioclase. Augite is frequently present. Now defined in the lamprophyre classification (Table B.3). *(Rosenbusch, 1887, p.319; Vosges, France; Trög. 249; Joh. v.3, p.37; Tomk. p.597)*

VOLCANIC. A loosely defined term pertaining to those igneous processes that occur on or very close to the surface of the earth. Volcanic rocks are usually fine-grained. *(Original reference uncertain)*

VOLCANITE. A name proposed for porphyritic material found in some of the bombs ejected from Vulcano in the Lipari Islands. The phenocrysts were described as anorthoclase, andesine and augite. *(Hobbs, 1893, p.602;*

Vulcano, Lipari Islands, Italy; Joh. v.1(2nd.Ed.), p.287; Tomk. p.599)

VOLHYNITE (WOLHYNITE). A local name, originally spelt wolhynite, for a variety of microgranodiorite which occurs as dykes and contains phenocrysts of labradorite, hornblende and biotite in a groundmass of andesine, orthoclase and quartz. *(Kroustchoff, 1885, p.441; Volhynian province, W. Ukraine, USSR; Trög. 339; Tomk. p.600)*

VREDEFORTITE. A porphyritic granogabbro containing phenocrysts of labradorite, hypersthene, and biotite in a matrix of potential quartz and K-feldspar, corresponding to the vredefortitic magma-type of Niggli (1936, p.369). *(Tröger, 1938, p.56; Vredefort, South Africa; Trög(38). 115½; Tomk. p.600)*

VULCANITE. A term proposed for all extrusive rocks. *(Scheerer, 1862, p.138; Joh. v.1(2nd.Ed.), p.288; Tomk. p.600)*

VULSINITE. A term originally defined as the volcanic equivalent of monzonite. The rock consists mainly of Na-orthoclase with smaller amounts of labradorite and minor clinopyroxene and biotite. *(Washington, 1896a, p.547; from the Etruscan tribe of Vulsinii, Italy; Trög. 253; Joh. v.4, p.42; Tomk. p.600)*

WEBSTERITE. A variety of pyroxenite consisting of equal amounts of orthopyroxene and clinopyroxene. Now defined modally in the ultramafic rock classification (Fig. B.8). *(Williams, 1890, p.44; Webster County, North Carolina, USA; Trög. 678; Joh. v.4, p.460; Tomk. p.604)*

WEHRLITE. An ultramafic plutonic rock composed of olivine and clinopyroxene often with minor brown hornblende. Now defined modally in the ultramafic rock classification (Fig. B.8). *(Kobell, 1838, p.313; named after Wehrle, who analysed the rock; Trög. 734; Joh. v.4, p.419; Tomk. p.604)*

WEIGELITE. An obsolete name for a variety of peridotite which occurs as dykes and consists essentially of enstatite, later altered to actinolite, olivine and hornblende. Cf. valbellite. *(Kretschmer, 1917, p.113; Weigelsberg, near Habartice, N. Moravia, Czechoslovakia; Trög. 1017; Joh. v.4, p.428; Tomk. p.604)*

WEILBURGITE. A local term proposed for a rock, previously called keratophyre spilite,

consisting mainly of alkali feldspar, chlorite and considerable carbonate. *(Lehmann, 1949, p.80; Weilburg, Lahn district, Hessen, West Germany; Tomk. p.604)*

WEISELBERGITE. An obsolete term originally used for a palaeovolcanic augite andesite which was usually altered. Later used for an altered glassy andesite (Wadsworth, 1884) – the fresh variety has been called shastalite. *(Rosenbusch, 1887, p.501; Weiselberg, St. Wendel, Saarland, West Germany; Trög. 155; Joh. v.3, p.170; Tomk. p.604)*

WENNEBERGITE. An obsolete local name for a variety of trachyandesite containing phenocrysts of sanidine, chloritised biotite and quartz in a groundmass of the same minerals with oligoclase. *(Schowalter, 1904, p.33; Wenneberg, Ries, Bavaria, West Germany; Trög. 104; Tomk. p.605)*

WERNERITE. An obsolete term for coarsegrained rock consisting of feldspar with minor ferromagnesian minerals. *(Pinkerton, 1811a, p.205; named after A.G. Werner; Tomk. p.605)*

WESSELITE. A local name for a melanocratic variety of nephelinite which occurs as dykes and consists of abundant phenocrysts of anomite (trioctahedral mica), syntagmatite (aluminous hornblende), barkevikite, and Ti-augite in a matrix of nepheline, analcime, and haüyne. *(Scheumann, 1922, p.505; Wesseln (now Veselí), N. Bohemia, Czechoslovakia; Trög. 624; Joh. v.4, p.385; Tomk. p.605)*

WESTERWALDITE. An obsolete name for a volcanic rock, previously called essexite-basalt, consisting of serpentinised olivine phenocrysts in a groundmass of olivine, augite, labradorite, sanidine, biotite, ore and interstitial nepheline. *(Johannsen, 1938, p.203; Stöffel, Marienberg, Westerwald, West Germany; Trög(38). 579½; Tomk. p.605)*

WIBORGITE (VIBORGITE). A local name for a variety of rapakivi granite. *(Wahl, 1925, p.42; Wiborg, now Vyborg, near Lenigrad, USSR; Trög. 80; Tomk. p.606)*

WICHTISITE. An obsolete name for a glassy dolerite dyke or selvage to a dyke. *(Hausmann, 1847, p.551; Wichtis, now Vihti, Finland; Trög. 1018; Joh. v.3, p.324; Tomk. p.606)*

WILSONITE. A term for a strongly welded rhyolitic or dacitic tuff later renamed owharoite. *(Henderson, 1913, p.70; Trög(38). 1018½; Tomk. p.606)*

WINDSORITE. A local name for an aplitic rock consisting essentially of alkali feldspar and oligoclase with smaller amounts of quartz and biotite. *(Daly, 1903, p.48; Windsor, Vermont, USA; Trög. 91; Joh. v.2, p.303; Tomk. p.606)*

WOLGIDITE. A local name for a lamproitic rock largely composed of leucite, mica and amphibole with subordinate clinopyroxene in a serpentine-rich matrix. *(Wade & Prider, 1940, p.50; Wolgidee Hills, Kimberley district, West Australia; Tomk. p.607)*

WOLHYNITE. See volhynite.

WOODENDITE. An obsolete name for a volcanic rock composed of phenocrysts of augite, enstatite and serpentinised olivine in an abundant matrix of glass. Although the rock is rich in alkalis, it is devoid of modal feldspar. *(Skeats & Summers, 1912, p.29; Woodend, Victoria, Australia; Trög. 272; Joh. v.3, p.120; Tomk. p.607)*

WYOMINGITE. A local name for a variety of leucitite composed of clinopyroxene, mica and leucite in a glassy matrix. Madupite is a more mafic variety. *(Cross, 1897, p.120; Leucite Hills, Wyoming, USA; Trög. 503; Joh. v.4, p.356; Tomk. p.607)*

XENITE. An obsolete name given to veins containing feldspar, some micas and garnets. *(Rozière, 1826, p.305; from the Greek xenos = stranger; Tomk. p.608)*

XENOPHYRE. An obsolete name for porphyritic rocks occurring as veins. *(Rozière, 1826, p.305; Tomk. p.609)*

YAMASKITE. A local name for a variety of pyroxenite, related to jacupirangite, composed of Ti-augite, alkali amphibole and small amounts of anorthite. *(Young, 1906, p.16; Mt. Yamaska, Montreal, Quebec, Canada; Trög. 690; Joh. v.3, p.341; Tomk. p.610)*

YATALITE. A local name for a coarse-grained uralitised mafic pegmatite composed of actinolite (after augite), albite, microcline and quartz. *(Benson, 1909, p.104; Hundred of Yatala, Houghton, South Australia; Trög. 230; Joh. v.3, p.144; Tomk. p.610)*

YENTNITE. A name given to a rock consisting of sodic plagioclase, biotite and scapolite. However, the scapolite was shown to be quartz and the name was withdrawn. *(Spurr, 1900b, p.315; Yentna River, Alaska, USA; Trög. 1019; Joh. v.3, p.157; Tomk. p.610)*

YOGOITE. A name originally defined as a variety of syenite containing equal amounts of orthoclase and augite. However, as the type rock also contained equal amounts of orthoclase and plagioclase the term was later withdrawn in favour of monzonite. *(Weed & Pirsson, 1895b, p.472; Yogo Peak, Little Belt Mts., Montana, USA; Trög. 279; Joh. v.3, p.63; Tomk. p.610)*

YOSEMITITE. A term for light-coloured granites that correspond to the yosemititic magmatype of Niggli (1923, p.111). *(Tröger, 1935, p.47; El Capitan, Yosemite Valley, California, USA; Trög. 84; Tomk. p.610)*

YUKONITE. A name suggested by J.E. Spurr for a leucocratic aplitic dyke rock of tonalitic composition consisting of oligoclase, quartz and biotite. *(Clarke, 1904, p.270; Yukon River, above Fort Hamlin, Alaska, USA; Trög. 136; Joh. v.2, p.401; Tomk. p.611)*

ZIRKELYTE (ZIRKELITE). An obsolete general name for altered basaltic glass. *(Wadsworth, 1887, p.30; named after Zirkel; Tomk. p.612)*

ZOBTENITE. A local name for an augen gabbro-gneiss with knots of augite set in streams of uralite embedded in a matrix of epidote and saussuritised plagioclase. *(Roth, 1887, p.611; Zobtenberg, Silesia, Poland; Trög. 1021; Joh. v.3, p.230; Tomk. p.612)*

ZUTTERITE. An erroneous spelling of rutterite. *(Tomkeieff et al., 1983, p.613)*

D. Bibliography

The references given in this section are either the source references for rock terms or are subsidiary references used in the petrological descriptions. Each reference is followed, in square brackets and italics, by the terms it contains.

The Subcommission took a great deal of effort to check as many as possible of the references, a task which took an enormous amount of time. Of the total of 787 references listed only 15, or 1.9%, have not been seen and checked – a truly remarkable effort. These references have all been marked with an "*" to indicate that they may not be entirely accurate.

To avoid any possible ambiguity and confusion in the future, it was also decided to give references in full in the bibliography, including the journal titles. This was done because many of the references cited in the early literature were so abbreviated that a considerable amount of time was wasted in trying to locate them in the various libraries when they were being checked.

The style of the journal titles is, wherever possible, that used in the British Museum (Natural History) list of publications.

D.1. FURTHER STATISTICS

Table D.1 shows which languages were used in the publication of new rock names. As can be seen, although ten different languages were involved, nearly 90% of all the new rock names were published in English, German or French.

Tables D.2 and D.3 give the more prolific authors of new rock names and publications containing new rock names, respectively. Although Johannsen was by far the most prolific contributor of new rock names with 133, Lacroix was by far the most prolific author of publications containing new rock names with 36 – 24 of which were published in the Compte Rendu. This is more than twice the number published by Washington with 16 and next on the list.

Similar data for some of the journals and publishing houses are shown in Tables D.4 and D.5. Worthy of mention are the Journal of Geology, Compte Rendu and the Neues Jahrbuch, for their contributions to igneous nomenclature.

Table D.1. Distribution of new rock names and references containing new rock names by the language in which the publication was written.

Language	Rocks	Acc%	Refs	Acc%
English	719	46.9	332	42.2
German	460	76.9	236	72.3
French	199	89.8	100	85.0
Russian	80	95.0	59	92.5
Italian	30	97.0	27	95.9
Scandinavian	28	98.8	23	98.9
Spanish	14	99.7	6	99.6
Bulgarian	2	99.9	1	99.7
Dutch	1	99.9	1	99.9
Slovakian	1	100.0	1	100.0
Totals	1534		786	

Table D.2. List of authors who introduced 10 or more new rock names. The number of references in which they were published is also given.

	Rocks	Acc%	Refs
Johannsen	133	8.7	11
Streckeisen	87	14.4	9
Lacroix	70	18.9	36
Brögger	65	23.2	10
Rosenbusch	32	25.2	9
Loewinson-Lessing	26	26.9	11
Pinkerton	24	30.1	2
Cordier	23	31.6	3
Tröger	22	33.0	5
Iddings	22	34.4	7
Washington	21	35.8	16
Belyankin	21	37.2	8
Gümbel	19	38.4	5
Brongniart	17	39.5	4
Shand	16	40.6	6
Duparc	13	41.4	10
Bořický	13	42.3	4
Vogelsang	12	43.1	2
Rittmann	11	43.8	3
Leonhard	11	44.5	4
Niggli	10	45.1	3

Table D.3. List of authors who published 5 or more publications containing new rock names. The total number of new rock names which they contained is also given.

	Refs.	Acc%	Rocks
Lacroix	36	4.6	70
Washington	16	6.6	21
Johannsen	11	8.0	133
Duparc	10	9.3	13
Brögger	10	10.6	65
Streckeisen	9	11.7	87
Rosenbusch	9	12.8	32
Loewinson-Lessing	9	14.0	12
Belyankin	8	15.0	21
Osann	7	15.9	8
Iddings	7	16.8	22
Tyrrell	6	17.6	7
Shand	6	18.3	16
Judd	6	19.1	6
Hibsch	6	19.8	6
Eckermann	6	20.6	9
Zirkel	5	21.2	8
Tröger	5	21.9	22
Pirsson	5	22.5	8
Gümbel	5	23.2	19

Table D.4. List of journals and publishers responsible for 20 or more new rock names. The total number of new rock names which they contained is also given.

Rocks	Acc%	Refs	
109	7.1	33	Journal of Geology
62	11.2	5	Chicago University Press
62	15.2	12	Skrifter utgitt av det Norske Videnskaps-Akademi i Oslo. Mat.-naturv.Kl
61	19.2	31	Neues Jahrbuch
47	22.2	36	Compte Rendu Hebdomadaire des Séances de l'Académie des Sciences
46	25.2	8	John Wiley
45	28.2	1	Geotimes
37	30.6	20	Tschermaks Mineralogische und Petrographische Mitteilungen
35	32.9	30	American Journal of Science
33	35.0	10	Schweizerbart
24	36.6	2	White
24	38.2	13	Zeitschrift der Deutschen Geologischen Gesellschaft
23	39.7	19	Geological Magazine
23	41.2	15	Geologiska Föreningens i Stockholm Förhandlingar
20	42.5	3	Journal des Mines

Table D.5. List of journals and publishers with 10 or more publications containing new rock names. The total number of new rock names which they contained is also given.

Refs	Acc%	Rocks	
36	4.6	47	Compte Rendu Hebdomadaire des Séances de l'Académie des Sciences
33	8.8	109	Journal of Geology
31	12.7	61	Neues Jahrbuch
30	16.5	35	American Journal of Science
20	19.1	37	Tschermaks Mineralogische und Petrographische Mitteilungen
19	21.5	23	Geological Magazine
15	23.4	23	Geologiska Föreningens i Stockholm Förhandlingar
13	25.1	24	Zeitschrift der Deutschen Geologischen Gesellschaft
12	26.6	62	Skrifter utgitt av det Norske Videnskaps-Akademi i Oslo. Mat.-naturv.Kl
10	27.9	33	Schweizerbart
10	29.1	17	Quarterly Journal of the Geological Society of London

D.2. REFERENCES

ABICH, H., 1839. Über Erhebungs Kratere und das Band inneren Zusammenhanges, welches in der Richtung bestimmter Linien, räumlich oft weit von einander getrennte vulkanische Erscheinungen und Gebilde zu ausgedehnten Zügen unter einander vereinigt [Abstract]. *Neues Jahrbuch für Mineralogie, Geognosie, Geologie und Petrefaktenkunde. Stuttgart.* Vol.10, p.334–337. [Haüynophyre]

ABICH, H., 1841. Geologische Beobachtungen über die vulkanischen Erscheinungen und Bildungen in Unter- und Mittelitalien. Vol.1. Ueber die Natur und den Zusammenhang der vulkanischen Bildungen. *Vieweg, Braunschweig.* 134pp. [Acid, Basic, Trachydolerite]

ADAMS, F.D., 1913. Excursion A7 – The Monteregian Hills. Guide Book No. 3. *Department of Mines & Geological Survey, Canada. Issued as part of the 12th. International Geological Congress, Toronto.* p.29–80. [Montrealite]

ADAMS, F.D. & BARLOW, A.E., 1908. The Nepheline and Associated Alkali Syenites of Eastern Ontario. *Transactions of the Royal Society of Canada.* Vol.2, Ser. 3, p.3–76. [Dungannonite, Raglanite]

ADAMS, F.D. & BARLOW, A.E., 1910. Geology of the Haliburton and Bancroft areas, Province of Ontario. *Memoirs. Geological Survey, Canada. Ottawa.* Vol.6, p.1–419. [Craigmontite, Monmouthite]

ADAMS, F.D. & BARLOW, A.E., 1913. Excursion A2 – The Haliburton–Bancroft area of Central Ontario. Guide Book No. 2. *Department of Mines & Geological Survey, Canada. Issued as part of the 12th. International Geological Congress, Toronto.* p.1–98. [Congressite]

ADAMSON, O.J., 1944. The petrology of the Norra Kärr District. An occurrence of alkaline rocks in Southern Sweden. *Geologiska Föreningens i Stockholm Förhandlingar. Stockholm.* Vol.66, p.113–255. [Grennaite, Kaxtorpite, Subphonolite]

ADAN DE YARZA, R., 1893. Roca eruptiva de Fortuna (Murcia). *Boletin de la Comision del Mapa Geológico de España. Madrid.* Vol.20, p.349–353. [Fortunite]

AINBERG, L.F., 1955. K voprosu genezisa charnokitov i porod charnokitovoi serii [transliterated from Russian]. *Izvestiya Akademii Nauk SSSR, Seriya Geologicheskaya.* No.4, p.102–120. [Kuzeevite]

AITCHISON, J., 1984. The Statistical Analysis of Geochemical Compositions. *Journal of the International Association for Mathematical Geology. New York.* Vol.16, p.531–564. [Boxite, Coxite, Hongite, Kongite]

ALDROVANDI, U., 1648. Bononiensis Musæi Metallici. *Ferronii [Bologna]*. 992pp. [Variolite]

ALLEN, J.A., 1914. Geology of Field Map-area, B.C. and Alberta. *Memoirs. Geological Survey, Canada. Ottawa.* Vol.55, p.1–312. [Sphenitite]

ALMEIDA, F.F.M. DE, 1961. Geologia e Petrologia da Ilha da Trindade. *Monografias. Divisão de Geologia e Mineralogia, Ministério das Minas e Energia, Brasil. Rio de Janeiro.* Mono.18, 197pp. [Grazinite]

AMSTUTZ, A., 1925. Les roches éruptives des environs de Dorgali et Orosei en Sardaine. *Schweizerische Mineralogische und Petrographische Mitteilungen. Zürich.* Vol.5, p.261–321. [Dorgalite]

ANDERSON, C.A., 1941. Volcanoes of the Medicine Lake Highland, California. *University of California Publications in Geological Sciences. Berkeley.* Vol.25, No.7, p.347–422. [Basaltic andesite]

ANTONOV, L.B., 1934. The Apatite Deposits of the Khibina Tundra [transliterated from Russian]. *The Khibina Apatite, Leningrad.* Vol.7, 196pp. [Lujavritite]

BABINGTON, W., 1799. A New System of Mineralogy, in the form of a catalogue, after the manner of Baron Born's Systematic Catalogue of the Collection of Fossils of Mlle. Éléonore de Raab. *Bensley, London.* 279pp. [Pitchstone]

BACKLUND, H.G., 1915. Nefelinovyj bazalt (onkilonit) s severnago ledovitago okeana [transliterated from Russian]. *Izvestiya Imperatorskoi Akademii Nauk. S. Peterburg.* Vol.9, p.289–308. [Onkilonite]

BAILEY, E.B. & MAUFE, H.B., 1916. The geology of Ben Nevis and Glen Coe, and the surrounding country. *Memoirs. Geological Survey of Scotland. Edinburgh.* Sheet 53, p.1–247. [Aplo-, Appinite, Calc-tonalite]

BARTH, T.F.W., 1927. Die Pegmatitgänge der Kaledonischen Intrusivgesteine im Seiland-Gebiete. *Skrifter utgitt av det Norske Videnskaps-Akademi i Oslo. Mat.-naturv.Kl.* No. 8, p.1–123. [Antifenitepegmatite]

BARTH, T.F.W., 1930. Pacificite, an anemousite basalt. *Journal of the Washington Academy of Sciences.* Vol.20, p.60–68. [Olivine pacificite, Pacificite]

BARTH, T.F.W., 1944. Studies on the igneous rock complex of the Oslo Region. II. Systematic petrology of the plutonic rocks. *Skrifter utgitt av det Norske Videnskaps-Akademi i Oslo. Mat.-naturv.Kl.* No. 9, p.1–104. [Apotroctolite, Hovlandite, Oslo-essexite]

BASCOM, F., 1893. The structures, origin, and nomenclature of the acid volcanic rocks of South Mountain. *Journal of Geology. Chicago.* Vol.1, p.813–832. [Apo-]

*BAYAN, 1866. *Katalog geogn. Mus. kaukas. Mineralwäss. u. Kurorte.* [Beschtauite]

BAYLEY, W.S., 1892. Eleolite-syenite of Litchfield, Maine and Hawes' Hornblende-Syenite from Red Hill, New Hampshire. *Bulletin of the Geological Society of America. New York.* Vol.3, p.231–252. [Litchfieldite]

BECK, R., 1907. Untersuchungen über einige südafrikanische Diamantenlagerstätten. *Zeitschrift der Deutschen Geologischen Gesellschaft. Berlin.* Vol.59, No.11, p.275–307. [Griquaite]

BECKE, F., 1899. Der Hypersthen-Andesit der Insel Alboran. *Tschermaks Mineralogische und Petrographische Mitteilungen. Wien.* Vol.18, 2nd.Ser., p.525–555. [Alboranite, Santorinite]

BECKER, G.F., 1888. Geology of the quicksilver deposits of the Pacific slope. *Monographs of the United States Geological Survey. Washington.* Vol.13, p.1–486. [Asperite]

BELL, J.M., CLARKE, E.C. & MARSHALL, P., 1911. The Geology of the Dun Mountain Subdivision, Nelson. *Bulletin. Geological Survey of New Zealand. Wellington.* Vol.12, p.1–71. [Rodingite]

BELYANKIN, D.S., 1911. Ob albitovom diabaze iz Krasnoi Polyany i o kontakte ego so slantsami [transliterated from Russian]. *Izvestiya Sankt-Peterburgskago Polytekhnicheskago Instituta Imperatora Petra Velikago. S. Peterburg.* Vol.15. [Anorthobase, Orthobase]

BELYANKIN, D.S., 1923. K petrografii perevala Shtuli-Atsek v Tsentralnom Kavkaze [transliterated from Russian]. *Izvestiya Rossiiskoi Akademii Nauk. Petrograd. VI Ser.* Vol.17, p.95–102. [Cryptodacite, Introdacite, Phanerodacite]

BELYANKIN, D.S., 1924. K voprosu o vozraste nekotorykh kavkazskikh intruzii [transliterated from Russian]. *Izvestiya Geologicheskogo Komiteta. Sankt Peterburg.* Vol.43, No.3, p.409–424. [Caucasite]

BELYANKIN, D.S., 1929. On the term "rock" and on petrographical classification and nomenclature. *Trudy Mineralogicheskogo Muzeya(imA.E.Fersmana).AkademiyaNauk SSSR. Leningrad.* Vol.3, p.12–24. [Aegineite, Aegisodite, Amneite, Anam-aegisodite, Barneite, Bineite, Binemelite, Dimelite, Pynosite, Tipyneite]

BELYANKIN, D.S., 1931. K voprosu ob anemuzite (po povodu patsifitov Barta) [transliterated from Russian]. *Doklady Akademii Nauk SSSR. Leningrad.* Ser.A, p.28–31. [Albasalt]

BELYANKIN, D.S. & KUPLETSKII, B.M., 1924. Gornye porody i poleznye iskopaemye Severnogo poberezhya i prilegayushchikh k nemu ostrovov Kandalakshskoi guby [transliterated from Russian].*Trudy SevernoiNauchnopromyshlennoi ekspedicii.* Vol.28, p.3–76. [Turjite]

BELYANKIN, D.S. & PETROV, V.P., 1945. Petrografiya Gruzii [transliterated from Russian]. *Petrografiya SSSR, Ser. I., Regionalnaya Petrografiya, No.11, Institut Geologicheskikh Nauk, Akademiya Nauk SSSR.* 394pp. [Rikotite]

BELYANKIN, D.S. & VLODAVETS, V.I., 1932. Shchelochnoi kompleks Turego mysa [transliterated from Russian]. *Trudy Petrograficheskogo Instituta. Akademiya Nauk SSSR. Leningrad.* Vol.2, p.45–71. [Aegiapite, Calcitite]

BENSON, W.N., 1909. Petrographical Notes on Certain Pre-Cambrian Rocks of the Mount Lofty Ranges, with special Reference to the Geology of the Houghton District. *Transactions of the Royal Society of South Australia. Adelaide.* Vol.33, p.101–140. [Yatalite]

BERTELS, G.A., 1874. Ein neues vulkanisches Gesteine. *Verhandlungen der Physikalisch-Medizinischen Gesellschaft zu Würzburg.* Vol.8, Neue Folge, p.149–178. [Isenite]

BERTOLIO, S., 1895. Sulle comenditi, nuovo gruppo di rioliti con aegirina. *Atti della Reale Accademia (Nazionale) dei Lincei. Rendiconti. Classe di Scienze Fisiche, Matematiche e Naturali. Roma.* Vol.4, Pt 2, p.48–50. [Comendite]

BÉTHUNE, P. DE, 1956. La busorite, une roche feldspathoïdale nouvelle, du Kivu. *Bulletin de la Société Belge de Géologie, de Paléontologie et d'Hydrologie. Bruxelles.* Vol.65, p.394–399. [Busorite]

BEUDANT, F.S., 1822. Voyage minéralogique et géologique en Hongrie pendant l'année 1818. *Verdière libraire, Paris.* Vol.3, 659pp. [Perlite]

BEUS, A.A., SEVEROV, E.A., SITNIN A.A. & SUBBOTIN, K.D., 1962. Albitizirovannye i greizenizirovannye granity (apogranity) [transliterated from Russian]. *Izadatel'stvo Akademii Nauk SSSR, Moskva.* 196pp. [Apogranite]

BEZBOROD'KO, N.I., 1931. K petrogenezisu temnotsvetnykh porod Podolii i sosednykh raionov [transliterated from Russian]. *Trudy Mineralogicheskogo Instituta. Akademiya Nauk SSSR. Leningrad.* Vol.1, p.127–159. [Bugite, Epibugite, Katabugite, Mesobugite, Sabarovite]

BEZBOROD'KO, N.I., 1935. Montsonitovyi ryad i montsonity Ukrainy [transliterated from Russian]. *Trudy Petrograficheskogo Instituta. Akademiya Nauk SSSR. Leningrad.* Vol.5, p.169–207. [Ukrainite]

BLUM, R., 1861. Foyait, ein neues Gestein aus Süd-Portugal. *Neues Jahrbuch für Mineralogie, Geognosie, Geologie und Petrefaktenkunde. Stuttgart.* Vol.32, p.426–433. [Foyaite]

BOASE, H.S., 1834. A Treatise on Primary Geology. *Longmans, London.* 399pp. [Felsparite]

BOETIUS DE BOOT, ANSELMUS, 1636. Gemmarvm et Lapidum Historia. *Joannis Maire, Lvgdvni Batavorvm [Rotterdam].* Ed. Adrianus Toll. 2nd. Edition, 576pp. [Bimsstein]

BOMBICCI, L., 1868. Sulla oligoclasite di Monte Cavaloro, presso Riola nel Bolognese e sulla composizione della pirite magnetica. *Memorie della Reale Accademia delle Scienze dell'Instituto di Bologna. Classe di Scienze Fisiche. Bologna.* Vol.8, p.77–114. [Oligoclasite]

BONNEY, T.G., 1899. The Parent-Rock of the

Diamond in South Africa. *Geological Magazine. London.* Vol.6, Decade 4, p.309–321. [Newlandite]

Bořický, E., 1873. Petrographische Studien an den Phonolithgesteinen Böhmens. *Archiv für die Naturwissenschaftliche Landesdurchforschung von Böhmen. Prag.* Vol.3, Pt.II, No.1, p.1–96. [Haüyne phonolite, Trachyphonolite]

Bořický, E., 1874. Petrographische Studien an den Basaltgesteinen Böhmens. *Archiv für die Naturwissenschaftliche Landesdurchforschung von Böhmen. Prag.* Vol.2, Pt.II, No.5, p.1–294. [Andesite-basalt, Leucitoid-basalt, Magmabasalt, Nephelinitoid, Noseanite, Oligoclase basalt, Peperin-basalt, Trachybasalt]

Bořický, E., 1878. Der Glimmerpikrophyr, eine neue Gesteinsart und die Libsicer Felswand. *Mineralogische und Petrographische Mitteilungen. Wien.* Vol.1, p.493–516. [Picrophyre]

Bořický, E., 1882. Petrologische Studien an den Porphyrgesteinen Böhmens. *Archiv für die Naturwissenschaftliche Landesdurchforschung von Böhmen. Prag.* Vol.4, No.4, p.1–177. [Radiophyre, Radiophyrite]

Borisov, I., 1963. Petrografski izuchvaniya na magmatitite severno ot gr. Burgas i sravnitelna petrokhimichna kharakteristika na gornokrednite vulkanity v Balgariya [transliterated from Bulgarian]. *Godishnik na Sofiiskiya Universitet, Geologo-geografski Fakultet, Kniga 1, Geologiya.* Vol.58, p.197–233. [Balgarite, Burgasite]

Born, T. von, 1790. Catalogue méthodique et raisonné de la collection des fossiles de Mlle Eléonore de Raab. *Vienne.* Vol.1, 500pp. [Basaltine]

Bowen, N.L., 1914. Art. XIX – The Ternary System: Diopside – Forsterite – Silica. *American Journal of Science. New Haven.* Vol.38, 4th.Ser., p.207–264. [Eutectite]

Bowen, N.L., 1915. Art. XVI – The Crystallization of Haplobasaltic, Haplodioritic and Related Magmas. *American Journal of Science. New Haven.* Vol.40, 4th.Ser., p.161–185. [Haplo-]

Bradley, F.H., 1874. Communication: On unakyte, an epidote rock from the Unaka range, on the borders of Tennessee and North Carolina. *American Journal of Science. New Haven.* Vol.7, 3rd.Ser., p.519–520. [Unakyte]

Brauns, R., 1922. Die phonolitischen Gesteine des Laacher Seegebietes und ihre Beziehungen zu anderen Gesteinen dieses Gebietes. *Neues Jahrbuch für Mineralogie, Geologie und Paläontologie. Stuttgart. Beilagebände.* Vol.46, p.1–116. [Riedenite, Schorenbergite, Selbergite]

Breithaupt, A., 1826. Tachylyt, sehr wahrscheunlich eine neue Mineral-Species. *Archiv für die Gesammte Naturlehre. Nürnberg.* Vol.7, p.112–113. [Tachylyte]

Breithaupt, J.F.A., 1861. Timazit, eine neue Gesteinsart, und Gamsigradit, ein neuer Amphibol. *Berg-und Hüttenmännische Zeitung. Freiberg, Leipzig.* Vol.20, p.51–54. [Timazite]

Brocchi, G.B., 1817. Catalogo Ragionato di una Raccolta di Rocce. *Dall'Imperiale regia stamperia, Milano.* Monograph, 346pp. [Necrolite]

Brögger, W.C., 1890. Die Mineralien der Syenitpegmatitgänge der Südnorwegischen Augit-und Nephelinsyenite. *Zeitschrift für Kristallographie und Mineralogie. Leipzig.* Vol.16, p.1–663. [Akerite, Grorudite, Lardalite, Larvikite, Lujavrite, Nordmarkite, Särnaite]

Brögger, W.C., 1894. Die Eruptivgesteine des Kristianiagebietes. I. Die Gesteine der Grorudit-Tinguait Serie. *Skrifter udgit av Videnskabsselskabet i Kristiania. I. Math.-Nat. Klasse.* No.4, p.1–206. [Hypabyssal, Lindöite, Sölvsbergite, Sussexite]

Brögger, W.C., 1895. Die Eruptivgesteine des Kristianiagebietes. II. Die Eruptionsfolge der triadischen Eruptivgesteine bei Predazzo in Südtyrol. *Skrifter udgit av Videnskabsselskabet i Kristiania. I. Math.-Nat. Klasse.* No.7, p.1–183. [Dellenite, Plauenite, Quartz monzonite, Quartz porphyrite, Trachyte-andesite]

Brögger, W.C., 1898. Die Eruptivgesteine des Kristianiagebietes. III Das Ganggefolge des Laurdalits. *Skrifter udgit av Videnskabsselskabet i Kristiania. I. Math.-Nat. Klasse.* No.6, p.1–377. [Alkaliplete, Calcioplete, Farrisite, Ferroplete, Hedrumite,

Heumite, Kalioplete, Leucocrate, Leucocratic, Maenaite, Melanocrate, Melanocratic, Natrioplete, Osloporphyry, Oxyplete, Tönsbergite]

BRÖGGER, W.C., 1904. "Krageröit". Oversigt over Videnskabs-Selskabets Møder. *Forhandlinger i Videnskabsselskabet i Kristiania.* p.1–74. [Krageröite]

BRÖGGER, W.C., 1906. Eine Sammlung der Wichtigsten Typen der Eruptivgesteine des Kristianiagebietes nach ihren geologischen Verwandtschaftsbeziehungen geordnet. *Nyt Magazin for Naturvidenskaberne. Christiania.* Vol.44, Pt.2, p.113–144. [Ekerite, Essexite-melaphyre, Essexite-porphyrite, Kvellite, Tjosite]

BRÖGGER, W.C., 1921. Die Eruptivgesteine des Kristianiagebietes. IV. Das Fengebiet in Telemark, Norwegen. *Skrifter udgit av Videnskabsselskabet i Kristiania. I. Math.-Nat. Klasse.* No.9, p.1–408. [Calcitfels, Carbonatite, Damkjernite, Fenite, Hollaite, Juvite, Kamperite, Kåsenite, Kristianite, Melteigite, Rauhaugite, Ringite, Sannaite, Silicocarbonatite, Sövite, Tveitåsite, Vibetoite]

BRÖGGER, W.C., 1931. Die Eruptivgesteine des Oslogebietes. V. Der grosse Hurumvulkan. *Skrifter utgitt av det Norske Videnskaps-Akademi i Oslo. Mat.-naturv.Kl.* No.6, p.1–146. [Hurumite, Pyritosalite, Töienite]

BRÖGGER, W.C., 1932. Die Eruptivgesteine des Oslogebietes. VI. Über verschiedene Ganggesteine des Oslogebietes. *Skrifter utgitt av det Norske Videnskaps-Akademi i Oslo. Mat.-naturv.Kl.* No.7, p.1–88. [Björnsjöite]

BRÖGGER, W.C., 1933. Die Eruptivgesteine des Oslogebietes. VII. Die chemische Zusammensetzung der Eruptivgesteine des Oslogebietes. *Skrifter utgitt av det Norske Videnskaps-Akademi i Oslo. Mat.-naturv.Kl.* No.1, p.1–147. [Essexite-akerite, Husebyite, Katnosite, Kjelsåsite, Modumite, Sörkedalite]

BRONGNIART, A., 1807. Traité elémentaire de minéralogie. *Deterville, Paris.* Vol.1, 564pp. [Diabase]

BRONGNIART, A., 1813. Essai d'une classification minéralogique des roches mélangées. *Journal des Mines, Paris.* Vol.34, p.5–48.

[Amphibolite, Euphotide, Eurite, Graphic granite, Hyalomicte, Melaphyre, Mimophyre, Mimose, Ophiolite, Pegmatite, Stigmite, Tephrine, Trachyte, Trappite]

BRONGNIART, A., 1824. Mémoire sur les terrains de sédiments supérieurs calcaréotrapéens du Vincentin. *Levrault, Paris.* 86pp. [Brecciole]

BRONGNIART, A., 1827. Classification et caractères minéralogiques des roches homogènes et hétérogènes. *Levrault, Paris.* 144pp. [Spilite]

BROUWER, H.A., 1909. Pienaarite, a melanocratic foyaite from Transvaal. *Proceedings of the Section of Sciences. Koninklijke (Nederlandse) Akademie van Wetenschappen te Amsterdam.* Vol.12, p.563–565. [Pienaarite]

BROUWER, H.A., 1917. On the Geology of the Alkali Rocks in the Transvaal. *Journal of Geology. Chicago.* Vol.25, p.741–778. [Leeuwfonteinite]

BRÜCKMANNS, U.F.B., 1778. Beiträge zu seiner Abhandlung von Edelsteinen. *Fürstl. Waisenhaus Buchhandlungen, Braunschweig.* 252pp. [Gestellstein, Granitello, Hornberg]

BUCH, L. VON, 1809. Geognostische Beobachtungen auf Reisen durch Deutschland und Italien mit ünum Anhange von mineralogischen Briefen aus Auvergne an den Geh. Ober-Bergrath Karsten von demsilben Verfasser. *Haude und Spener, Berlin.* Vol.1 & 2, p.1–224 & p.225–318. [Domite, Peperino]

BUCH, L. VON, 1810. Reise durch Norwegen und Lappland. *Nauck, Berlin.* Vol.1, 406pp. [Rhombporphyry]

BUCH, L. VON, 1824. VII. Ueber geognostische Erscheinungen im Fassa-Thale. *Taschenbuch für die Gesammte Mineralogie. Frankfurt am Main.* Vol.24, p.343–396. [Monzosyenite]

BUCH, L. VON, 1836. Ueber Erhebungscrater und Vulkane. *Annalen der Physik und Chemie. Leipzig.* Vol.37, p.169–190. [Andesite]

BÜCKING, H., 1881. Basaltische Gesteine aus der Gegend südwestlich vom Thüringer Walde und aus der Rhön. *Jahrbuch der Königlich Preussischen Geologischen Landesanstalt und Bergakademie. Berlin.* Vol.1, for 1880, p.149–189. [Basanitoid, Tephritoid]

BUGGE, J.A.W., 1940. Geological and petrographical investigations in the Arendal District. *Norsk Geologisk Tidsskrift. Oslo.* Vol.20, p.71–116. [Arendalite]

BUNSEN, R.W., 1851. Ueber die Processe der vulkanischen Gesteinsbildungen Islands. *Annalen der Physik und Chemie. Leipzig.* Vol.83, p.197–272. [Baulite]

BURRI, C. & PARGA-PONDAL, I., 1937. Die Eruptivgesteione der Insel Alborán (Provinz Almería, Spanien). *Schweizerische Mineralogische und Petrographische Mitteilungen. Zürich.* Vol.17, p.230–270. [Peralboranite]

BURT, D.M., SHERIDAN, M.F., BIKUN, J.V. & CHRISTIANSEN, E.H., 1982. Topaz rhyolites – Distribution, Origin, and Significance for Exploration. *Economic Geology and Bulletin of the Society of Economic Geologists. Lancaster, Pa.* Vol.77, p.1818–1836. [Topaz rhyolite]

BUSZ, K., 1904. Heptorit, ein Hauyn-Monchiquit aus dem Siebengebirge am Rhein. *Neues Jahrbuch für Mineralogie, Geologie und Paläontologie. Stuttgart. Jahrgang.* Bd. 2, p.86–92. [Heptorite]

BUTAKOVA, E.L., 1956. K petrologii maimechakotuiskogo kompleksa ultraosnovnykh i shchelochnykh porod [transliterated from Russian]. *Trudy Nauchno-Issledovatel'skogo Instituta Geologii Arktiki. Leningrad.* Vol.89, p.201–249. [Kotuite]

CAESALPINO, A., 1596. De metallicis. *ex Typographia A. Zannetti, Rome.* Vol.2, Ch 11, 222pp. [Granite]

CAPELLINI, G., 1878. Inclusioni di apatite nella roccia di Monte Cavaloro. *Rendiconti delle Sessioni dell'Accademia delle Scienze dell'Instituto di Bologna.* p.122–125. [Cavalorite]

CARMICHAEL, I.S.E., 1964. The Petrology of Thingmuli, a Tertiary Volcano in Eastern Iceland. *Journal of Petrology. Oxford.* Vol.5, p.435–460. [Icelandite]

CATHREIN, A., 1890. Zur Dünnschliffsammlung der Tiroler Eruptivgesteine. *Neues Jahrbuch für Mineralogie, Geologie und Paläontologie. Stuttgart. Jahrgang.* Bd. 1, p.71–82. [Adamellite]

CATHREIN, A., 1898. Dioritische Gang- und Stockgesteine aus dem Pusterthal. *Zeitschrift*

der Deutschen Geologischen Gesellschaft. *Berlin.* Vol.50, p.257–278. [Klausenite]

CEDERSTRÖM, A., 1893. Om berggrunden på Norra delen af Ornön. *Geologiska Föreningens i Stockholm Förhandlingar. Stockholm.* Vol.15, p.103–118. [Ornöite]

CHAPPELL, B.W. & WHITE, A.J.R., 1974. Two contrasting granite types. *Pacific Geology. Tokyo.* Vol.8, p.173–174. [I-type granite, S-type granite]

CHAYES, F., 1966. Alkaline and subalkaline basalts. *American Journal of Science. New Haven.* Vol.264, p.128-145. [Subalkali basalt]

CHELIUS, C., 1892. Das Granitmassiv des Melibocus und seine Ganggesteine. *Notizblatt des Vereins für Erdkunde zu Darmstadt, des Mittelrheinischen Geologischen Vereins (und des Naturwissenschaftlichen Vereins zu Darmstadt).* Ser.4, Pt.13, p.1–13. [Alsbachite, Gabbrophyre, Luciite, Odinite, Orbite]

CHESTERMAN, C.W., 1956. Pumice, Pumicite, and Volcanic Cinders in California. *Bulletin. California Division of Mines and Geology. San Francisco.* Vol.174, p.1–119. [Pumicite]

CHIRVINSKII, P.N., AFANASEV, M.S. & USHAKOVA, E.G., 1940. Massiv ultraosnovnykh porod pri stantsii Afrikanda [transliterated from Russian]. *Trudy Kolskoi Bazy Akademii Nauk SSSR.* Vol.5, p.31–70. [Afrikandite]

CHRUSTSCHOFF, K., 1888. Beiträge zur Petrographie Volhyniens und Russlands. *Mineralogische und Petrographische Mitteilungen. Wien.* Vol.9, p.470–527. [Perthitophyre]

CHRUSTSCHOFF, K., 1894. Über eine Gruppe eigenthümlicher Gesteine vom Taimyr-Lande aus der Middendorff'schen Sammlung. *Izvestiya Imperatorskoi Akademii Nauk. S. Peterburg.* Vol.35, p.421–431. [Taimyrite]

CHUDOBA, K., 1930. "Brandbergit", ein neues aplitisches Gestein aus dem Brandberg (SW - Afrika). *Centralblatt für Mineralogie, Geologie und Paläontologie. Stuttgart. Abt.A.* p.389–395. [Brandbergite]

CHUMAKOV, A.A., 1946. O nekotorykh kislykh differentsiatakh gabbro Mugodzhar [transliterated from Russian]. *Akademiku D.S. Belyankinu, Izdatel'stvo Akademii Nauk SSSR.* p.293–297. [Mugodzharite]

CLARKE, F.W., 1904. Analyses of Rocks from the Laboratory of the United States Geological Survey 1880 to 1903. *Bulletin of the United States Geological Survey, Washington.* Vol.228, p.1–375. [Yukonite]

CLARKE, F.W., 1910. Analyses of Rocks and Minerals from the Laboratory of the United States Geological Survey 1880 to 1908. *Bulletin of the United States Geological Survey, Washington.* Vol.419, p.1–323. [Devonite]

CLOIZEAUX, A. DES, 1863. Note sur la classification des roches dites hypérites et euphotides. *Bulletin de la Société Géologique de France. Paris.* Vol.21, p.105–109. [Diallagite]

COGNÉ, J. & GIOT, P., 1961. Observations à propos d'un Lamprophyre micacé à microcline et apatite, riche en enclaves granitiques, au flanc sud du Cap Sizun (Finistère). *Compte Rendu Hebdomadaire des Séances de l'Académie des Sciences. Paris.* Vol.252, p.2569-2571. [Sizunite]

COHEN, E., 1885. Berichtigung bezüglich des "Olivin-Diallag-Gesteins" von Schriesheim im Odenwald. *Neues Jahrbuch für Mineralogie, Geologie und Paläontologie. Stuttgart. Jahrgang.* Bd. 1, p.242–243. [Hudsonite]

COLEMAN, A.P., 1899. A New Analcite Rock from Lake Superior. *Journal of Geology. Chicago.* Vol.7, p.431–436. [Heronite]

COLEMAN, A.P., 1914. The Pre-cambrian rocks north of Lake Huron with special reference to the Sudbury Series. *Ontario Bureau of Mines. Toronto.* Report 23, Pt.1, No.4, p.204–236. [Sudburite]

COLEMAN, R.G. & PETERMAN, Z.E., 1975. Oceanic plagiogranite. *Journal of Geophysical Research. Richmond, Va.* Vol.80, p.1099–1108. [Oceanic plagiogranite]

COLONY, R.J. & SINCLAIR, J.H., 1928. The Lavas of the Volcano Sumaco, Eastern Ecuador, South America. *American Journal of Science. New Haven.* Vol.16, 5th.Ser., p.299–312. [Andesite-tephrite]

COMUCCI, P., 1937. Di una singolare roccia del ghiacciaio Baltoro (Karakorum). *Atti della Reale Accademia (Nazionale) dei Lincei. Rendiconti. Classe di Scienze Fisiche, Matematiche e Naturali. Roma.* Vol.25, p.648–652 & p.734–738. [Baltorite]

CONYBEARE, J.J., 1817. XXII. Memoranda relative to the Porphyritic veins, &c. of St.Agnes, in Cornwall. *Transactions of the Geological Society. London.* Vol.4, p.401–403. [Elvan]

COOMBS, D.S. & WILKINSON, J.F.G., 1969. Lineages and Fractionation Trends in Undersaturated Volcanic Rocks from the East Otago Province (New Zeaand) and Related Rocks. *Journal of Petrology. Oxford.* Vol.10, p.440–501. [Nepheline benmoreite, Nepheline hawaiite, Nepheline mugearite, Nepheline trachyandesite, Nepheline trachybasalt, Nepheline tristanite]

COQUAND, H., 1857. Traité des roches. *Baillière, Paris.* 421pp. [Albitophyre, Labradophyre, Oligophyre, Orthophyre, Pyroxenite]

CORDIER, P.L.A., 1816. Sur les substances minérales dites en masses qui entrent dans la composition des roches volcaniques de tous les ages. *Journal de Physique, de Chimie et d'Histoire Naturelle. Paris.* Vol.83, p.352–386. [Alloite, Asclerine, Cinerite, Pépérite, Phonolite, Pumite, Spodite, Trassoite, Tufaite]

CORDIER, P.L.A., 1842. Dictionnaire universel d'histoire naturelle (1842–1848) Ed. Ch. d'Orbigny. *Savy, Paris.* 25 Volumes. [Amphigenite, Coccolite, Lhercoulite, Mimosite, Nephelinite, Ophitone, Peridotite]

CORDIER, P.L.A., 1868. Description des roches composant l'écorce terrestre et des terrains cristallins. Ed. Ch. d'Orbigny. *Savy, Paris.* 553pp. [Cecilite, Cristulite, Eurynite, Feldspathine, Leucostite, Lherzoline, Syenilite, Tusculite]

COTELO NEIVA, J.M., 1947a. Deux nouvelles roches éruptives de la famille des péridotites. *Estudos. Notas e Trabalhos do Serviço de Fomento Mineiro.* Vol.3, p.105–117. [Abessedite, Bragançaite]

COTELO NEIVA, J.M., 1947b. Nouvelles roches éruptives de la famille des pyroxénolites. *Estudos. Notas e Trabalhos do Serviço de Fomento Mineiro.* Vol.3, p.118–129. [Bogueirite, Conchaite, Regadite]

COTTA, B. VON, 1855. Die Gesteinslehre. *Engelhardt, Freiberg.* 255pp. [Basalt-wacke, Schalstein]

COTTA, B. VON, 1864a. Letter in "Mittheilungen an Professor Leonhard". *Neues Jahrbuch für*

Mineralogie, Geologie und Paläontologie. Stuttgart. Jahrgang. Bd. 35, p.822–827. [Acidite, Basite]

COTTA, B. VON, 1864b. Erzlagerstätten im Banat und in Serbien. *Braumüller, Wien.* 108pp. [Banatite]

COTTA, B. VON, 1866. Rocks Classified and Described. Trans. P.H. Lawrence. *Longmans, London.* 425pp. [Elvanite, Eucrite, Napoleonite, Scoria]

CRONSTEDT, A.F., 1758. Försök til Mineralogie, eller Mineral-Rikets Upställning. *Wildiska Tryckeriet, Stockholm.* 251pp. [Amygdaloid]

CROSS, W., 1897. Art. XVI – Igneous Rocks of the Leucite Hills and Pilot Butte, Wyoming. *American Journal of Science. New Haven.* Vol.4, 4th.Ser., p.115–141. [Madupite, Orendite, Wyomingite]

CROSS, W., IDDINGS, J.P., PIRSSON, L.V. & WASHINGTON, H.S., 1902. A Quantitative Chemico-Mineralogical Classification and Nomenclature of Igneous Rocks. *Journal of Geology. Chicago.* Vol.10, p.555–690. [Femic, Salic]

CROSS, W., IDDINGS, J.P., PIRSSON, L.V. & WASHINGTON, H.S., 1906. The Texture of Igneous Rocks. *Journal of Geology. Chicago.* Vol.14, p.692–707. [Graphiphyre]

CROSS, W., IDDINGS, J.P., PIRSSON, L.V. & WASHINGTON, H.S., 1912. Modifications of the quantitative system of classification of igneous rocks. *Journal of Geology. Chicago.* Vol.20, p.550–561. [Felsic, Mafic]

D'AUBUISSON DE VOISINS, J.F., 1819. Traité de géognosie. *Levrault, Strasbourg.* Vol.2, 665pp. [Aphanite, Diorite, Dolerite]

DALY, R.A., 1903. Geology of Ascutney Mountain, Vermont. *Bulletin of the United States Geological Survey, Washington.* Vol.209, p.1–122. [Windsorite]

DALY, R.A., 1912. Geology of the North American Cordillera at the Forty-Ninth Parallel. *Memoirs. Geological Survey, Canada. Ottawa.* Vol.38, p.1–546. [Shackanite]

DANA, E.S., 1872. Art. 10 – Contributions from the Laboratory of the Sheffield Scientific School No. XXIII – On the composition of the Labradorite rocks of Waterville, New Hampshire. *American Journal of Science.*

New Haven. Vol.3, 3rd.Ser., p.48–50. [Ossipyte]

DANNENBERG, A., 1898. Die Trachyte, Andesite und Phonolithe des Westwaldes. *Tschermaks Mineralogische und Petrographische Mitteilungen. Wien.* Vol.17, 2nd.Ser., p.421–484. [Isenite]

DANNENBERG, A., 1900. Beiträge zur Petrographie der Kaukasusländer. *Tschermaks Mineralogische und Petrographische Mitteilungen. Wien.* Vol.19, 2nd.Ser., p.218–272. [Dacite-andesite]

DAVID, T.E.W., GUTHRIE, F.B. & WOOLNOUGH, W.G., 1901. The Occurrence of a Variety of Tinguaite at Kosciusko, N.S.Wales. *Journal of the Proceedings of the Royal Society of New South Wales. Sydney.* Vol.35, p.347–382. [Muniongite]

DAWSON, J.B. & GALE, N.H., 1970. Uranium and thorium in alkalic rocks from the active carbonatite volcano Oldoinyo Lerngai (Tanzania). *Chemical Geology.* Vol.6, p.221–231. [Lengaite]

DE-ANGELIS, M., 1925. Note di petrografia dancala (II). *Atti della Società Italiana di Scienze Naturali e del Museo Civico di Storia Naturale. Milano.* Vol.64, p.61–84. [Dancalite]

DECHEN, H. VON, 1845. Die Feldspath-Porphyre in den Lenne-Gegenden. *Karston's Archiv für Mineralogie, Geognosie, Bergbau und Hüttenkunde. Berlin.* Vol.19, p.367–452. [Lenneporphyry]

DECHEN, H. VON, 1861. Geognostischer Führer in das Siebengebirge am Rhein. *Henry & Cohen, Bonn.* 431pp. [Sanidophyre]

DELAMÉTHERIE, J.C., 1795. Théorie de la Terre. *Maradan, Paris.* Vol.3, 471pp. [Lherzolite, Pissite]

DELESSE, A., 1851. Mémoire sur la composition minéralogique et chimique des roches des Vosges. *Annales des Mines ou Recueil de Mémoires sur l'Exploration des Mines, et sur les Sciences qui s'y rapportent; rédigés par le Conseil Général des Mines. Paris.* Vol.19, p.149–183. [Kersantite]

DENAEYER, M.E., 1965. La rushayite, lave ultrabasique nouvelle du Nyiragongo (Virunga, Afrique centrale). *Compte Rendu Hebdomadaire des Séances de l'Académie des Sciences. Paris.* Vol.261, p.2119–2122. [Rushayite]

DERBY, O.A., 1891. Art. XXXVII – On the Magnetite Ore Districts of Jacupiranga and Ipanema, São Paulo, Brazil. *American Journal of Science. New Haven.* Vol.41, 3rd.Ser., p.311–321. [Jacupirangite]

DERWIES, V. DE, 1905. Recherches géologiques et pétrographiques sur les laccolithes des environs de Piatigorsk (Caucase du Nord). *Kündig, Genèva.* 84pp. [Trachyliparite]

DEWEY, H., 1910. The Geology of the country around Padstow and Camelford. *Memoirs of the Geological Survey. England & Wales. London.* Sheet 335, p.1–119. [Minverite]

DOELTER, C., 1902. Der Monzoni und seine Gesteine. *Sitzungsberichte der Kaiserlichen Akademie der Wissenschaften. Mathematisch-Naturwissenschaftliche Classe. Wien.* Vol.111, p.929–986. [Rizzonite]

DOLOMIEU, D. DE, 1794. Distribution méthodique de toutes les matières dont l'accumulation forme les montagnes volcaniques ou tableau systématique dans lequel peuvent se placer toutes les substances qui ont des relations avec les feux souterrains. *Journal de Physique, de Chimie et d'Histoire Naturelle. Paris.* p.102–125. [Retinite]

DU BOIS, C.G.B., FURST, J., GUEST, N.J. & JENNINGS D.J., 1963. Fresh Natro Carbonatite Lava from Oldoinyo L'engai. *Nature. London.* Vol.197, p.445–446. [Natrocarbonatite]

DUMAS, E., 1846. Notice sur la constitution géologique de la région supérieure ou cévennique du départment du Gard. *Bulletin de la Société Géologique de France. Paris.* Vol.3, p.572–573. [Fraidronite]

DUPARC, L., 1913. Sur l'Ostraïte, une pyroxénite riche en spinelle. *Bulletin de la Société Française de Minéralogie. Paris.* Vol.36, p.18–20. [Ostraite]

DUPARC, L., 1926. Contribution à la connaissance de la pétrographie et des gîtes minéraux du Maroc. *Annales de la Sociéte Géologique de Belgique.* Vol.49, p.b114–b139. [Aiounite, Mestigmerite]

DUPARC, L. & GROSSET, A., 1916. Recherches géologiques et pétrographiques sur le district minier de Nikolai Pawda. *Kündig, Genèva.*

294pp. [Kazanskite, Pawdite]

DUPARC, L. & JERCHOFF, S., 1902. Sur les plagiaplites quartzifères du Kosswinsky. Intervention reportée dans le "Compte rendu des Séances du 6 Fevrier 1902". *Archives des Sciences Physiques et Naturelles. Genève.* Vol.13, p.307–310. [Plagiaplite]

DUPARC, L. & MOLLY, E., 1928. Sur la tokeïte, une nouvelle roche d'Abyssinie. *Compte Rendu des Séances de la Société de Physique et d'Histoire Naturelle de Genève.* Vol.45, p.24–25. [Tokéite]

DUPARC, L. & PAMPHIL, G., 1910. Sur l'issite, une nouvelle roche filonienne dans la dunite. *Compte Rendu Hebdomadaire des Séances de l'Académie des Sciences. Paris.* Vol.151, p.1136–1138. [Issite]

DUPARC, L. & PEARCE, F., 1900. Sur les plagioliparites du Cap Marsa (Algérie). *Compte Rendu Hebdomadaire des Séances de l'Académie des Sciences. Paris.* Vol.130, p.56–58. [Plagioliparite]

DUPARC, L. & PEARCE, F., 1901. Sur la koswite, une nouvelle pyroxénite de l'Oural. *Compte Rendu Hebdomadaire des Séances de l'Académie des Sciences. Paris.* Vol.132, p.892–894. [Koswite]

DUPARC, L. & PEARCE, F., 1904. Sur la garéwaïte, une nouvelle roche filonienne basique de l'Oural du Nord. *Compte Rendu Hebdomadaire des Séances de l'Académie des Sciences. Paris.* Vol.139, p.154–155. [Garewaite]

DUPARC, L. & PEARCE, F., 1905. Sur la gladkaïte, nouvelle roche filonienne dans la dunite. *Compte Rendu Hebdomadaire des Séances de l'Académie des Sciences. Paris.* Vol.140, p.1614–1616. [Gladkaite, Tilaite]

DUROCHER, J., 1845. Sur l'origine des roches granitiques. *Compte Rendu Hebdomadaire des Séances de l'Académie des Sciences. Paris.* Vol.20, p.1275–1284. [Quartz porphyry]

EBERT, H. VON, 1968. Suprakrustale Glieder der Charnockit-Familie in Nordwestsachsen. *Geologie. Zeitschrift für das Gesamtgebiet der Geologie und Mineralogie sowie der Angewandten Geophysik. Beiheft. Berlin.* Vol.17, p.1031–1050. [Grimmaite]

EBRAY, TH., 1875. Quelques remarques sur les

granulites et les minettes. Nouvelle classification des roches éruptives. *Bulletin de la Société Géologique de France. Paris.* Vol.3, p.287–291. [Anthraphyre, Carbophyre, Kohliphyre, Triaphyre]

ECKERMANN, H. VON, 1928a. Dikes belonging to the Alnö formation in the cuttings of the East Coast Railway. *Geologiska Föreningens i Stockholm Förhandlingar. Stockholm.* Vol.50, p.381–412. [Beforsite, Stavrite]

ECKERMANN, H. VON, 1928b. Hamrongite, a new Swedish alkaline mica lamprophyre. *Fennia: Bulletin de la Société de Géographie de Finlande.* Vol.50, p.1–21. [Hamrongite]

ECKERMANN, H. VON, 1938a. The Anorthosite and Kenningite of the Nordingrå-Rödö Region. *Geologiska Föreningens i Stockholm Förhandlingar. Stockholm.* Vol.60, p.243–284. [Kenningite]

ECKERMANN, H. VON, 1938b. A Contribution to the Knowledge of the late Sodic Differentiates of Basic Eruptives. *Journal of Geology. Chicago.* Vol.46, p.412–437. [Svartvikite]

ECKERMANN, H. VON, 1942. Ett preliminärt meddelande om nya forskningsrön inom Alnö alkalina område. *Geologiska Föreningens i Stockholm Förhandlingar. Stockholm.* Vol.64, p.399–455. [Alkorthosite, Alvikite, Hartungite]

ECKERMANN, H. VON, 1960. Borengite – A new ultra-potassic rock from Alnö Island. *Arkiv för Mineralogi och Geologi. Stockholm.* Vol.2, No.39, p.519–528. [Borengite]

EDEL'SHTEIN, YA.,S., 1930. O novoi oblasti razvitiya shchelochnykh (nefelino-egirinovykh) porod v Yuzhnoi Sibiri [transliterated from Russian]. *Geologicheskii vestnik.* Vol. UP, No.1-3, p.15-23. [Saibarite]

EGOROV, L.S., 1969. Melilitovie porodi Maimecha-Kotuiskoi provintzii [transliterated from Russian]. *Trudy Nauchno-Issledovatel' skogo Instituta Geologii Arktiki. Leningrad.* Vol.159, 247pp. [Kugdite]

EIGENFELD, R., 1965. Raqqait, ein holo-melanokrates Lavagestein von pyroxenitischem Magmacharakter. *43. Jahrestagung der Deutschen Mineralogischen Gesellschaft, Kurzreferate der Vorträge (Hannover).* p.46. [Raqqaite]

ELIE DE BEAUMONT, M.L., 1822. Sur les mines de fer et les forges de Framont et de Rothau. *Annales des Mines ou Recueil de Mémoires sur l'Exploration des Mines, et sur les Sciences qui s'y rapportent; rédigés par le Conseil Général des Mines. Paris.* Vol.7, p.521–554. [Minette]

ELISEEV, N.A., 1937. Struktury rudnykh polei v pervichno rassloennykh plutonakh Kolskogo poluostrova [transliterated from Russian]. *Izvestiya Akademii Nauk SSSR, Seriya Geologicheskaya.* No. 6, p.1085-1104. [Eudialytite]

*ELLIS, W., 1825. Journal of a tour around Hawaii. *Crocker & Brewster, Boston.* [Pélé's hair]

EMERSON, B.K., 1902. Holyokeite, a purely Feldspathic Diabase from the Trias of Massachusetts. *Journal of Geology. Chicago.* Vol.10, p.508–512. [Holyokeite]

EMERSON, B.K., 1915. Art. XIX – Northfieldite, Pegmatite, and Pegmatite Schist. *American Journal of Science. New Haven.* Vol.40, 4th.Ser., p.212–217. [Northfieldite]

ENDLICH, F.M., 1874. Report of F.M. Endich S.N.D. *Report of the United States Geological (and Geographical) Survey of the Territories. Washington.* p.275–361. [Trachorheite]

ERDMANN, A., 1849. Letter in "Mittheilungen an den Geheimenrath v. Leonhard gerichtet". *Neues Jahrbuch für Mineralogie, Geognosie, Geologie und Petrefaktenkunde. Stuttgart.* Vol.20, p.837–838. [Eulysite]

ERDMANNSDÖRFFER, O.H., 1907. Über Vertreter der Essexit-Theralitheihe unter den diabasartigen Gesteinen der Deutschen Mittelgebirge. *Zeitschrift der Deutschen Geologischen Gesellschaft. Berlin.* Vol.59, Monatsberichte No.2, p.16–23. [Essexite-diabase]

ERDMANNSDÖRFFER, O.H., 1928. Ueber ein sibirisches Nephelingestein. *Festschrift Victor Goldschmidt, Heidelberg.* p.85–88. [Bjerezite]

ESCHWEGE, W.L.C. VON, 1832. Beiträge zur Gebirgskunde Brasiliens. *Reimer, Berlin.* 488pp. [Carvoeira]

ESKOLA, P., 1921. On the igneous rocks of Sviatoy Noss in Transbaikalia. *Ofversigt av Finska Vetenskaps-Societetens Förhandlingar. Helsingfors.* Vol.63, afd.A,1, p.1–99.

[Sviatonossite]

ESMARK, J., 1823. Om Norit-Formationen. *Magazin for Naturvidenskaberne, Christiania.* Vol.1, No.2, p.205–215. [Norite]

FAESSLER, C., 1939. The Stock of "Suzerite" in Suzor Township, Quebec. *University of Toronto Studies. Geological Series. Toronto.* Vol.42, p.47–52. [Suzorite]

FAUJAS DE SAINT-FOND, B.F., 1778. Recherches sur les Volcans Eteints du Vivarais et du Velay. *Cuchet, Grenoble.* 460pp. [Gallinace]

FEDERICO, M., 1976. On a Kalsilitolite from the Alban Hills, Italy. *Periodico di Mineralogia. Rome.* Vol.45, p.5–12. [Kalsilitolite]

*FEDOROV, E., 1896. O novoi gruppe izverzhennykh porod [transliterated from Russian]. *Izvestiya Moskovskago Sel' skokhozyaistvennago Instituta. Moskva.* No.1. [Drusite]

FEDOROV, E.S., 1901. Geologicheskiya izsledovaniya letom 1900 g [transliterated from Russian]. *Ezhegodnik po Geologii i Mineralogii Rossii. Varshava.* Vol.4, No.6, p.135–140. [Kedabekite]

FEDOROV, E.S., 1903. Kratkoe soobshchenie o rezultatakh mineralogicheskago i petrograficheskago izsledovaniya beregov Belago morya [transliterated from Russian]. *Zapiski Imperatorskogo (S. Petersburgskogo) Mineralogicheskogo Obshchestva.* Vol.40, p.211–220. [Kovdite]

FENNER, C.N., 1948. Incandescent tuff flows in southern Peru. *Bulletin of the Geological Society of America. New York.* Vol.59, p.879–893. [Sillar]

FERSMAN, A., 1929. Geochemische Migration der Elemente und deren wissenschaftliche und wirtschaftliche Bedeutung. *Abhandlungen zur praktischen Geologie und Bergwirtschaftslehre. Halle (Salle).* Vol.18, p.1–116. [Miaskitic]

FICHTEL, J.E. VON, 1791. Mineralogische Bemerkungen von den Karpathen. *Edlen von Kurzbeck, Wien.* Vol.1, 411pp. [Perlstein]

FIEDLER, A., 1936. Über Verflößungserscheinungen von Amphibolit mit diatektischen Lösungen im östlichen Erzgebirge. *Zeitschrift für Kristallographie, Mineralogie und Petrographie. Leipzig. Abt. B.* Vol.47, p.470–516. [Diatectite]

FLETT, J.S., 1907. Petrography of the Igneous Rocks. In: W.A.E. Ussher, The Geology of the country around Plymouth and Liskeard. *Memoirs of the Geological Survey. England & Wales. London.* Sheet 348, p.1–147. [Spilite]

FLETT, J.S., 1911. In: B.N. Peach, J.S.G. Wilson, J.B. Hill, E.B. Bailey & G.W. Grabham, The Geology of Knapdale, Jura and North Kintyre. *Memoirs. Geological Survey of Scotland. Edinburgh.* Sheet 28, p.1–149. [Crinanite]

FLETT, J.S., 1935. Petrography. In: Wilson, G.V., Edwards, W., Knox, J., Jones, R.C.V. and others. The Geology of the Orkneys. *Memoirs. Geological Survey of Great Britain. Scotland.* p.173–186. [Holmite]

FOERSTNER, E., 1881. Nota preliminare sulla geologia dell'Isola di Pantelleria secondo gli studii fatti negli anni 1874 e 1881, dal dottor Enrico Foerstner. *Bollettino del Reale Comitato Geologico d'Italia. Firenze.* Vol.12, p.523–556. [Pantellerite]

FOURNET, J., 1845. Note sur l'état actuel des connaissances touchant les roches éruptives des environs de Lyon. *Bulletin de la Société Géologique de France. Paris.* Vol.2, p.495–506. [Ilvaite, Miarolite]

FOURNET, J., 1861. Géologie lyonnaise. *Lyon.* Vol.1, 744pp. [Vaugnerite]

FRECHEN, J., 1971. Siebengebirge am Rhein – Laacher Vulkangebiet – Maargebiet der Westeifel. Sammlung Geologischer Führer 56. *Gebr. Borntraeger, Berlin.* 2nd. Edition, 195pp. [Boderite, Rodderite]

FRENTZEL, A., 1911. Das Passauer Granitmassiv. *Geognostische Jahreshefte. München.* Vol.24, p.105–192. [Engelburgite, Ilzite]

FRITSCH, K. VON, 1865. Notizen über geologische Verhältnisse im Hegau. *Neues Jahrbuch für Mineralogie, Geologie und Paläontologie. Stuttgart. Jahrgang.* Bd. 36, p.651–673. [Tephrite]

FUSTER, J.M., GASTESI, P., SAGREDO, J. & FERMOSO, M.L., 1967. Las rocas lamproíticas del S.E. de España. *Estudios Geológicos. Instituto de Investigaciones Geológicas "Lucas Mallada". Madrid.* Vol.23, p.35–69. [Cancalite]

GAGEL, C., 1913. Studien über den Aufbau und die Gesteine Madeiras. *Zeitschrift der*

Deutschen Geologischen Gesellschaft. Berlin. Vol.64, No.7, p.344–448. [Madeirite]

GAPEEVA, G.M., 1959. Ussurit – osobaya raznovidnost' shchelochnykh bazaltovykh porod [transliterated from Russian]. *Doklady Akademii Nauk SSSR. Leningrad.* Vol.126, No.1, p.157–159. [Ussurite]

GAVELIN, A., 1915. Den geologiska byggnaden inom Ruoutevare-området i Kvikkjokk, Norbottens Län. *Geologiska Föreningens i Stockholm Förhandlingar. Stockholm.* Vol.37, p.17–32. [Vallevarite]

GAVELIN, A., 1916. Über Högbomit. *Bulletin of the Geological Institution of the University of Upsala.* Vol.15, p.289–316. [Magnetite Högbomitite]

*GEIKIE, A., 1903. Textbook of Geology. *Macmillan, London.* 4th. Edition. [Plateau basalt]

GEMMELLARO, C., 1845. Sul basalto decomposto dell'isola de'ciclopi. *Atti dell' Accademia Gioenia di Scienze Naturali in Catania.* Vol.2, 2nd. Ser., p.309–319. [Analcimite]

GERHARD, A., 1815. Beiträge zur Gesdrichte des Weissteins, des Felsit und anderer verwandber Arten. *Abhandlungen der Königlichen Akademie der Wissenschaften. Berlin.* p.12–26. [Felsite]

GILL, J, 1981. Orogenic Andesites and Plate Tectonics. *Springer-Verlag, Berlin.* 390pp. [Orogenic andesite]

GIRAUD, P., 1964. Essai de classification modale des roches à caractère charnockitique. *Bulletin du Bureau de Recherches Géologiques et Minières. Paris.* Vol.4, p.36–53. [Ankaranandite]

GLOCKER, E.F., 1831. Handbuch der Mineralogie. *Schrag, Nürnberg.* Vol.2. [Fluolite]

*GMELIN, L., 1814. Observatisner Oryctognostica et Chemiae de Haüyne. *Mohr & Zimmer, Heidelberg.* 58pp. [Asprone, Sperone]

GOLDSCHMIDT, V.M., 1916. Geologisch-Petrographische Studien im Hochgebierge des Südlichen Norwegiens. IV. Übersicht der Eruptivgesteine im Kaledonischen Gebirge zwischen Stavanger und Trondhjem. *Skrifter udgit av Videnskabsselskabet i Kristiania. I. Math.-Nat. Klasse.* p.1–140. [Jotun-norite,

Opdalite, Trondhjemite]

GORDON, C.H., 1896. Syenite-gneiss (Leopard Rock) from the Apatite Region of Ottawa County, Canada. *Bulletin of the Geological Society of America. New York.* Vol.7, p.95–134. [Leopard rock]

GORNOSTAEV, N. N., 1933. Differentsirovannyi ekstruzivnyi lakkolit Kyz-Emchik v gorakh Semeitau bliz g. Semipalatinska [transliterated from Russian]. *Sbornik po Geologii Sibiri posvyashchennyi 25-letiyu nauchnopedagogicheskoi deyatelnosti prof. M.A.Usova. Tomsk.* p.153–223. [Semeitavite, Terektite]

GRAHMANN, R., 1927. Blatt Riesa-Strehla. *Erläuterungen zur Geologischen Karte von Sachsen. Leipzig.* No. 16, 2nd. Edition, p.1–82. [Gröbaite]

GRANGE, L.I., 1934. Rhyolite sheet flows of the North Island, New Zealand. *New Zealand Journal of Science & Technology. Wellington.* Vol.16, p.57–58. [Owharoite]

GREGORY, J.W., 1900. Contributions to the Geology of British East Africa – Part II, The Geology of Mount Kenya. *Quarterly Journal of the Geological Society of London.* Vol.56, p.205–222. [Kenyte]

GREGORY, J.W., 1902. Article XVI – The Geology of Mount Macedon, Victoria. *Proceedings of the Royal Society of Victoria. Melbourne.* Vol.14, 2nd.Ser., p.185–217. [Geburite-dacite]

GROOM, T.T., 1889. On the Occurrence of a New Form of Tachylyte in Association with the Gabbro of Carrock Fell, in the Lake District. *Geological Magazine. London.* Vol.6, Decade 3, p.43. [Carrockite]

GROTH, P., 1877. Das Gneissgebiet von Markirch im Ober-Elsass. *Abhandlungen zur Geologischen Spiezialkarte von Elsass-Lothringen. Strassburg.* Vol.1, p.395–489. [Kammgranite]

GRUBENMANN, U., 1908. Der Granatolivinfels des Gordunotales und seine Begleitgesteine. *Vierteljahrsschrift der Naturforschenden Gesellschaft in Zürich.* Vol.53, p.129–156. [Gordunite]

GRUSS, K., 1900. Beiträge zur Kenntnis der Gesteine des Kaiserstuhlgebirges. Tephritische Strom-und Ganggesteine. *Mittheilungen* der Grossherzoglich Badischen Geologischen Landesanstalt. Heidelberg. Vol.4, p.83–144. [Mondhaldeite]

GÜMBEL, C.W. VON, 1861. Geognostische Beschreibung des bayerischen Alpengebirges und series Vorlandes. Geognostische Beschreibung des Königreiches Bayern. Gotha. p.1–948. [Sillite]

GÜMBEL, C.W. VON, 1865. Geognostische Beschreibung der Fränkischen Alb (Frankenjura). Geognostische Beschreibung des Königreiches Bayern. Gotha. Vol.4, p.1–763. [Aschaffite]

GÜMBEL, C.W. VON, 1868. Geognostische Beschreibung des Ostbayerischen Grenzgebirges oder des Bayerischen und oberpfälzer Waldgebirges. Geognostische Beschreibung des Königreiches Bayern. Gotha. p.1–968. [Nadeldiorite, Regenporphyry]

GÜMBEL, C.W. VON, 1874. Die paläolithischen Eruptivgesteine des Fichtelgebirges. Weiss, Munich. 50pp. [Epidiorite, Keratophyre, Lamprophyre, Leucophyre, Palaeophyre, Palaeopicrite, Proterobase]

GÜMBEL, C.W. VON, 1888. Geologie von Bayern. Erster Theil, Grundzuge der Geologie. Fischer, Kassel. Vol.1, 1144pp. [Dioritoid, Gabbroid, Granitophyre, Hyalopsite, Kokkite, Peridotoid, Phonolitoid, Trachytoid]

HACKL, O. & WALDMANN, L., 1935. Ganggesteine der Kalireihe aus dem neiderösterreichischen Waldviertal. Jahrbuch der Geologischen Bundesanstalt. Wein. Vol.85, p.259–285. [Karlsteinite, Raabsite, Thuresite]

HAIDINGER, W., 1861. Meeting/session report from 28.5.1861, without title. Verhandlungen der Geologischen Reichsanstalt (Staatanstalt-Landesanstalt). Wien. Vol.12, p.64. [Haüynfels]

HALL, A.L., 1922. On the Marundites and allied Corundum-bearing rocks in the Leydsdorp District of the Eastern Transvaal. Transactions of the Geological Society of South Africa. Johannesburg. Vol.25, p.43–67. [Marundite]

HALL, T.C., 1915. Summary of Progress of the Geological Survey of Great Britain and the Museum of Practical Geology for 1914. Memoirs of the Geological Survey. England & Wales. London. p.1–84. [Lundyite]

HARKER, A., 1904. The Tertiary Igneous Rocks of Skye. Memoirs of the Geological Survey of the United Kingdom. p.1–481. [Marscoite, Mugearite]

HARKER, A., 1908. The geology of the small Isles of Inverness-shire (Rum, Canna, Eigg, Muck, etc.). Memoirs. Geological Survey of Scotland. Edinburgh. Sheet 60, p.1–210. [Allivalite, Harrisite]

HATCH, F.H., 1909. Text-book of Petrology. Sonnenschein, London. 5th. Edition, 404pp. [Markfieldite]

HATCH, F.H., WELLS, A.K. & WELLS M.K., 1949. Petrology of the Igneous Rocks. Thomas Murby, London. Vol.1, 10th. Edition, 469pp. [Orthogabbro]

HATCH, F.H., WELLS, A.K. & WELLS M.K., 1961. Petrology of the Igneous Rocks. Thomas Murby, London. Vol.1, 12th. Edition, 515pp. [Kalmafite, Leumafite, Melmafite, Nemafite]

HAUER F. VON & STACHE, G., 1863. Geologie Siebenbürgens (Ed. Graeser). Braumüller, Wien. 636pp. [Dacite, Quartz trachyte]

HAUSMANN, J., 1847. Handbuch der Mineralogie. Vandenhoek & Ruprecht, Göttingen. Theil II, Band I, 2nd. Edition, 896pp. [Hyalomelane, Wichtisite]

HAÜY, A., 1822. Traité de Minéralogie. Bachelier, Huzard, Paris. Vol.4, 2nd. Edition, 604pp. [Basaltoid, Eclogite, Γogonite, Resinite, Selagite, Trematode]

HEDDLE, M.F., 1897. On the crystalline forms of riebeckite. Transactions of the Edinburgh Geological Society. Vol.7, p.265–267. [Ailsyte]

HENDERSON, J.A.L., 1898. Petrographical and Geological Investigations of Certain Transvaal Norites, Gabbros and Pyroxenites. Dulau, London. 56pp. [Hatherlite, Pilandite]

HENDERSON, J.A.L., 1913. The Geology of the Aroha Subdivision, Hauraki, Auckland. Bulletin. Geological Survey of New Zealand. Wellington. Vol.16, p.1–127. [Wilsonite]

HENNIG, A., 1899. Kullens kristalliniska bergarter. II. Den postsiluriska gångformationen. Acta Universitatis Lundensis. Lund. Vol.35, Afdeln.2, No.5., p.1–34. [Kullaite]

HESS, H.H., 1937. Island arcs, gravity anomalies and serpentine intrusions; a contribution to the ophiolite problem. *International Geological Congress, Report of the XVII Session, Moscow.* Vol.2, p.263–283. [Ultramafic]

HIBSCH, J.E., 1896. Erläuterungen zur geologischen Specialkarte des böhmischen Mittelgebirges. Blatt I. Umgebung von Tetschen. *Tschermaks Mineralogische und Petrographische Mitteilungen. Wien.* Vol.15, 2nd.Ser., p.201–290. [Ash tuff]

HIBSCH, J.E., 1898. Erläuterungen zur geologischen Karte des böhmischen Mittelgebirges. Blatt III. Bensen. *Tschermaks Mineralogische und Petrographische Mitteilungen. Wien.* Vol.17, 2nd.Ser., p.1–96. [Gauteite]

HIBSCH, J.E., 1899. Die Tiefengesteine des böhmischen Mittelgebirges. *Sitzungsberichte des deutschen naturwissenschaftlichmedicinischen Vereines für Böhmen "Lotos" in Prag.* Vol.19, 2nd.Ser., p.68–72. [Analcime syenite]

HIBSCH, J.E., 1902. Geologische Karte des böhmischen Mittelgebirges. Blatt V. Grosspriesen. *Tschermaks Mineralogische und Petrographische Mitteilungen. Wien.* Vol.21, 2nd.Ser., p.465–588. [Sodalitophyre]

HIBSCH, J.E., 1910. Geologische Karte des böhmischen Mittelgebirges. Blatt VI. Wernstadt-Zinkenstein. *Tschermaks Mineralogische und Petrographische Mitteilungen. Wien.* Vol.29, 2nd.Ser., p.381–438. [Alkali basalt]

HIBSCH, J.E., 1920. Geologische Karte des böhmischen Mittelgebirges. Blatt XIV. Meronitz - Trebnitz. *Prag.* 120pp. [Analcime basanite]

HILL, J.B. & KYNASTON, H., 1900. On Kentallenite and its relations to other Igneous Rocks in Argyllshire. *Quarterly Journal of the Geological Society of London.* Vol.56, p.531–558. [Kentallenite]

HILLS, E.S., 1958. Cauldron Subsidence, Granitic rocks, and crustal fracturing in S.E. Australia. *Geologische Rundschau. Internationale Zeitschrift für Geologie. Stuttgart.* Vol.47, p.543–561. [Snobite]

HJÄRNE, U., 1694. En kort Anledning till Atskillge Malm och Bergarters, Mineraliers, Wäxters och Jordeslags, samt flere sällsamme Tings effterspöriande och angiftvande. *Stock-holm.* [Rapakivi]

HOBBS, W.H., 1893. Volcanite, an Anorthoclase-augite Rock chemically like Dacites. *Bulletin of the Geological Society of America. New York.* Vol.5, p.598–602. [Volcanite]

HOCHSTETTER, F., 1859. Lecture on the geology of the Province of Nelson. *New Zealand Government Gazette.* No.39, p.269–281. [Dunite]

HØDAL, J., 1945. Rocks of the anorthosite kindred in Vossestrand (Norway). *Norsk Geologisk Tidsskrift. Oslo.* Vol.24, p.129–243. [Jotunite]

HÖGBOM, A.G., 1905. Zur Petrographie der kleinen Antillen. *Bulletin of the Geological Institution of the University of Upsala.* Vol.6, p.214–233. [Plagioclase granite]

HOHENEGGER, L., 1861. Die geognostischen Verhältnisse der Nordkarpathen in Schlesien und den angrenzenden Theilen von Mähren und Galizien. *Perthes, Gotha.* 50pp. [Teschenite]

HOLLAND, T.H., 1900. The charnockite series, a group of Archaean hypersthenic rocks in Peninsular India. *Memoirs of the Geological Survey of India. Calcutta.* Vol.28, p.119–249. [Charnockite]

HOLLAND, T.H., 1907. General Report of the Geological Survey of India for the year 1906. *Records of the Geological Survey of India. Calcutta.* Vol.35, p.1–61. [Kodurite]

HOLMES, A., 1937. The Petrology of Katungite. *Geological Magazine. London.* Vol.74, p.200–219. [Katungite]

HOLMES, A., 1942. A Suite of Volcanic Rocks from south-west Uganda containing Kalsilite (a polymorph of $KAlSiO_4$). *Mineralogical Magazine and Journal of the Mineralogical Society. London.* Vol.26, p.197–217. [Mafurite, Protokatungite]

HOLMES, A. & HARWOOD, H.F., 1937. The Petrology of the Volcanic Area of Bufumbira. *Memoirs. Geological Survey of Uganda. Entebbe.* Vol.3, Pt.2, p.1–300. [Lutalite, Murambite, Murambitoid, Ugandite]

HOLMQUIST, P.J., 1908. Utkast till ett bergartsschema för urbergsskiffrarna. *Geologiska Föreningens i Stockholm Förhandlingar. Stockholm.* Vol.30, p.269–293. [Peridotitoid]

HOLST, N.O. & EICHSTÄDT, F., 1884. Klotdiorit från Slättmossa, Järeda socken, Kalmar län. *Geologiska Föreningens i Stockholm Förhandlingar. Stockholm.* Vol.7, p.134–142. [Klotdiorite, Klotgranite]

HOLTEDAHL, O., 1943. Studies on the igneous rock Complex of the Oslo Region. I. Some structural features of the district near Oslo. *Skrifter utgitt av det Norske Videnskaps-Akademi i Oslo. Mat.-naturv. Kl.* No.2, p.1–71. [Lathus porphyry, Østern porphyry]

HORNE, J. & TEALL, J.J.H., 1892. On Borolanite – an Igneous Rock intrusive in the Cambrian Limestone of Assynt, Sutherlandshire, and the Torridon Sandstones of Ross-shire. *Transactions of the Royal Society of Edinburgh.* Vol.37, p.163–178. [Borolanite]

HUMBOLDT, A. VON, 1823. Essai Géognostique sur le gisement des roches dans les deux hemisphères. *Levrault, Paris.* 379pp. [Lozero, Texontli]

HUMBOLDT, H.A. VON, 1837. Geognostische und physikalische Beobachtungen über die Vulkane des Hochlandes von Quito. *Neues Jahrbuch für Mineralogie, Geognosie, Geologie und Petrefaktenkunde. Stuttgart.* Vol.8, p.253–284. [Leucitophyre]

HUME, W.F., HARWOOD, H.F. & THEOBALD, L.S., 1935. Notes on some analyses of Egyptian Igneous and Metamorphic Rocks. *Geological Magazine. London.* Vol.72, p.8–32. [Baramite]

*HUNT, T.S., 1852. *Annual Report. Geological Survey of Canada. Montreal and Ottawa.* [Parophite]

HUNT, T.S., 1862. Descriptive catalogue of a collection of the crystalline rocks of Canada: Descriptive catalogue of a collection of the economic minerals of Canada and of its crystalline rocks sent to the London International Exhibition Montreal. *Geological Survey of Canada.* p.61–83. [Anorthosite]

HUNT, T.S., 1864. On the chemical and mineralogical relations of metamorphic rocks. *Journal of the Geological Society of Dublin.* Vol.10, p.85–95. [Anortholite]

HUNTER, C.L., 1853. Art. 42 – Notices of the rarer minerals and New localities in western North Carolina. *American Journal of Science. New Haven.* Vol.15, 2nd.Ser., p.373–378. [Leopardite]

HUNTER, M. & ROSENBUSCH, H., 1890. Ueber Monchiquit, ein camptonitisches Ganggestein aus der Gefolgschaft der Eläolithsyenite. *Tschermaks Mineralogische und Petrographische Mitteilungen. Wien.* Vol.11, 2nd.Ser., p.445–466. [Bostonite, Monchiquite]

IDDINGS, J.P., 1895a. Absarokite – Shoshonite – Banakite Series. *Journal of Geology. Chicago.* Vol.3, p.935–959. [Absarokite, Banakite, Shoshonite]

IDDINGS, J.P., 1895b. The Origin of Igneous Rocks. *Bulletin of the Philosophical Society of Washington.* Vol.12, p.89–213. [Alkali, Subalkali]

IDDINGS, J.P., 1904. Quartz-Feldspar-Porphyry (Graniphyro Liparose-Alaskose) from Llano, Texas. *Journal of Geology. Chicago.* Vol.12, p.225–231. [Llanite]

IDDINGS, J.P., 1909. Igneous Rocks. *Wiley, New York.* Vol.1, 464pp. [Cumulophyre, Linophyre, Planophyre, Skedophyre]

IDDINGS, J.P., 1913. Igneous Rocks. *Wiley, New York.* Vol.2, 685pp. [Andesine basalt, Bandaite, Hawaiite, Kauaiite, Kohalaite, Laugenite, Marosite, Oligoclase andesite, Shastaite, Ungaite]

IDDINGS, J.P. & MORLEY, E.W., 1917. A Contribution to the Petrography of Southern Celebes. *Proceedings of the National Academy of Sciences of the United States of America. Washington.* Vol.3, p.592–597. [Batukite]

IDDINGS, J.P. & MORLEY, E.W., 1918. A Contribution to the Petrography of the South Sea Islands. *Proceedings of the National Academy of Sciences of the United States of America. Washington.* Vol.4, p.110–117. [Tautirite]

IPPEN, J.A., 1903. Ueber den Allochetit vom Monzoni. *Verhandlungen der Geologischen Reichsanstalt (Staatanstalt-Landesanstalt). Wien.* No.7 & 8, p.133–143. [Allochetite]

IRVINE, T.N. & BARAGAR, W.R.A., 1971. A Guide to the Chemical Classification of the Common Volcanic Rocks. *Canadian Journal of Earth Sciences, Ottawa.* Vol.8, p.523–548. [Commendite]

ISSEL, A., 1880. Osservazioni intorno a carte rocce anfiboliche della Liguria, a proposito d'una nota del Prof. Bonney concernente alcune Serpentine della Liguria e della Toscana. *Bollettino del Reale Comitato Geologico d'Italia. Firenze.* Vol.11, p.183–192. [Borzolite, Coschinolite]

ISSEL, A., 1892. Liguria Geologica e Preistorica. *Donath, Genoa.* Vol.1, 440pp. [Epidiabase]

ISSEL, A., 1916. Prime linee di un ordinamento sistematico delle Pietre figurate. *Memorie della Reale Accademia (Nazionale) dei Lincei. Rendiconti. Classe di Scienze Fisiche, Matematiche e Naturali. Roma.* Vol.11, p.631–667. [Peleiti]

IVES, R.L., 1956. Tezontli rock. *Rocks and Minerals, Peekskill, New York.* Vol.31, p.122–124. [Tezontli]

JENNER, G.A., 1981. Geochemistry of the High-Mg andesites from Cape Vogel, Papua New Guinea. *Chemical Geology.* Vol.33, p.307–332. [High-Mg andesite]

JENZSCH, G., 1853. Amygdalophyr, ein Felsit-Gestein mit Weissigit, einem neuen Minerale in Blasen-Räumen. *Neues Jahrbuch für Mineralogie, Geognosie, Geologie und Petrefaktenkunde. Stuttgart.* Vol.58, p.385–398. [Amygdalophyre]

JEVONS, H.S., JENSEN, H.I., TAYLOR, T.G. & SÜSSMILCH, C.A., 1912. The Geology and Petrology of the Prospect Intrusion. *Journal of the Proceedings of the Royal Society of New South Wales. Sydney.* Vol.45, p.445–553. [Pallioessexite, Palliogranite]

JOHANNSEN, A., 1911. Petrographic Terms for Field Use. *Journal of Geology. Chicago.* Vol.19, p.317–322. [Amphiboleid, Anameseid, Anameseid porphyry, Aphaneid, Dioreid, Dolereid, Felseid, Felseid porphyry, Gabbreid, Graneid, Leuco-aphaneid, Leucophyreid, Melano-aphaneid, Melanophyreid, Peridoteid, Phanereid, Poikileid, Pyriboleid, Pyroxeneid, Quartz dioreid, Quartz dolereid, Quartz gabbreid, Syeneid, Vitrophyreid]

JOHANNSEN, A., 1917. Suggestions for a Quantitative Mineralogical Classification of Igneous Rocks. *Journal of Geology. Chicago.* Vol.25, p.63–97. [Granogabbro, Syenodiorite, Syenogabbro]

JOHANNSEN, A., 1920a. A Quantitative Mineralogical Classification of Igneous Rocks – Revised. Part I. *Journal of Geology. Chicago.* Vol.28, p.38–60. [Bytownitite, Leuco-,

Mela-, Meso-, Orthogranite, Orthoshonkinite, Orthotarantulite, Tarantulite, Topazite, Tourmalite]

JOHANNSEN, A., 1920b. A Quantitative Mineralogical Classification of Igneous Rocks – Revised. Part II. *Journal of Geology. Chicago.* Vol.28, p.158–177. [Adam-gabbro, Adam-diorite, Adam-tonalite, Andelatite, Beloeilite, Monzodiorite, Moyite, Nepheline diorite, Nepheline monzodiorite, Orthofoyaite, Orthorhyolite, Orthosyenite, Toryhillite]

JOHANNSEN, A., 1920c. A Quantitative Mineralogical Classification of Igneous Rocks – Revised. Part III. *Journal of Geology. Chicago.* Vol.28, p.210–232. [Basalatite, Chromitite, Magnetitite, Monzogabbro, Monzonorite, Nepheline monzogabbro, Rhyobasalt, Ricolettaite]

JOHANNSEN, A., 1926. A Revised Field Classification of Igneous Rocks. *Journal of Geology. Chicago.* Vol.34, p.181–182. [Amphibolide, Aphanide, Dioride, Felside, Gabbride, Granide, Leucophyride, Melanophyride, Peridotide, Perknide, Phaneride, Pyroxenide, Syenide, Trappide, Vitrophyride]

JOHANNSEN, A., 1931. A Descriptive Petrography of the Igneous Rocks. *Chicago University Press.* Vol.1, 267pp. [Biquahororthandite, Diaschistite, Hobiquandorthite, Quartz diorite, Quartz gabbride, Quartz leucophyride, Topatourbiolilepiquorthite]

JOHANNSEN, A., 1932. A Descriptive Petrography of the Igneous Rocks. *Chicago University Press.* Vol.2, 428pp. [Haplite, Hraftinna, Pearlstone, Quartz gabbro, Quartz norite]

JOHANNSEN, A., 1937. A Descriptive Petrography of the Igneous Rocks. *Chicago University Press.* Vol.3, 360pp. [Calciclasite, Clinopyroxene norite, Elkhornite, Felsoandesite, Labradoritite, Masafuerite, Melanephelinite, Orthopyroxene gabbro]

JOHANNSEN, A., 1938. A Descriptive Petrography of the Igneous Rocks. *Chicago University Press.* Vol.4, 523pp. [Analcimolith, Baldite, Barshawite, Bogusite, Caltonite, Columbretite, Cuyamite, Deldoradite, Feldspathoidite, Fiasconite, Glenmuirite, Gooderite, Haüynitite, Haüynolith, Highwoodite, Hilairite, Holmite, Leucitite basanite, Leuci-

tite tephrite, Leucitolith, Linosaite, Marien-
bergite, Martinite, Melilitholith, Nepheline
andesite, Nephelinite basanite, Nephelinite
tephrite, Nordsjöite, Noseanolith, Noselitite,
Parchettite, Penikkavaarite, Pikeite, Rafaelite,
Sodalite basalt, Sumacoite, Tannbuschite,
Tasmanite, Tutvetite, Ventrallite, Vetrallite,
Westerwaldite]

JOHANNSEN, A., 1939. A Descriptive Petrography of the Igneous Rocks. *Chicago University Press*. Vol.1, 2nd. Edition, 318pp.

JOPLIN, G.A., 1964. A Petrography of Australian Igneous Rocks. *Angus and Robertson,
Sydney*. 253pp. [Shoshonite]

JUAN, V.C., TAI, H. & CHANG, F.H., 1953.
Taiwanite, a new basaltic glassy rock of East
Coastal Range, Taiwan, and its bearing on
parental magma-type. *Acta Geologica Tai-
wanica. Taipei*. Vol.5, p.1–25. [Taiwanite]

JUDD, J.F., 1886a. On the Gabbros, Dolerites,
and Basalts, of Tertiary Age, in Scotland and
Ireland. *Quarterly Journal of the Geological
Society of London*. Vol.42, p.49–97. [Inter-
mediate]

JUDD, J.W., 1876. On the Ancient Volcano of
the District of Schemnitz, Hungary. *Quarterly Journal of the Geological Society of
London*. Vol.32, p.292–325. [Matraite]

JUDD, J.W., 1881. Volcanoes. *Kegan Paul &
Trench, London*. 2nd. Edition, 381pp. [Ultra-
basic]

JUDD, J.W., 1885. On the Tertiary and Older
Peridotites of Scotland. *Quarterly Journal of
the Geological Society of London*. Vol.41,
p.354–422. [Scyelite]

JUDD, J.W., 1886b. On Marekanite and its
allies. *Geological Magazine. London*. Vol.3,
Decade 3, p.241–248. [Marekanite]

JUDD, J.W., 1897. On the Petrology of Rockall.
*Transactions of the Royal Irish Academy.
Dublin*. Vol.31, p.48–58. [Rockallite]

JUNG, D., 1958. Untersuchungen am Tholeyit
von Tholey (Saar). *Beiträge zur Mineralogie
und Petrographie. Berlin, etc*. Vol.6,
p.147–181. [Tholeiite]

JUNG, J. & BROUSSE, R., 1959. Classification
Modale des Roches Eruptives. *Masson &
Cie., Paris*. 122pp. [Leucitonephelinite, Tra-
chylabradorite]

JURINE, L., 1806. Lettre de M. le Professeur

Jurine, de Genève, à M. Gillet-Laumont,
Membre du Conseil des Mines, Correspon-
dant de l'Institut. *Journal des Mines, Paris*.
Vol.19, p.367–378. [Arkesine, Dolerine,
Notite, Protogine, Spurine]

KAISER-GIESSEN, E., 1913. Neue nephil-
ingesteine aus Deutsch-Südwestafrika. *Verhandlungen der Gesellschaft Deutscher
Naturforscher und Ärtze*. Vol.86, Part 2,
p.595–598. [Klinghardtite]

KALB, G., 1936. Beiträge zur Kenntnis der
Auswürflinge des Laacher Seegebietes. II.
Zwei Arten von Umbildungen kristalliner
Schiefer zu sanidiniten. *Zeitschrift für Kris-
tallographie, Mineralogie und Petrographie.
Leipzig. Abt. B*. Vol.47, p.185–210. [Laachite]

KALB, G. & BENDIG, M., 1938. Beiträge zur
Kenntnis der Auswürflinge des Laacher
Seegebietes. *Decheniana. Verhandlungen des
Naturhistorischen Vereins der Rheinlande
und Westfalens. Bonn*. Vol.98, p.1–12. [Gle-
esites]

KARPINSKII, A., 1869. Avgitovye porody der-
evni Muldakaevoi i gory Kachkanar [translit-
erated from Russian]. *Gornyi zhurnal, St.
Petersburg*. Vol.2, No.5, p.225–255.
[Muldakaite]

KARPINSKII, A., 1903. O zamechatelnoi tak
nazyvaemoi groruditovoi gornoi porode iz
Zabaikalskoi oblasti [transliterated from
Russian]. *Izvestiya Imperatorskoi Akademii
Nauk. S. Peterburg*. Vol.19, No.2, p.1–32.
[Karite]

KATO, T., 1920. A contribution to the knowledge of the cassisterite veins of pneumato-
hydatogenetic or hydrothermal origin. A study
of the copper–tin veins of the Akénobé district in the Province of Tajima, Japan. *Journal of the College of Science, Imperial University of Tokyo*. Vol.43, Art.5, p.17–18.
[Akenobeite]

KEILLER, A., 1936. Two axes of Presely stone
from Ireland. *Antiquity. A Quarterly Review
of Archaeology. Gloucester*. Vol.10,
p.220–221. [Preselite]

KEMP, J.F., 1890. The basic dikes occurring
outside of the syenite areas of Arkansas.
*Annual Report of the Geological Survey of
Arkansas. Little Rock*. Vol.2, p.392–406.
[Ouachitite]

KEYES, C.R., 1895. The origin and relations of Central Maryland granites. *Report of the United States Geological Survey. Washington.* p.651–755. [Binary granite]

KHAZOV, R.A., 1983. Ladogality – novye apatitonosnye shchelochnye ultraosnovnye porody [transliterated from Russian]. *Doklady Akademii Nauk SSSR. Leningrad.* Vol.268, p.1199–1203. [Ladogalite, Ladogite, Nevoite]

KINAHAN, G.H., 1875. On the nomenclature of rocks. *Geological Magazine. London.* Vol.2, Decade 2, p.425–426. [Bottleite]

KIRWAN, R., 1784. Elements of Mineralogy. *Elmsly, London.* 412pp. [Granitell, Granitone]

KIRWAN, R., 1794. Elements of Mineralogy. *Elmsly, London.* Vol.1, 2nd. Edition, 510pp. [Ferrilite, Granatine, Granilite, Igneous rock, Plutonic]

KLIPSTEIN, A. VON, 1843. Beiträge zur geologischen Kenntnis der östlichen Alpen. *Heyer, Giessen.* Vol.1, 311pp. [Mulatoporphyry]

KLOOS, J.H., 1885. Ein Uralitgestein von Ebersteinburg im nördlichen Schwarzwald. *Neues Jahrbuch für Mineralogie, Geologie und Paläontologie. Stuttgart. Jahrgang.* Bd. 2, p.82–88. [Uralitite]

KNIGHT, C.W., 1905. Analcite-trachyte tuffs and breccias from southeast Alberta, Canada. *The Canadian Record of Science. Montreal.* Vol.9, p.265–278. [Blairmorite]

KNIGHT, W.C., 1898. Bentonite. *Engineering and Mining Journal.* Vol.66, p.491. [Bentonite]

KOBELL, F. VON, 1838. Grundzüge der Mineralogie. *Schrag, Nürnberg.* 348pp. [Wehrlite]

KOCH, A., 1858. Paläozoische Schichten und Grünsteine in den Herzogisch Nassauischen Aemtern Dillenburg und Herborn, unter Berücksichtigung allgemeiner Lagerungsverhältnisse in angränzenden Ländertheilen. *Jahrbuch des Vereins für Naturkunde im Herzogthum Nassau. Wiesbaden.* Vol.13, p.85–326. [Lahnporphyry]

KOLDERUP, C.F., 1896. Die Labradorfelse des westlichen Norwegens. I. Das Labradorfelsgebiet bei Ekersund und Soggendal. *Bergens Museums Aarbog (Årbok) Afhandlinger og Årsberetning.* No.5, p.1–222. [Ilmenitite, Soggendalite]

KOLDERUP, C.F., 1898. Lofotens og Vesteraalens gabbrobergarter. *Bergens Museums Aarbog (Årbok) Afhandlinger og Årsberetning.* No.7, p.1–54. [Oligoclasite]

KOLDERUP, C.F., 1903. Die Labradorfelse des westlischen Norwegens. II. Die Labradorfelse und die mit denselben verwandten Gesteine in dem Bergensgebiete. *Bergens Museums Aarbog (Årbok) Afhandlinger og Årsberetning.* Vol.12, p.1–129. [Birkremite, Farsundite, Mangerite]

KOLENEC, F., 1904. Über einige leukokrate Gang-Gesteine vom Monzoni und Predazzo. *Mitteilungen des Naturwissenschaftlichen Vereins für Steiermark. Graz.* Vol.40, p.161–212. [Feldspathite]

KOTÔ, B., 1909. Journeys through Korea (First contribution). *Journal of the College of Science, Imperial University of Tokyo.* Vol.26, Art.2, p.1–207. [Eutectofelsite, Granomasanite, Masanite, Masanophyre]

KOTO, B., 1916a. The great eruption of Sakurajima in 1914. *Journal of the College of Science, Imperial University of Tokyo.* Vol.38, Art.3, p.1–237. [Keramikite]

KOTO, B., 1916b. On the volcanoes of Japan. V. *Journal of the Geological Society of Toyko.* Vol.23, p.95–127. [Orthoandesite, Orthobasalt, Sanukitoid]

KOVALENKO, V.I., KUZMIN, M.I., ANTIPIN, V.S. & PETROV, L.L., 1971. Topassoderzhashchii kvartsevyi keratofyr (ongonit) – novaya raznovidnost subvulkanicheskikh zhilnykh magmaticheskikh porod [transliterated from Russian]. *Doklady Akademii Nauk SSSR. Leningrad.* Vol.199, No.2, p.430–433. [Ongonite]

KRANCK, E.H., 1939. The rock-ground of the coast of Labrador and the connection between the Pre-cambrian of Greenland and North America. *Bulletin de la Commission Géologique de la Finlande. Helsingfors.* Vol.22, No.125, p.65–86. [Aillikite]

KRENNER, M.J., 1910. Über tephrite in Ungarn. *Compte Rendu de la XI:e Session du Congrès Géologique International (Stockholm).* Vol.1, p.740. [Danubite]

KRETSCHMER, F., 1917. Der metamorphe Dioritgabbrogang nebst seinen Peridotiten und Pyroxeniten im Spieglitzer Schnee- und Bielengebirge. *Jahrbuch der Kaiserlich-Königlichen Geologischen Reichsanstalt. Wien.* Vol.67, p.1–201. [Aegirinolith, Bielenite, Marchite, Minettefels, Niklesite, Titanolite, Weigelite]

KROUSTCHOFF, K. DE, 1885. Note préliminaire sur la Wolhynite de M. d Ossowski. *Bulletin de la Société Française de Minéralogie. Paris.* Vol.8, p.441–451. [Volhynite]

KTÉNAS, C.A., 1928. Sur la présence de laves alcalines dans la mer Egée. *Compte Rendu Hebdomadaire des Séances de l'Académie des Sciences. Paris.* Vol.186, p.1631–1633. [Eustratite]

KUNO, H., 1960. High-alumina Basalt. *Journal of Petrology. Oxford.* Vol.1, p.121–145. [High-alumina basalt, Transitional basalt]

KUNZ, G.F., 1908. Precious Stones. *Mineral Industry, its Statistics, Technology and Trade (in the United States and other Countries). New York.* Vol.16, p.790–812. [Jadeolite]

KUPLETSKII, B.M., 1932. Kukisvumchorr i prilegayushchie k nemu massivy tsentralnoi chasti Khibinskikh tundr [transliterated from Russian]. *Trudy Soveta po Izucheniyu Proizvoditel'nykh Sil, Seriya Kolskaya. Materialy po Petrografii i Geokhimii Kolskogo poluostrova.* Vol.2, p.5–72. [Rischorrite]

LACROIX, A., 1893. Les Enclaves des Roches Volcaniques. *Macon, Paris.* 710pp. [Ittnerite rock, Melilitite]

LACROIX, A., 1894. Etude minéralogique de la lherzolite des Pyrénées et de ses phénomènes de contact. *Nouvelles Archives du Muséum d'Histoire Naturelle. Paris.* Vol.6, 3rd.Ser., p.209–308. [Amphibololite]

LACROIX, A., 1895. Sur les roches basiques constituant des filons minces dans la lherzolite des Pyrénées. *Compte Rendu Hebdomadaire des Séances de l'Académie des Sciences. Paris.* Vol.120, p.752–755. [Diopsidite, Pyroxenolite]

LACROIX, A., 1900. Sur un nouveau groupe d'enclaves homoeogènes des roches volcaniques, les microtinites des andésites et des téphrites. *Compte Rendu Hebdomadaire des Séances de l'Académie des Sciences. Paris.* Vol.130, p.348–351. [Microtinite]

LACROIX, A., 1901a. Sur un nouveau groupe de roches très basiques. *Compte Rendu Hebdomadaire des Séances de l'Académie des Sciences. Paris.* Vol.132, p.358–360. [Ariegite]

LACROIX, A., 1901b. Les roches basiques accompagnant les lherzolites et les ophites des Pyrénées. *8th. International Geological Congress, Paris.* Vol.8, p.826–829. [Avezacite]

LACROIX, A., 1902. Matériaux pour la Minéralogie de Madagascar – Les roches alcalines caractérisant la province pétrographique d'Ampafindrava. *Nouvelles Archives du Muséum d'Histoire Naturelle. Paris.* Vol.4, p.1–214. [Nepheline gabbro, Nepheline monzonite]

LACROIX, A., 1905. Sur un nouveau type pétrographique représentant la forme de profondeur de certaines leucotéphrites de la Somma. *Compte Rendu Hebdomadaire des Séances de l'Académie des Sciences. Paris.* Vol.141, p.1188–1193. [Sommaite]

LACROIX, A., 1907. Etude minéralogique des produits silicatés de l'éruption du Vésuve (Avril 1906). *Nouvelles Archives du Muséum d'Histoire Naturelle. Paris.* Vol.9, p.1–172. [Pollenite]

LACROIX, A., 1909. Sur l'existence de roches grenues intrusives pliocènes dans le massif volcanique du Cantal. *Compte Rendu Hebdomadaire des Séances de l'Académie des Sciences. Paris.* Vol.149, p.540–546. [Essexitegabbro]

LACROIX, A., 1910. Sur l'existence à la Côte d'Ivoire d'une série pétrographique comparable à celle de la charnockite. *Compte Rendu Hebdomadaire des Séances de l'Académie des Sciences. Paris.* Vol.150, p.18–22. [Ivoirite]

LACROIX, A., 1911a. Le cortège filonien des péridotites de Nouvelle-Calédonie. *Compte Rendu Hebdomadaire des Séances de l'Académie des Sciences. Paris.* Vol.152, p.816–822. [Ouenite]

LACROIX, A., 1911b. Les syénites néphéliniques de l'Archipel de Los et leurs minéraux. *Nouvelles Archives du Muséum d'Histoire Naturelle. Paris.* Vol.3, p.1–131. [Topsailite]

LACROIX, A., 1914. Les roches basiques non

volcaniques de Madagascar. *Compte Rendu Hebdomadaire des Séances de l'Académie des Sciences. Paris.* Vol.159, p.417–422. [Anabohitsite]

LACROIX, A., 1915. Sur un type nouveau de roche granitique alcaline renfermant une eucolite. *Compte Rendu Hebdomadaire des Séances de l'Académie des Sciences. Paris.* Vol.161, p.253–258. [Fasibitikite]

LACROIX, A., 1916a. Sur quelques roches volcaniques mélanocrates des Possessions françaises de l'Océan Indien et du Pacifique. *Compte Rendu Hebdomadaire des Séances de l'Académie des Sciences. Paris.* Vol.163, p.177–183. [Ankaramite]

LACROIX, A., 1916b. La constitution des roches volcaniques de l'Estrême Nord de Madagascar et de Nosy bé; les ankaratrites de Madagascar en général. *Compte Rendu Hebdomadaire des Séances de l'Académie des Sciences. Paris.* Vol.163, p.253–258. [Ankaratrite, Fasinite, Olivine nephelinite]

LACROIX, A., 1916c. Les syénites à riébéckite d'Alter Pedroso (Portugal), leurs formes mésocrates (lusitanites) et leur transformation en leptynites et en gneiss. *Compte Rendu Hebdomadaire des Séances de l'Académie des Sciences. Paris.* Vol.163, p.279–283. [Lusitanite]

LACROIX, A., 1917a. Les laves à haüyne d'Auvergne et leurs enclaves homeogènes: importance théorique de ces dernières. *Compte Rendu Hebdomadaire des Séances de l'Académie des Sciences. Paris.* Vol.164, p.581–588. [Mareugite, Ordanchite, Tahitite]

LACROIX, A., 1917b. Les roches grenues d'un magma leucitique étudiées à l'aide des blocs holocristallins de la Somma. *Compte Rendu Hebdomadaire des Séances de l'Académie des Sciences. Paris.* Vol.165, p.207–211. [Campanite, Ottajanite, Puglianite, Sebastianite]

LACROIX, A., 1917c. Les péridotites des Pyrénées et les autres roches intrusives non feldspathiques qui les accompagnent. *Compte Rendu Hebdomadaire des Séances de l'Académie des Sciences. Paris.* Vol.165, p.381–387. [Lherzite]

LACROIX, A., 1917d. Les laves leucitiques de la Somma. *Compte Rendu Hebdomadaire des Séances de l'Académie des Sciences. Paris.* Vol.165, p.481–487. [Vesuvite]

LACROIX, A., 1917e. Les formes grenues du magma leucitique du volcan Laziale. *Compte Rendu Hebdomadaire des Séances de l'Académie des Sciences. Paris.* Vol.165, p.1029–1035. [Braccianite]

LACROIX, A., 1918. Sur quelques roches filoniennes sodiques de l'Archipel de Los, Guinée française. *Compte Rendu Hebdomadaire des Séances de l'Académie des Sciences. Paris.* Vol.166, p.539–545. [Kassaite, Tamaraite]

LACROIX, A., 1919. Dacites et dacitoïdes à propos des laves de la Martinique. *Compte Rendu Hebdomadaire des Séances de l'Académie des Sciences. Paris.* Vol.168, p.297–302. [Dacitoid]

LACROIX, A., 1920. La systématique des roches grenues à plagioclases et feldspathoïdes. *Compte Rendu Hebdomadaire des Séances de l'Académie des Sciences. Paris.* Vol.170, p.20–25. [Berondrite, Luscladite, Mafraite]

LACROIX, A., 1922. Minéralogie de Madagascar. *Challamel, Paris.* Vol.2, 694pp. [Algarvite, Ampasimenite, Antsohite, Dissogenite, Finandranite, Itsindrite, Lindinosite]

LACROIX, A., 1923. Minéralogie de Madagascar. *Challamel, Paris.* Vol.3, 450pp. [Dellenitoid, Doreite, Etindite, Kivite, Oceanite, Rhyolitoid, Sakalavite, Sancyite]

LACROIX, A., 1924. Les laves analcimiques de l'Afrique du Nord et, d'une façon générale, la classification des laves renfermant de l'analcime. *Compte Rendu Hebdomadaire des Séances de l'Académie des Sciences. Paris.* Vol.178, p.529–535. [Scanoite]

LACROIX, A., 1925. Sur un nouveau type de roche éruptive alcaline récente. *Compte Rendu Hebdomadaire des Séances de l'Académie des Sciences. Paris.* Vol.180, p.481–484. [Ordosite]

LACROIX, A., 1926. La systématique des roches leucitiques; les types de la famille syénitique. *Compte Rendu Hebdomadaire des Séances de l'Académie des Sciences. Paris.* Vol.182, p.597–601. [Gaussbergite, Kajanite]

LACROIX, A., 1927a. Les rhyolites et les trachytes hyperalcalins quartzifères, à propos de ceux de la Corée. *Compte Rendu Hebdo-madaire des Séances de l'Académie des Sciences. Paris.* Vol.185, p.1410–1415. [Hakutoite]

LACROIX, A., 1927b. La constitution lithologique des îles volcaniques de la Polynésie Australe. *Mémoires de l'Académie des Sciences(de l'Institut de France). Paris.* Vol.59, p.107. [Murite]

LACROIX, A., 1928a. La composition minéralogique et chimique des roches éruptives et particulièrement des laves mésozoiques et plus récentes de la Chine orientale. *Bulletin of the Geological Society of China. Peking.* Vol.7, p.13–59. [Mandchurite]

LACROIX, A., 1928b. Les pegmatitoïdes des roches volcaniques à faciès basaltiques. *Compte Rendu Hebdomadaire des Séances de l'Académie des Sciences. Paris.* Vol.187, p.321–326. [Pegmatitoid]

LACROIX, A., 1933. Contribution à la connaissance de la composition chimique et minéralogique des roches éruptives de l'Indochine. *Bulletin du Service Géologique de l'Indochine. Saïgon.* Vol.20, No 3, 208pp. [Argeinite, Cocite, Florinite, Jerseyite, Kemahlite, Melfite, Melilitolite, Mikenite, Nemite, Niligongite, Plagioclasolite]

LAGORIO, A., 1897. Itinéraire Géologique d'Alouchta à Sébastopol par Yalta, Bakhtchissaraï et Mangoup-Kalé. *7th. International Geological Congress, St. Petersburg. Excursions Guide.* Vol.33, p.1–28. [Taurite]

LAITAKARI, A., 1918. Einige Albitepidotgesteine von Südfinnland. *Bulletin de la Commission Géologique de la Finlande. Helsingfors.* Vol.9, No.51, p.1–13. [Helsinkite]

LANG, H.O., 1877. Grundriss der Gesteinskunde. *Haessel, Leipzig.* 289pp. [Hungarite, Plädorite]

LANG, H.O., 1891. Versuch einer Ordnung der Eruptivgesteine nach ihren chemischen Bestande. *Tschermaks Mineralogische und Petrographische Mitteilungen. Wien.* Vol.12, 2nd.Ser., p.199–252. [Aetna-basalt, Amiatite, Bolsenite, Christianite, Puys-andesite, Rhön basalt]

LANGIUS, C.N., 1708. Historia Lapidum Figuratorum Helvetiae ejusqui Viciniae. *Jacobi*

Tomasini, Venice. 165pp. [Grammite]

LAPPARENT, A. DE, 1885. Traité de Géologie. *Savy, Paris.* Vol.3, 2nd. Edition, p.1289–2015. [Granulophyre]

LAPPARENT, A. DE, 1906. Traité de Géologie. Géologie proprement dite. *Savy, Paris.* 5th. Edition, p.593–1288. [Ortholite]

LAPPARENT, A.A.C. DE, 1864. Mémoire sur la constitution géologique du Tyrol méridional. *Annales des Mines ou Recueil de Mémoires sur l'Exploration des Mines, et sur les Sciences qui s'y rapportent; rédigés par le Conseil Général des Mines. Paris.* Vol.6, p.245–314. [Monzonite]

LAPPARENT, M. DE, 1900. Traité de Géologie. *Savy, Paris.* 4th. Edition, 1912pp. [Granoliparite]

LAPWORTH, C., 1898. Long excursion to the Birmingham district. *Proceedings of the Geologists' Association, London.* Vol.15, p.417–428. [Anchorite]

LARSEN, E.S. & HUNTER, J.F., 1914. Melilite and other minerals from Gunnison County, Colorado. *Journal of the Washington Academy of Sciences.* Vol.4, p.473–479. [Uncompahgrite]

LARSEN, E.S. & PARDEE, J.T., 1929. The Stock of Alkaline Rocks near Libby, Montana. *Journal of Geology. Chicago.* Vol.37, p.97–112. [Glimmerite]

LASAULX, A. VON, 1875. Elemente der Petrographie. *Strauss, Bonn.* 486pp. [Olivine gabbro, Orthophonite, Troctolite]

LASPEYRES, H., 1869. Über das Zusammenvorkommen von Magneteisen und Titaneisen in Eruptivgesteinen und über die sogenannten petrographischen Gesetze. *Neues Jahrbuch für Mineralogie, Geologie und Paläontologie. Stuttgart. Jahrgang.* Bd. 40, p.513–531. [Palatinite]

LAWSON, A.C., 1893. The Geology of Carmelo Bay. *University of California Publications. Bulletin of the Department of Geology. Berkeley.* Vol.1, p.1–59. [Carmeloite]

LAWSON, A.C., 1896. On malignite, a family of basic plutonic orthoclase rocks rich in alkalis and lime intrusive in the Coutchiching schists of Poohbah Lake. *University of California Publications. Bulletin of the Department of Geology. Berkeley.* Vol.1, No.12, p.337–362.

[Malignite]
LAWSON, A.C., 1903. Plumasite, an oligoclase – corundum rock near Spanish Peak, California. *University of California Publications. Bulletin of the Department of Geology. Berkeley.* Vol.3, No.8, p.219–229. [Plumasite]
LE BAS, M.J., 1977. Carbonatite-nephelinite volcanism. *Wiley, London.* 347pp. [Ferrocarbonatite]
LE BAS, M.J., LE MAITRE, R.W., STRECKEISEN, A. & ZANETTIN, B., 1986. A Chemical Classification of Volcanic Rocks Based on the Total Alkali–Silica Diagram. *Journal of Petrology. Oxford.* Vol.27, p.745–750. [Basaltic trachyandesite]
LE MAITRE, R.W., 1984. A proposal by the IUGS Subcommission on the Systematics of Igneous Rocks for a chemical classification of volcanic rocks based on the total alkali silica (TAS) diagram. *Australian Journal of Earth Science, Melbourne.* Vol.31, p.243–255. [Picrobasalt, Potassic trachybasalt]
LE PUILLON DE BOBLAYE, E. & VIRLET, TH., 1833. Expédition Scientifique de Morée: Géologie et Minéralogie. *Levrault, Paris.* Vol.2, Part 2, p.1–375. [Prasophyre]
LEA, F.M. & DESCH, C.H., 1935. The Chemistry of Cement and Concrete. *Arnold, London.* 429pp. [Tetin]
LEBEDEV, A.P. & VAKHRUSHEV, V.A., 1953. Yavleniya kontaminatsii v zhilnykh giperbazitakh yuzhnoi Fergany [transliterated from Russian]. *Izvestiya Akademii Nauk SSSR, Seriya Geologicheskaya.* No.1, p.114–131. [Ferganite]
LEBEDINSKY, V.I. & CHU TSZYA-SYAN, 1958. Ob anortoklaze v shchelochnykh bazaltakh yuzhnoi okrainy mongolskogo plato (KNR) [transliterated from Russian]. *Zapiski Vsesoyuznogo Mineralogicheskogo Obshchestva. Moskva.* Vol.87, p.14–22. [Spumulite]
LEHMANN, E., 1924. Das Vulkangebiet am Nordende des Nyassa als magmatische Provinz. *Ergänzungsband 4, zur Zeitschrift für Vulkanologie. Reimer, Berlin.* 209pp. [Atlantite, Essexite-basalt]
LEHMANN, E., 1949. Das Keratophyr –Weilburgit–Problem. *Heidelberger Beiträge zur Mineralogie und Petrographie. Berlin.*

Vol.2, p.1–166. [Weilburgite]
LEONHARD, K.C. VON, 1821. Handbuch der Oryktognosie. *Mohr & Winter, Heidelberg.* 720pp. [Cantalite]
LEONHARD, K.C. VON, 1823a. Charakteristik der Felsarten. *Engelmann, Heidelberg.* Vol.1, p.1–230. [Aplite, Blatterstein, Felstone, Granitite, Greisen, Kugeldiorite, Tholerite]
LEONHARD, K.C. VON, 1823b. Charakteristik der Felsarten. *Engelmann, Heidelberg.* Vol.3, p.599–772. [Agglomerate, Trass]
LEONHARD, K.C. VON, 1832. Die Basalt-Gebilde in ihren Beziehungen zu normalen und abnormen Felsmassen. *Schweizerbart, Stuttgart.* Vol.1, 498pp. [Anamesite]
LEPSIUS, R., 1878. Das Westliche Süd-Tyrol. *Hertz, Berlin.* 372pp. [Nonesite]
LEWIS, H.C., 1888. The matrix of the diamond. *Geological Magazine. London.* Vol.5, Decade 3, p.129–131. [Kimberlite]
LINDGREN, W., 1886. Eruptive Rocks: In W.M. Davis, Relation of the coal of Montana to the older rocks: In R. Pumpelly, Report on the Mining Industries of the United States (exclusive of the precious metals), with special investigation into the iron resources of the Republic and into the Cretaceous coal of the Northwest. *10th Decennial Census of the United States: 1880.* Vol.15, p.727. [Analcime basalt]
LINDGREN, W., 1893. Art. XXX – The Auriferous Veins of Meadow Lake, California. *American Journal of Science. New Haven.* Vol.46, 3rd.Ser., p.201–206. [Granodiorite]
†LOEWINSON-LESSING, F. YU., 1896. K voprosu o khimicheskoi klassifikatsii izverzhennykh gornykh porod [transliterated from Russian]. *Trudy Imperatorskago Sankt-Petersburgskago Obshchestva Estestvoispytatelei. S.-Peterburg.* Vol.27, No.1, p.175–176. [Hyperacidite]
†LOEWINSON-LESSING, F. YU., 1898. Issledovaniya po teoreticheskoi petrografii v svyazi s izucheniem izverzhennykh porod Central-nogo Kavkaza [transliterated from Russian].

---

† The current transliteration of this name would be Levinson-Lessing. However, to maintain uniformity with previous publications the older transliteration of Loewinson-Lessing has been used.

*Trudy Imperatorskago Sankt-Peters-
burgskago Obshchestva Estestvoispytatelei.
S.-Peterburg.* Vol.26, No.5, p.1–104. [Hypobasite, Mesite]

LOEWINSON-LESSING, F. YU., 1900a. Kritische Beiträge zur Systematik der Eruptivgesteine II. *Tschermaks Mineralogische und Petrographische Mitteilungen. Wien.* Vol.19, 2nd.Ser., p.169–181. [Anorthophyre, Anorthosyenite]

†LOEWINSON-LESSING, F. YU., 1900b. Geologicheskii ocherk Yuzhno-Zaozerskoi dachi i Denezhkina kamnya na severnom Urale [transliterated from Russian]. *Trudy Imperatorskago Sankt-Petersburgskago Obshchestva Estestvoispytatelei. S.-Peterburg.* Vol.30, No.5, p.1–257. [Alkaliptoche]

LOEWINSON-LESSING, F. YU., 1901. Kritische Beiträge zur Systematik der Eruptivgesteine. IV. *Tschermaks Mineralogische und Petrographische Mitteilungen. Wien.* Vol.20, 2nd.Ser., p.110–128. [Anorthoclasite, Calciptoche, Feldspathidolite, Feldspatholite, Leucolite, Leucoptoche, Melanolite, Melanoptoche, Melilithite, Microclinite, Nephelinolith, Pikroptoche]

†LOEWINSON-LESSING, F. YU., 1905a. Petrograficheskaya ekskursiya po r. Tagilu [transliterated from Russian]. *Izvestiya Sankt-Peterburgskago Polytekhnicheskago Instituta Imperatora Petra Velikago. S. Peterburg.* Vol.3, p.1–40. [Camptovogesite]

†LOEWINSON-LESSING, F. YU., 1905b. Petrographische Untersuchungen im Centralen Kaukasus [transliterated from Russian]. *Zapiski Imperatorskogo (S. Petersburgskogo) Mineralogicheskogo Obshchestva.* Vol.42, p.237–280. [Dumalite]

LOEWINSON-LESSING, F. YU., 1928. Some queries on rock classification and nomenclature. *Doklady Akademii Nauk SSSR. Leningrad.* Ser.A., p.139–142. [Keratophyrite]

†LOEWINSON-LESSING, F. YU., 1934. Problema genezisa magmaticheskikh porod i puti k ee razresheniyu [transliterated from Russian]. *Izadatel' stvo Akademii Nauk SSSR, Moskva.* 58pp. [Anatectite, Prototectite, Syntectite]

†LOEWINSON-LESSING, F. YU., 1936. O nesilikatnykh magmakh [transliterated from Russian]. *Akademiku V.I.Vernadskomu k pyatidesyatiletiyu nauchnoi i pedagogicheskoi deyatel' nosti. Akademiya Nauk SSSR.* Vol.2, p.989–997. [Apatitolite]

†LOEWINSON-LESSING, F. YU., GINZBERG, A.S. & DILAKTORSKII, N.P., 1932. Trappy Tuluno-Udinskogo i Bratskogo raionov v Vostochnoi Sibiri [transliterated from Russian]. *Trudy Soveta po Izucheniyu Proizvoditel' nykh Sil. Seriya Sibirskaya.* Vol.1, p.1–82. [Angarite]

LOISELLE, M.C. & WONES, D.R., 1979. Characteristics and Origin of Anorogenic Granites. *Abstracts of papers to be presented at the Annual Meetings of the Geological Society of America and Associated Societies, San Diego, California, November 5-8, 1979.* Vol.11, p.468. [A-type granite]

LORENZO, G. DE, 1904. The History of Volcanic Action in the Phlegræan Fields. *Quarterly Journal of the Geological Society of London.* Vol.60, p.296–314. [Mappamonte, Piperno]

LOSSEN, K.A., 1886. Über die verschiedene Bedeutung, welche dem Worte Palatinit von den Petrographen beigelegt worden ist, und knüpfte daran einige Mittheilungen über seine Stellung zur Mealphyr-Frage. *Zeitschrift der Deutschen Geologischen Gesellschaft. Berlin.* Vol.38, p.921–926. [Hysterobase]

LOSSEN, K.A., 1892. Vergleichende Studien über die Gesteine des Spiemonts und des Bosenbergs bei St. Wedel und verwandte benachbarte Eruptivtypen aus der Zeit des Rothliegenden. *Jahrbuch der Königlich Preussischen Geologischen Landesanstalt und Bergakademie. Berlin.* Vol.10, for 1889, p.258–321. [Pegmatophyre]

LOUGHLIN, G.F., 1912. The gabbros and associated rocks at Preston, Connecticut. *Bulletin of the United States Geological Survey, Washington.* Vol.492, p.1–158. [Quartz anorthosite]

LUCHITSKII, I.V., 1960. Vulkanizm i tektonika devonskikh vpadin Minusinskogo mezhgornogo progiba [transliterated from Russian]. *Izadatel' stvo Akademii Nauk SSSR, Moskva.* 176pp. [Goryachite]

LYELL, C., 1835. Principles of Geology. *Murray, London.* Vol.1, 4th. Edition, 406pp. [Lapilli]

MACDONALD, G.A., 1949. Hawaiian Petrographic Province. *Bulletin of the Geological Society of America. New York.* Vol.60, p.1541–1596. [Mimosite]

MACHATSCHKI, F., 1927. Enstatit-Hornblendit von Grönland. *Centralblatt für Mineralogie, Geologie und Paläontologie. Stuttgart. Abt.A.* p.172–175. [Grönlandite]

MARCET RIBA, J., 1925. El método notural en Petrografía. Rocas eruptivas intrusivas de la serie calco-alcalina. *Memorias de la Real Academia de Ciencias y Artes de Barcelona.* Vol.19, No.10, p.245–419. [Alentegite, Andersonite, Eichstädtite, Finlandite, Ivreite, Kolderupite, Langite, Lossenite, Rotenburgite]

MARSHALL, P., 1906. The Geology of Dunedin (New Zealand). *Quarterly Journal of the Geological Society of London.* Vol.62, p.381–424. [Kaiwekite, Ulrichite]

MARSHALL, P., 1932. Notes on some volcanic rocks of the North Island of New Zealand. *New Zealand Journal of Science & Technology. Wellington.* Vol.13, p.198–200. [Ignimbrite]

MARSHALL, P., 1935. Acid Rocks of the Taupo-Rotorua Volcanic District. *Transactions of the Royal Society of New Zealand. Dunedin.* Vol.64, p.323–366. [Lapidite, Lenticulite, Pulverulite]

MARTIN, R., & DE SITTER-KOOMANS, C., 1955. Pseudotectites from Colombia and Peru. *Leidse Geologische Mededelingen. Leiden.* Vol.20, p.151–164. [Macusanite]

*MARZARI-PENCATI, G., 1819. *Osservatore Veneziano.* p.118–127. [Sievite]

MAWSON, D., 1906. The Minerals and Genesis of the Veins and Schlieren Traversing the Aegirine-Syenite in the Bowral Quarries. *Proceedings of the Linnean Society of New South Wales. Sydney.* Vol.31, p.579–607. [Bowralite]

MAZZUOLI, L. & ISSEL, A., 1881. Relazione degli studi fatti per un rilievo delle masse ofiolitiche nella riviera di Levante (Liguria). *Bollettino del Reale Comitato Geologico d'Italia. Firenze.* Vol.12, p.313–349. [Calcogranitone, Ophigranitone, Silicogranitone]

MCHENRY, A. & WATT, W.W., 1895. Guide to the Collection of Rocks and Fossils Belonging to the Geological Survey of Ireland. *Ireland Geological Survey. Museum of Science and Art, Dublin.* 155pp. [Ivernite]

MENNELL, F.P., 1929. Some Mesozoic and Tertiary Igneous Rocks from Portuguese East Africa. *Geological Magazine. London.* Vol.66, p.529–540. [Lupatite]

MICHEL-LEVY, A., 1874. Note sur une classe de roches éruptives intermédiaires entre les granites porphyroïdes et les porphyres granitoïdes groupe des granulites. *Bulletin de la Société Géologique de France. Paris.* Vol.2, p.177–189. [Granulite]

MICHEL-LEVY, A., 1894. Etude sur la Détermination des Feldspaths. *Baudry, Paris.* 107pp. [Trachyandesite]

MICHEL-LEVY, A., 1897. Mémoire sur le porphyre bleu de l'Esterel. *Bulletin des Services de la Carte Géologique de la France et des Topographies Souterraines. Paris.* Vol.9, No.57, p.1–40. [Esterellite]

MILCH, L., 1927. A. Johannsen: A revised Field Classification of Igneous Rocks. (Journ. of Geol. 34. 1926. 181-182.). *Neues Jahrbuch für Mineralogie, Geologie und Paläontologie. Stuttgart. Referate Abt. A.* Band II, p.62–63. [Melanide]

MILLER, W.J., 1919. Pegmatite, Silexite, and Aplite of Northern New York. *Journal of Geology. Chicago.* Vol.27, p.28–54. [Silexite]

MILLOSEVICH, F., 1908. Studi sulle rocce vulcaniche di Sardegna. I – Le rocce di Sassari e di Porto Torres. *Atti della Reale Accademia (Nazionale) dei Lincei. Rendiconti. Classe di Scienze Fisiche, Matematiche e Naturali. Roma.* Vol.6, p.403–438. [Trachydacite]

MILLOSEVICH, F., 1927. Le rocce a corindone della Val Sessera (Prealpi Biellesi). *Atti della Reale Accademia (Nazionale) dei Lincei. Rendiconti. Classe di Scienze Fisiche, Matematiche e Naturali. Roma.* Vol.5, p.22–31. [Sesseralite]

MONTEIRO, M, 1814. Du pyroméride globaire ou de la roche connue sous le nom de Porphyre globuleux de Corse. *Journal des Mines, Paris.* Vol.35, p.347–360. [Pyromeride]

MOOR, G.G. & SHEINMAN, YU. M., 1946. Porody iz severnoi okrainy Sibirskoi platformy [transliterated from Russian]. *Doklady Akademii Nauk SSSR. Leningrad.* Vol.51,

No.2, p.141–144. [Meimechite]

MOROZEWICZ, J., 1899. Experimentelle Untersuchungen über die Bildung der Minerale im Magma. *Tschermaks Mineralogische und Petrographische Mitteilungen. Wien.* Vol.18, 2nd.Ser., p.105–240. [Kyschtymite]

MOROZEWICZ, J., 1901. Gora Magnitnaya i eya blizhaishiya okrestnosti [transliterated from Russian]. *Trudy Geologicheskago Komiteta. S.-Peterburg.* Vol.18, No.1, p.1–104. [Atatschite]

MOROZEWICZ, J., 1902. Ueber Mariupolit, ein extremes Glied der Elaeolithsyenite. *Tschermaks Mineralogische und Petrographische Mitteilungen. Wien.* Vol.21, 2nd.Ser., p.238–246. [Mariupolite]

MOROZOV, A.I., 1938. Zametka o kamchatskoi novoi gornoi porode [transliterated from Russian]. *Byulleten Vulkanologicheskoi Stantsii na Kamchatke.* No.3, p.17–20. [Kamchatite]

MOSEBACH, R., 1956. Khewraite vom Khewra Gorge, Pakistan, ein neuer Typus kalireicher Effusivgesteine. *Neues Jahrbuch für Mineralogie. Stuttgart. Abhandlungen.* Vol.89, p.182–209. [Khewraite]

MUELLER, G., 1964. Lepaigite, a new porphyritic glass form Northern Chile. *International Geological Congress. Report of the Twenty-Second Session India. New Delhi.* Vol.16, p.374–385. [Lepaigite]

MÜGGE, O., 1893. Untersuchungen über die "Lenneporphyre" in Westfalen und den angrenzenden Gebieten. *Neues Jahrbuch für Mineralogie, Geologie und Paläontologie. Stuttgart. Beilagebände.* Vol.8, p.535-721. [Lenneporphyry]

MYIASHIRO, A. & SHIDO, F., 1975. Tholeiitic and calc-alkali series in relation to the behaviours of titanium, vanadium, chromium, and nickel. *American Journal of Science. New Haven.* Vol.275, p.165-277. [Abyssal tholeiite]

NARICI, E., 1932. Contributo alla petrografia chimica della provincia magmatica compana e del Monte Vulture. *Zeitschrift für Vulkanologie. Berlin.* Vol.14, p.210–239. [Guardiaite]

NAUMANN, C.F., 1850. Lehrbuch der Geognosie. *Engelmann, Leipzig.* Vol.1,

1000pp. [Hyperite, Hypersthenite, Leucilite, Nepheline basalt]

NAUMANN, C.F., 1854. Lehrbuch der Geognosie. *Engelmann, Leipzig.* Vol.2, 1222pp. [Porphyrite]

NEMOTO, T., 1934. Preliminary note on alkaline rhyolites from Tokati, Hokkaido. *Journal of the Faculty of Science, Hokkaido University. Sapporo.* Vol.2, Ser. 4, p.299–321. [Okawaite]

NIGGLI, P., 1923. Gesteins- und Mineralprovinzen. *Borntraeger, Berlin.* Vol.1, 602pp. [Coloradoite, Greenhalghite, Lamproite, Lamprosyenite, Nosykombite, Orvietite]

NIGGLI, P., 1931. Die quantitive mineralogische Klassifikation der Eruptivgesteine. *Schweizerische Mineralogische und Petrographische Mitteilungen. Zürich.* Vol.11, p.296–364. [Alkali diorite, Alkali gabbro, Pheno-]

NIGGLI, P., 1936. Die Magmatypen. *Schweizerische Mineralogische und Petrographische Mitteilungen. Zürich.* Vol.16, p.335–399. [Achnahaite]

NIKITIN, V. & KLEMEN, R., 1937. Diorit-pirokseniti iz okolice Cizlaka na Pohorju. *Geoloshki Anali Balkanskoga Poluostrva.* Vol.14, p.149–198. [Cizlakite]

NOCKOLDS, S.R., 1954. Average chemical composition of some igneous rocks. *Bulletin of the Geological Society of America. New York.* Vol.65, p.1007–1032. [Nepheline latite]

NORDENSKJÖLD, N., 1820. Bidrag till Närmare Kännedom af Finlands Mineralier och Geognosie. *Första Häftet, Stockholm.* 103pp. [Sordawalite]

NORDENSKJÖLD, O., 1893. Ueber archaeische Ergussgesteine aus Småland. *Bulletin of the Geological Institution of the University of Upsala.* Vol.1, p.133–255. [Eorhyolite]

*NOSE, C.W., 1808. Mineralogische Studien über die Gebirge am Niederrhein. Ed. J. Nöggerath. *Hermann, Frankfurt am Main.* [Sanidinite]

O'NEILL, J.J., 1914. St. Hilaire (Beloeil) and Rougemont Mountains, Quebec. *Memoirs. Geological Survey, Canada. Ottawa.* Vol.43, p.1–108. [Rougemontite, Rouvillite]

OEBBEKE, K., 1881. Beiträge zur Petrographie

der Philippinen und Palau-Inseln. *Neues Jahrbuch für Mineralogie, Geologie und Paläontologie. Stuttgart. Beilagebände.* Vol.1, p.451–501. [Orthoandesite]

OGURA, T., MATSUDA, K., NAKAGAWA, T., MATSUMOTO, M. & MURATA, K., 1936. Volcanoes of the Wu Ta Lien Chih district, Lung Chiang Province, Manchuria (In Japanese with English summary). *Survey Reports of Volcanoes in Manchuria. Ryojun.* No.1, p.1–96. [Shihlunite]

OSANN, A., 1889. Beiträge zur Kenntniss der Eruptivgesteine des Cabo de Gata (Prov. Almeria). *Zeitschrift der Deutschen Geologischen Gesellschaft. Berlin.* Vol.41, p.297–311. [Verite]

OSANN, A., 1892. Über dioritische Ganggesteine im Odenwald. *Mittheilungen der Grossherzoglich Badischen Geologischen Landesanstalt. Heidelberg.* Vol.2, p.380–388. [Malchite]

OSANN, A., 1893. Report on the rocks of Trans-Pecos Texas. *Report of the Geological Survey of Texas. Austin.* Vol.4, p.123–138. [Paisanite]

OSANN, A., 1896. Beiträge zur Geologie und Petrographie der Apache (Davis) Mts, Westtexas. *Tschermaks Mineralogische und Petrographische Mitteilungen. Wien.* Vol.15, 2nd.Ser., p.394–456. [Apachite]

OSANN, A., 1922. H. Rosenbusch: Elemente der Gesteinslehre. *Schweizerbart, Stuttgart.* 4th. Edition, 779pp. [Kasanskite, Pedrosite]

OSANN, A.H., 1902. Versuch einer chemischen Classification der Eruptivgesteine. III. Die Ganggesteine. *Tschermaks Mineralogische und Petrographische Mitteilungen. Wien.* Vol.21, 2nd.Ser., p.365–448. [Katzenbuckelite]

OSANN, A.H., 1906. Über einige Alkaligesteine aus Spanien. In: E.A. Wülfing (Ed.) Festschrift Harry Rosenbusch. *Schweizerbart, Stuttgart.* p.263–310. [Jumillite]

OSBORNE, F.F. & WILSON, N.L., 1934. Some Dike Rocks from Mount Johnson, Quebec. *Journal of Geology. Chicago.* Vol.42, p.180–187. [Monnoirite]

OSTADAL, C., 1935. Über ein calcitführendes Tiefengestein aus dem nordwestlichen Waldviertel. *Verhandlungen der Geologis-chen Reichsanstalt (Staatanstalt-Landesanstalt). Wien.* No.8/9, p.117–126. [Hörmannsite]

OYAWOYE, M.O., 1965. Bauchite: a new variety in the quartz monzonite series. *Nature. London.* Vol.205, p.689. [Bauchite]

PALMIERI, L. & SCACCHI, A., 1852. Della regione vulcanica del Monte Vulture. *Nobile, Naples.* Monograph, 160pp. [Augitophyre]

PAPASTAMATIOU, J., 1939. Sur quelques minéraux types de roches à corindon de l'Ile de Naxos (Archipel Grec). *Compte Rendu Hebdomadaire des Séances de l'Académie des Sciences. Paris.* Vol.208, p.2088–2090. [Naxite]

PARGA-PONDAL, I., 1935. Quimismo de las manifestaciones magmáticas cenozoicas de la Península Ibérica. *Trabajos Museo Nacional de Ciencias Naturales. Madrid. Serie Geologica.* Vol.39, p.1–174. [Cancarixite]

PASOTTI, P., 1954. Sobre una roca filoneana adamellítica del Cerro Tandileofú, Prov. de Buenos Aires, Argentina. *Publicaciones del Instituto de Fisiografia y Geologia. Universidad Nacional del Litoral. Rosario.* Vol.9, No.41, p.1–25. [Tandileofite]

PELIKAN, A., 1902. Dahamit, ein neues Ganggestein aus der Gefolgschaft des Alkaligranit (a review of another paper). *American Journal of Science. New Haven.* Vol.14, 4th.Ser., p.397. [Dahamite]

PELIKAN, A., 1906. Über zwei Gesteine mit primärem Analcim nebst Bemerkungen über die Entstehung der Zeolithe. *Tschermaks Mineralogische und Petrographische Mitteilungen. Wien.* Vol.25, 2nd.Ser., p.113–126. [Analcime phonolite]

PERRET, F.A., 1913. Art. 52 – Some Kilauean Ejectamenta. *American Journal of Science. New Haven.* Vol.35, 4th.Ser., p.611–618. [Pélé's tears]

PETERSEN, J., 1890a. Beiträge zur Petrographie von Sulphur Island, Peel Island, Hachijo und Mijakeshima. *Jahrbuch der Hamburgischen Wissenschaftlichen Anstalten. Hamburg.* Vol.8, p.1–58. [Boninite, Mijakite]

PETERSEN, J., 1890b. Der Boninit von Peel island. *Jahrbuch der Hamburgischen Wissenschaftlichen Anstalten. Hamburg.* Vol.8, p.341–349. [Boninite]

PETERSEN, TH., 1869. Über den Basalt und Hydrotachylyt von Rossdorf bei Darmstadt. *Neues Jahrbuch für Mineralogie, Geologie und Paläontologie. Stuttgart. Jahrgang.* Bd. 40, p.32–41. [Hydrotachylyte]

PETTERSEN, K., 1883. Sagvandit, eine neue enstatitführende Gebirgsart. *Neues Jahrbuch für Mineralogie, Geologie und Paläontologie. Stuttgart. Jahrgang.* Bd. 2, p.247. [Sagvandite]

PHEMISTER, J., 1926. The geology of Strath Oykell and Lower Loch Shin (South Sutherlandshire and North Ross-shire). *Memoirs. Geological Survey of Scotland. Edinburgh.* Sheet 102, p.1–220. [Perthosite]

PHILLIPS, J., 1848. The Malvern Hills compared with the Palaeozoic districts of Abberley, &c. *Memoirs of the Geological Survey of Great Britain, and of the Museum of Practical Geology in London.* Vol.2, Part 1, p.1–330. [Hornblendite]

PHILLIPS, W., 1815. An Outline of Mineralogy and Geology, intended for the use of those who may desire to become acquainted with the elements of those sciences; especially of young persons. *Phillips, London.* 193pp. [Tuff]

PICHLER, A., 1875. Beiträge zur Geognosie Tirols. *Neues Jahrbuch für Mineralogie, Geologie und Paläontologie. Stuttgart. Jahrgang.* Bd. 46, p.926–936. [Ehrwaldite, Töllite, Vintlite]

PINKERTON, J., 1811a. Petralogy: A treatise on Rocks. *White, Cochrane & Co., London.* Vol.1, 599pp. [Amygdalite, Basaltin, Basalton, Granitel, Granitin, Granitoid, Graniton, Intrite, Lehmanite, Patrinite, Porphyrin, Porphyron, Toadstone, Wernerite]

PINKERTON, J., 1811b. Petralogy: A treatise on Rocks. *White, Cochrane & Co., London.* Vol.2, 654pp. [Bergmanite, Blacolite, Bornite, Corsilite, Firmicite, Miagite, Niolite, Rhazite, Runite, Torricellite]

PIRSSON, L.V, 1915. The Microscopical Characters of Volcanic Tuffs — a Study for Students. *American Journal of Science. New Haven.* Vol.40, 4th.Ser., p.191–211. [Crystal tuff, Lithic tuff, Vitric tuff]

PIRSSON, L.V., 1895. Art. XIII – Complementary Rocks and Radial Dikes. *American Journal of Science. New Haven.* Vol.50, 3rd.Ser., p.116–121. [Oxyphyre]

PIRSSON, L.V., 1896. On the Monchiquites or Analcite group of Igneous Rocks. *Journal of Geology. Chicago.* Vol.4, p.679–690. [Analcitite]

PIRSSON, L.V., 1905. Petrology and Geology of the Igneous Rocks of the Highwood Mountains, Montana. *Bulletin of the United States Geological Survey, Washington.* Vol.237, p.1–208. [Cascadite, Fergusite]

PIRSSON, L.V., 1914. Art. XXVII – Geology of Bermuda Island; Petrology of the lavas. *American Journal of Science. New Haven.* Vol.38, 4th.Ser., p.331–344. [Bermudite]

PISANI, F., 1864. Sur quelques nouveaux minéraux du Cornouailles. *Compte Rendu Hebdomadaire des Séances de l'Académie des Sciences. Paris.* Vol.59, p.912–913. [Luxulianite]

PLINY, C., 77AD. Naturalis historiae. *[various printed versions and translations are available]*. [Basalt, Ophite, Syenite, Tephrite]

POLENOV, B.K., 1899. Massivnye gornye porody severnoi chasti Vitimskogo ploskogorya [transliterated from Russian]. *Trudy Imperatorskago Sankt-Petersburgskago Obshchestva Estestvoispytatelei. S.-Peterburg.* Vol.27, p.89–483. [Diabasite, Dioritite, Dioritophyrite, Gabbrite, Gabbrophyrite, Syenitite]

POLKANOV, A.A., 1940. Egirinity plutona Gremyakha-Vyrmes na Kolskom poluostrove [transliterated from Russian]. *Zapiski Vserossiiskogo Mineralogicheskogo Obshchestva. Moskva.* Vol.69, p.303–307. [Aegirinite]

POLKANOV, A.A. & ELISEEV, N.A., 1941. Petrologiya plutona Gremyakha-Vyrmes [transliterated from Russian]. *Leningradskii Gosudarstvennyi Universitet.* 244pp. [Hortonolitite]

PRATT, J.H. & LEWIS, J.V., 1905. Corundum and the Peridotites of Western North Carolina. *Report North Carolina Geological & Economic Survey. Chapel Hill, N.C.* Vol.1, p.1–464. [Enstatolite]

PRELLER, C.S. DU RICHE, 1924. Italian Mountain Geology. Parts I & II, Northern Italy and

Tuscany. *Wheldon & Wesley, London.* 3rd. Edition, 195pp. [Romanite]

PREOBRAZHENSKII, I.A., 1956. O nekotorykh nazvaniyakh v oblasti petrografii osadochnykh porod [transliterated from Russian]. *Geologicheskii Sbornik. L'vovskoe Geologicheskoe Obshchestvo pri L'vovskom Gosudarstvennom Universitete im. Iv. Franko. L'vov.* No.2-3, p.320–322. [Pyrolite]

PREYER, E. & ZIRKEL, F., 1862. Reise nach Island im Sommer 1860. *Brockhaus, Leipzig.* 497pp. [Krablite]

QUENSEL, P., 1913. The alkaline rocks of Almunge. *Bulletin of the Geological Institution of the University of Upsala.* Vol.12, p.9–200. [Canadite]

QUENSEL, P.D., 1912. Die Geologie der Juan Fernandezinseln. *Bulletin of the Geological Institution of the University of Upsala.* Vol.11, p.252–290. [Picrite basalt]

QUIRKE, T.T., 1936. New Nepheline Syenites from Bigwood Township, Ontario. *Transactions of the Illinois State Academy of Science. Springfield.* Vol.29, p.179–185. [Bigwoodite, Rutterite]

RACHKOVSKY, J., 1911. Über Alkaligesteine aus dem Südwesten des Gouvernements Yenissej. I. Der Teschenit und seine Beziehung zu den Ergussgesteinen. *Trudy Geologicheskago (i Mineralogicheskago) Muzeya imeni Petra Velikago Imperatorskoi Akademii Nauk. S.Peterburg.* Vol.5, p.217–283. [Ijussite]

RAMSAY, W., 1896. Urtit, ein basisches Endglied der Augitsyenit–Nephelinsyenit –Serie. *Geologiska Föreningens i Stockholm Förhandlingar. Stockholm.* Vol.18, p.459–468. [Urtite]

RAMSAY, W., 1921. En melilitförande djupbergart från Turja på sydsidan av Kolahalvön. *Geologiska Föreningens i Stockholm Förhandlingar. Stockholm.* Vol.43, p.488–489. [Turjaite]

RAMSAY, W. & BERGHELL, H., 1891. Das Gestein vom Iiwaara in Finnland. *Geologiska Föreningens i Stockholm Förhandlingar. Stockholm.* Vol.13, p.300–312. [Ijolite]

RAMSAY, W. & HACKMAN, V., 1894. Das Nephelinsyenitgebiet auf der Halbinsel Kola

I. *Fennia: Bulletin de la Société de Géographie de Finlande.* Vol.11, No.2, p.1–225. [Imandrite, Khibinite, Tawite, Umptekite]

RANSOME, F.L., 1898. Art. XLV – Some Lava Flows of the Western Slope of the Sierra Nevada, California. *American Journal of Science. New Haven.* Vol.5, 4th.Ser., p.355–375. [Latite, Quartz latite]

RATH, G. VOM, 1855. Chemische Untersuchung einiger Grünsteine aus Schlesien. *Annalen der Physik und Chemie. Leipzig.* Vol.95, p.533–561. [Forellenstein]

RATH, G. VOM, 1864. Beiträge zur Kenntnis der Eruptiven Gesteine der Alpen. *Zeitschrift der Deutschen Geologischen Gesellschaft. Berlin.* Vol.16, p.249–266. [Tonalite]

RATH, G. VOM, 1868. Geognostisch mineralogische Fragmente aus Italien. *Zeitschrift der Deutschen Geologischen Gesellschaft. Berlin.* Vol.20, p.265–364. [Leucite trachyte, Petrisco]

RAUMER, K. VON, 1819. Das Gebirge Niederschlesiens, der Grafschaft Glatz und eines Theiles von Böhmen und der Ober-Lausitz. *Reimer, Berlin.* 184pp. [Basaltite, Schillerfels]

REINISCH, R., 1912. Petrographisches Praktikum. *Borntraeger, Berlin.* Vol.2, 217pp. [Arsoite, Drakontite, Ponzaite]

REINISCH, R., 1917. Blatt Wiesenthal-Weipert. *Erläuterungen zur Geologischen Spezialkarte des Königreichs Sachsen. Leipzig.* No.147, p.1–84. [Haüynite]

REYNOLDS, D.L., 1937. Augite-biotite-diorite of the Newry complex. *Geological Magazine. London.* Vol.74, p.476-477. [Garronite]

REYNOLDS, D.L., 1958. Granite: Some Tectonic, Petrological, and Physico-Chemical Aspects. *Geological Magazine. London.* Vol.95, p.378–396. [Haplo-pitchstone, Minimite]

RICHTHOFEN, F. BARON, 1868. The Natural System of Volcanic Rocks. *Memoirs of the California Academy of Sciences. San Francisco.* Vol.1, Pt.2, p.1–94. [Nevadite, Propylite]

RICHTHOFEN, F. VON, 1860. Studien aus den u n g a r i s c h - s i e b e n b ü r g i s c h e n Trachytgebirgen. *Jahrbuch der Kaiserlich-Königlichen Geologischen Reichsanstalt.*

Wien. Vol.11, p.153–278. [Lithoidite, Rhyolite]

RINGWOOD, A.E., 1962. A model for the Upper Mantle. *Journal of Geophysical Research. Richmond, Va.* Vol.67, No.2, p.857–867. [Pyrolite]

RINMAN, A., 1754. Anmärkningar angående Järnhaltiga Jord-och Stenarter. *Handlingar Kongliga Svenska Vetenskaps-Akademiens, Stockholm.* Vol.15, p.282–297. [Trapp]

RINNE, F., 1904. Beitrag zur Gesteinskunde des Kiautschou-Schutz-Gebietes. *Zeitschrift der Deutschen Geologischen Gesellschaft. Berlin.* Vol.56, p.122–167. [Tsingtauite]

RINNE, F., 1921. Gesteinskunde. *Jänecke, Leipzig.* 6th. & 7th. Edition, 365pp. [Kiirunavaarite, Peracidite, Silicotelite]

RIO, N. DA, 1822. Traité sur la Structure extérieure du globe. *Giegler, Milan.* Vol.3, 557pp. [Masegna]

RITTER, E.A., 1908. The Evergreen Copper-Deposit, Colorado. *Transactions of the American Institute of Mining Engineers. New York.* Vol.38, p.751–765. [Evergreenite]

RITTMANN, A., 1960. Vulkane und ihre Tätigkeit. *Enke, Stuttgart.* 2nd. Edition, 336pp. [Etnaite]

RITTMANN, A., 1962. Volcanoes and their Activity. Translated from the Second German Edition by E.A. Vincent. *Wiley, New York.* 305pp. [Hyaloclastite]

RITTMANN, A., 1973. Stable Mineral Assemblages of Igneous Rocks. *Springer-Verlag Berlin, Heidelberg.* 262pp. [Latiandesite, Latibasalt, Phonoleucitite, Phononephelinite, Phonotephrite, Plagidacite, Tephrileucitite, Tephrinephelinite, Tephriphonolite]

RIVIÈRE, A., 1844. Mémoire minéralogique et géologique sur les roches dioritiques de la France occidentale. *Bulletin de la Société Géologique de France. Paris.* Vol.1, p.528–569. [Kersanton]

ROEVER, W.P. DE, 1940. Geological Investigations in the Southwestern Moetis Region (Netherlands Timor). In: Brouwer, H.A. (Ed.). Geological expedition of the University of Amsterdam to the Lesser Sunda Islands in the southeastern part of the Netherlands East Indies 1937 (Bouwer, H.A. Ed.). *N.V. Noord-Hollandsche Uitgevers Maatschappij, Amsterdam.* Vol.2, p.97–344. [Poeneite]

ROGERS, G., SAUNDERS, A.D., TERRELL, D.J., VERMA, S.P. & MARRINER, G.F., 1985. Geochemistry of the Holocene volcanic rocks associated with ridge subduction in Baja California, Mexico. *Nature. London.* Vol.315, p.389–392. [Bajaite]

ROSE, G., 1837. Mineralogisch-geognostische Reise nach dem Ural, dem Altai und dem Kaspischen Meere. Vol 1. Reise nach dem nördlichen Ural und dem Altai. *Sandersche Buchhandlung, Berlin.* 641pp. [Beresite]

ROSE, G., 1839. Ueber die mineralogische und geognostische Beschaffenheit des Ilmengebirges. *Annalen der Physik und Chemie. Leipzig.* Vol.47, p.373–384. [Miaskite]

ROSE, G., 1863. Beschreibung und Eintheilung der Meteoriten auf Grund der Sammlung im mineralogischen Museum zu Berlin. *Abhandlungen der Königlichen Akademie der Wissenschaften. Berlin.* p.23–161. [Eucrite]

ROSENBUSCH, H., 1872. Petrographische Studien an den Gesteinen des Kaiserstuhls. *Neues Jahrbuch für Mineralogie, Geologie und Paläontologie. Stuttgart. Jahrgang.* Bd. 43, p.35–65. [Limburgite]

ROSENBUSCH, H., 1877. Mikroskopische Physiographie der Mineralien und Gesteine. II. Massigen Gestein. *Schweizerbart, Stuttgart.* 596pp. [Leucite phonolite, Leucite tephrite, Nepheline syenite, Nepheline tephrite]

ROSENBUSCH, H., 1883. In review of C. Dölter: Die Vulkane der Capeverden und ihre Produkte. *Neues Jahrbuch für Mineralogie, Geologie und Paläontologie. Stuttgart. Jahrgang.* Bd. 1, p.396–405. [Augitite]

ROSENBUSCH, H., 1887. Mikroskopische Physiographie der Mineralien und Gesteine. II. Massigen Gestein. *Schweizerbart, Stuttgart.* 2nd. Edition, 877pp. [Alnöite, Camptonite, Cuselite, Harzburgite, Leucite basanite, Navite, Nepheline basanite, Olivine tholeiite, Theralite, Tholeiite, Tinguaite, Vogesite, Weiselbergite]

ROSENBUSCH, H., 1896. Mikroskopische Physiographie der Mineralien und Gesteine. II. Massigen Gestein. *Schweizerbart, Stuttgart.* 3rd. Edition, 1360pp. [Alkali granite, Allalinite, Lestiwarite, Olivine basalt, Schriesheimite, Spessartite]

ROSENBUSCH, H., 1898. Elemente der Gesteinslehre. *Schweizerbart, Stuttgart.* 546pp. [Gabbroporphyrite]

ROSENBUSCH, H., 1899. Über Euktolith, ein neues Glied der Theralithschen Effusivmagmen. *Sitzungsberichte der Preussischen Akademie der Wissenschaften zu Berlin.* No.7, p.110–115. [Euktolite]

ROSENBUSCH, H., 1907. Mikroskopische Physiographie der Mineralien und Gesteine. II. Massigen Gestein. *Schweizerbart, Stuttgart.* Pt.1, 4th. Edition, p.1–716. [Alkali syenite, Bekinkinite]

ROSENBUSCH, H., 1908. Mikroskopische Physiographie der Mineralien und Gesteine. II. Massigen Gestein. *Schweizerbart, Stuttgart.* Pt.2, 4th. Edition, p.717–1592. [Alkali trachyte, Orthotrachyte, Phonolitoid tephrite, Prowersite]

ROTH, J., 1861. Die Gesteins-Analysen in tabellarischer Übersicht und mit kritischen Erläuterungen. *Hertz, Berlin.* 68pp. [Liparite]

ROTH, J., 1887. Über den Zobtenit. *Sitzungsberichte der Preussischen Akademie der Wissenschaften zu Berlin.* No.32, p.611–630. [Zobtenite]

ROZIÈRE, F.M. DE, 1826. Description de l'Egypte. Histoire Naturelle, Minéralogie, Zoologie. *Paris.* Vol.21, 2nd. Edition, 482pp. [Chlorophyre, Iophyre, Sinaite, Syenitelle, Xenite, Xenophyre]

RUSSELL, H.D., HEIMSTRA, S.A. & GROENEVELD, D., 1955. The Mineralogy and Petrology of the Carbonatite at Loolekop, Eastern Transvaal. *Transactions and Proceedings of the Geological Society of South Africa. Johannesburg.* Vol.57, p.197–208. [Phoscorite]

SABATINI, V., 1899. Relazione sul lavoro eseguito nel triennio 1896–97–98 sui vulcani dell'Italia centrale e i loro prodotti. *Bollettino del Reale Comitato Geologico d'Italia. Firenze.* Vol.30, p.30–60. [Venanzite]

SABATINI, V., 1903. La pirossenite melilitica di Coppaeli (Cittaducale). *Bollettino del Reale Comitato Geologico d'Italia. Firenze.* Vol.34, p.376–378. [Coppaelite]

SÆTHER, E., 1957. The Alkaline rock province

of the Fen area in southern Norway. *Det Kongelige Norske Videnskabers Selskabs Skrifter, Trondheim.* No.1, p.1–148. [Damtjernite, Vipetoite]

SAGGERSON, E.P. & WILLIAMS, L.A.J., 1963. Ngurumanite, a New Hypabyssal Alkaline Rock from Kenya. *Nature. London.* Vol.199, p.479. [Ngurumanite]

SAHAMA, T.G., 1974. Potassium-Rich Alkaline Rocks. In: Sørensen, H. (Ed.). The Alkaline Rocks. *Wiley, New York.* 622pp. [Kamafugite]

SAKSELA, M., 1948. Das pyroklastische Gestein von Lappajärvi und seine Verbreitung als Geschiebe. *Bulletin de la Commission Géologique de la Finlande. Helsingfors.* Vol.25, No.144, p.19–30. [Kärnäite]

SANDBERGER, F., 1872. Vorläufige Bemerkungen über den Buchonit, ein Felsart aus der Gruppe der Nephelingesteine. *Sitzungsberichte der Mathematisch-Physikalischen Classe der Königlich Bayerischen Akademie der Wissenschaften zu München.* Vol.2, p.203–208. [Buchonite]

SAUER, A., 1892. Der Granitit von Durbach im nördlichen Schwarzwalde und seine Grenzfacies von Glimmersyenit (Durbachite). *Mittheilungen der Grossherzoglich Badischen Geologischen Landesanstalt. Heidelberg.* Vol.2, p.231–276. [Durbachite]

SAUER, A., 1919. Atlasblatt Bopfingen. *Begleitworte zu der geognostischen Spezialkarte von Württemberg. Stuttgart.* No.20, 2nd. Edition, p.1–31. [Suevite]

SCHAEFER, R.W., 1898. Der basische Gesteinszug von Ivrea im Gebiet des Mastallone-Thales. *Tschermaks Mineralogische und Petrographische Mitteilungen. Wien.* Vol.17, 2nd. Ser., p.495–517. [Valbellite]

SCHEERER, TH., 1862. Die Geneuse des Sächsischen Erzgebirges und verwandte Gesteine, nach ihrer chemischen Constitution und geologischen Bedeutung. *Zeitschrift der Deutschen Geologischen Gesellschaft. Berlin.* Vol.14, p.23–150. [Plutonite, Vulcanite]

SCHEERER, TH., 1864. Vorläufiger Bericht über Krystallinische Silikatgesteine des Fassathales und benachbarter Gegenden Südtyrols. *Neues Jahrbuch für Mineralogie, Geologie und Paläontologie. Stuttgart.*

Jahrgang. Bd. 35, p.385–411. [Plutovolcanite]

SCHEIBE, R., 1933. Compilacion de los estudios Geologicos officiales en Colombia 1917 a 1933. *Biblioteca de Departmento de Minas y Petroleo. Ministerio de Industria. Bogota.* Vol.1, p.1–470. [Corcovadite]

SCHEUMANN, K.H., 1913. Petrographische Untersuchungen an Gesteinen des Polzengebietes in Nord-Böhmen. *Abhandlungen der Königlich-Sächsischen Gesellschaft der Wissenschaften. Mathematisch-Physische Classe. Leipzig.* Vol.32, p.605–776. [Polzenite]

SCHEUMANN, K.H., 1922. Zur Genese alkalisch-lamprophyrischer Ganggesteine. *Centralblatt für Mineralogie, Geologie und Paläontologie. Stuttgart.* Abt.A. p.495–520, 521–545. [Luhite, Modlibovite, Vesecite, Wesselite]

SCHEUMANN, K.H., 1925. Ausländische Systematik, Klassifikation und Nomenklatur der Magmengesteine. *Fortschritte der Mineralogie, Kristallographie und Petrographie. Jena.* Vol.10, p.187–310. [Mienite]

SCHMID, E.E., 1880. Die quartzfreien Porphyre des Centralen Thüringer Waldgebirges und ihre Begleiter. *Fischer, Jena.* 106pp. [Paramelaphyre]

SCHMID, R., 1981. Descriptive nomenclature and classification of pyroclstic deposits and fragments: Recommendations of the IUGS Subcommission on the Systematics of Igneous Rocks. *Geology. The Geological Society of America. Boulder.* Vol.9, p.41–43. [Ash, Ash grain, Dust grain, Dust tuff, Pyroclastic deposit, Tuffite]

SCHOWALTER, E., 1904. Chemisch-geologische Studien im vulkanichen Ries bei Nördlingen. *Inaugural Dissertation. Jacob, Erlangen.* 65pp. [Wennebergite]

SCHÜLLER, A., 1949. Ein Plagioklas-Charnockit vom Typus Akoafim und seine Stellung innerhalb der Charnockit-Serie. *Heidelberger Beiträge zur Mineralogie und Petrographie. Berlin.* Vol.1, p.573–592. [Akoafimite]

*SCHULZ, C.F. & POETSCH, C.G., 1759. Neue Gesellschaftliche Erzählungen. *Leipzig.* Vol.2. [Pechstein]

SCHUSTER, M.E., 1907. Beiträge zur mikroskopischen Kenntnis der basischen Eruptivgesteine aus der bayerischen Rheinpfalz. *Geognostische Jahreshefte. München.* Vol.18, p.1–70. [Rhenopalite]

SEARS, J.H., 1891. Elaeolite-Zircon-Syenites and associated Granitic Rocks in the vicinity of Salem, Essex County, Massachusetts. *Bulletin of the Essex Institute. Salem, Mass.* Vol.23, p.145–155. [Essexite]

SEDERHOLM, J.J., 1891. Ueber die finnländischen Rapakiwigesteine. *Tschermaks Mineralogische und Petrographische Mitteilungen. Wien.* Vol.12, 2nd.Ser., p.1–31. [Anoterite]

SEDERHOLM, J.J., 1926. On migmatites and associated Pre-Cambrian rocks of South-western Finland. II. The region around the Barösundsfjärd W. of Helsingfors and neighbouring areas. *Bulletin de la Commission Géologique de la Finlande. Helsingfors.* Vol.12, No.77, p.1–139. [Epibasite]

SEDERHOLM, J.J., 1928. On orbicular granites. Spotted and nodular granites etc. and on the rapakivi texture. *Bulletin de la Commission Géologique de la Finlande. Helsingfors.* Vol.13, No.83, p.1–105. [Esboite, Orbiculite]

SEDERHOLM, J.J., 1934. On migmatites and associated Pre-cambrian rocks of South-western Finland. III. The Åland Islands. *Bulletin de la Commission Géologique de la Finlande. Helsingfors.* Vol.17, No.107, p.1–68. [Kleptolith]

SENFT, F., 1857. Classification und Beschreibung der Felsarten. *Korn, Breslau.* 442pp. [Alabradorite, Labradorite, Leucitite, Melaporphyre, Orthoclasite, Pyroxenite]

SHAND, S.J., 1910. On borolanite and its associates in Assynt. *Transactions of the Edinburgh Geological Society.* Vol.9, p.376–416. [Aploid, Assyntite, Cromaltite, Dioroid, Doleroid, Ledmorite, Pegmatoid, Syenoid]

SHAND, S.J., 1913. On Saturated and Unsaturated Igneous Rocks. *Geological Magazine. London.* Vol.10, Decade 5, p.508–514. [Oversaturated, Saturated, Undersaturated]

SHAND, S.J., 1916. A Recording Micrometer for Geometrical Rock Analysis. *Journal of Geology. Chicago.* Vol.24, p.394–404. [Colour index]

SHAND, S.J., 1917. A System of Petrography. *Geological Magazine. London.* Vol.4, Decade 6, p.463–469. [Granodolerite]

SHAND, S.J., 1927. Eruptive Rocks. *Murby, London.* 360pp. [Gabbroid, Metaluminous, Peraluminous]

SHAND, S.J., 1945. Coronas and coronites. *Bulletin of the Geological Society of America.New York.* Vol.56, p.247–266. [Coronite]

SHARASKIN, A.Y., DOBRETSOV, N.L. & SOBOLEV, N.V., 1980. Marianites: the clinoenstatite-bearing pillow lavas associated with the ophiolite assemblage of the Mariana Trench. In: A, Panayiotou. Ophiolites. Proceedings International Ophiolite Conference. Cyprus, 1979. *Ministry of Agriculture and Natural Resources. Geological Survey Department. Republic of Cyprus.* p.473–479. [Marianite]

SHILIN, L.L., 1940. Perovskit-shpinelevyi magnetitit iz Praskovye-Evgenevskoi kopi Shishimskikh gor na Yuzhnom Urale [transliterated from Russian]. *Doklady Akademii Nauk SSSR. Leningrad.* Vol.28, p.347–350. [Shishimskite]

SIGMUND, A., 1902. Die Eruptivgesteine bei Gleichenberg. *Tschermaks Mineralogische und Petrographische Mitteilungen. Wien.* Vol.21, 2nd.Ser., p.261–306. [Andesitoid]

SILVESTRI, O., 1888. Sopra alcune lave antiche e moderne del vulcano Kilauea nelle isole Sandwich. III – Le lave preistoriche stratificate le quali costituiscono le pareti all'intorno del grande bacino del Kilauea. *Bollettino del Reale Comitato Geologico d'Italia. Firenze.* Vol.19, p.168–196. [Kilaueite]

SJÖGREN, A., 1876. Om förekomsten af Tabergs jernmalmsfyndighet i Småland. *Geologiska Föreningens i Stockholm Förhandlingar. Stockholm.* Vol.3, p.42–62. [Olivinite]

SJÖGREN, H., 1893. En ny jernmalmstyp representerad af Routivare malmberg. *Geologiska Föreningens i Stockholm Förhandlingar. Stockholm.* Vol.15, p.55–63. [Magnetite Spinellite, Routivarite]

SKEATS, E.W., 1910. The Volcanic Rocks of Victoria (Presidential Address, Section C). *Report of the Twelfth Meeting of the Australian Association for the Advancement of Science, Brisbane 1909.* p.173–235.

SKEATS, E.W. & SUMMERS, H.S., 1912. The Geology and Petrology of the Macedon District. *Bulletin of the Geological Survey of Victoria. Melbourne.* Vol.24, p.1–58. [Anorthoclase basalt, Woodendite]

SOBOLEV, N.D., 1959. Neivit – novaya gornaya poroda iz gruppy zhilnykh [transliterated from Russian]. *Izvestiya Akademii Nauk SSSR, Seriya Geologicheskaya.* No.10, p.115–120. [Neivite]

SOBRAL, J.M., 1913. Contributions to the Geology of Nordingrå Region. *Almquist & Wiksell's boktrycherii, Uppsala.* 179pp. [Värnsingite]

SÖLLNER, J., 1913. Über Bergalith ein neues melilithreiches Ganggestein aus dem Kaiserstuhl. *Mittheilungen der Grossherzoglich Badischen Geologischen Landesanstalt.Heidelberg.* Vol.7, p.413–466. [Bergalite]

SØRENSEN, H., 1960. On the agpaitic rocks. *International Geological Congress. Report of the Twenty-First Session Norden.* Vol.13, p.319-327. [Agpaite]

SØRENSEN, H. (EDITOR), 1974. The Alkaline Rocks. *Wiley, London.* 622pp. [Fitzrodyite]

SOROTCHINSKY, S., 1934. Types pétrographiques nouveaux (kanzibite, kahusite, et quartz porphyre siliceux) provenant de l'édifice volcanique du Kahusi et du Biega (Kivu). *Bulletin de l'Académie Royale de Belgique. Classe des Sciences. Bruxelles.* Vol.20, p.183–195. [Kahusite, Kanzibite]

SOUZA-BRANDÃO, V., 1907. Les espichellites. Une nouvelle famille de roches de filons, au Cap Espichel. *Annals da Academia Polytechnica do Porto.* Vol.2, p.1–70. [Espichellite]

SPURR J.E. & WASHINGTON, H.S., 1917. In: Washington H.S., 1917, Chemical analyses of igneous rocks. *Professional Paper. United States Geological Survey. Washington.* Vol.99, p.1–120̇1. [Arizonite]

SPURR, J.E., 1900a. A reconnaissance in southwestern Alaska in 1898. *Report of the United States Geological Survey.Washington.* Pt.VII, p.31–264. [Alaskite, Aleutite, Belugite, Tordrillite]

SPURR, J.E., 1900b. Art. XXXI – Scapolite Rocks from Alaska. *American Journal of*

Science. New Haven. Vol.10, 4th.Ser., p.310–315. [Kuskite, Yentnite]

SPURR, J.E., 1906. The Southern Klondike District, Esmeralda County, Nevada – A Study in Metalliferous Quartz Veins of Magmatic Origin. *Economic Geology and Bulletin of the Society of Economic Geologists. Lancaster, Pa.* Vol.1, p.369–382. [Esmeraldite]

STACHE, G. & JOHN, C. VON, 1877. Geologische und petrographische Beiträge zur Kenntniss der älteren Eruptiv- und Massengesteine der Mittel- und Ostalpen. I. Die Gesteine der Zwölferspitzgruppe in Westtirol. *Jahrbuch der Kaiserlich-Königlichen Geologischen Reichsanstalt. Wien.* Vol.27, p.143–242. [Haplophyre]

STACHE, G. & JOHN, C. VON, 1879. Geologische und petrographische Beiträge zur Kenntniss der älteren Eruptiv- und Massengesteine der Mittel- und Ostalpen. II. Das Cevedale-Gebiet als Hauptdistrict älterer dioritischer Porphyrite (Palaeophyrite). *Jahrbuch der Kaiserlich-Königlichen Geologischen Reichsanstalt. Wien.* Vol.29, p.317–404. [Ortlerite, Palaeophyrite, Protopylite, Suldenite]

STANSFIELD, J., 1923a. Extensions of the Monteregian Petrographical Province to the West and North-West. *Geological Magazine. London.* Vol.60, p.433–453. [Okaite]

STANSFIELD, J., 1923b. Nomenclature and Relations of the Lamprophyres. *Geological Magazine. London.* Vol.60, p.550–554. [Bizardite]

STARZYNSKI, Z., 1912. Ein Beitrag zur Kenntnis der pazifischen Andesite und der dieselben bildenden Mineralien. *Bulletin International de l'Académie des Sciences et des Lettres de Cracovie. Cracovie.* Ser.A, p.657–681. [Beringite]

STEENSTRUP, K.J.V., 1881. Bemaerkninger til et geognostisk Oversigtskaart over en del af Julianehaabs distrikt. *Meddelelser om Grønland, Kjøbehavn.* Vol.2, p.27–41. [Sodalite syenite]

STEININGER, J., 1840. Geognostische Beschreibung des Landes zwischen der untern Saar und dem Rheine. *Lintz, Trier.* 149pp. [Tholeiite]

STEININGER, J., 1841. Geognostische

Beschreibung des Landes zwischen der untern Saar und dem Rheine. *Lintz, Trier.* Nachträge, p.48. [Spiemontite]

STELZNER, A., 1882. Vorläufige Mittheilungen über Melilithbasalte. *Neues Jahrbuch für Mineralogie, Geologie und Paläontologie. Stuttgart. Jahrgang.* Bd. 1, p.229–231. [Melilite basalt]

STELZNER, A.W., 1885. Beiträge zur Geologie und Palaeontologie der Argentinischen Republik. *Fischer, Cassel & Berlin.* Vol.1, 329pp. [Andendiorite, Andengranite]

STRECKEISEN, A., 1938. Das Nephelinsyenite-Massiv von Ditro in Rumänien als Beispiel einer kombinierten Differentiation und Assimilation. *Verhandlungen der Schweizerischen Naturforschenden Gesellschaft.* p.159–161. [Orotvite]

STRECKEISEN, A., 1952. Das Nephelinsyenit-Massiv von Ditro (Siebenbürgen). *Schweizerische Mineralogische und Petrographische Mitteilungen. Zürich.* Vol.32, p.251–308. [Ditro-essexite]

STRECKEISEN, A., 1954. Das Nephelinsyenit-Massiv von Ditro (Siebenbürgen) – Part II. *Schweizerische Mineralogische und Petrographische Mitteilungen. Zürich.* Vol.34, p.336–409. [Ditro-essexite]

STRECKEISEN, A., 1965. Die Klassifikation der Eruptivgesteine. *Geologische Rundschau. Internationale Zeitschrift für Geologie. Stuttgart.* Vol.55, p.478–491. [Foidite]

STRECKEISEN, A., 1967. Classification and Nomenclature of Igneous Rocks (Final Report of an Inquiry). *Neues Jahrbuch für Mineralogie. Stuttgart. Abhandlungen.* Vol.107, p.144–214. [Latite andesite, Latite basalt, Mafitite, Monzogranite, Syenogranite]

STRECKEISEN, A., 1973. Plutonic Rocks. Classification and nomenclature recommended by the IUGS Subcommission on the Systematics of Igneous Rocks. *Geotimes.* Vol.18, No.10, p.26–30. [Alkali feldspar granite, Alkali feldspar syenite, Alpha granite, Beta granite, Foid diorite, Foid dioritoid, Foid gabbro, Foid gabbroid, Foid monzodiorite, Foid monzogabbro, Foid monzosyenite, Foid plagisyenite, Foid syenite, Foid syenitoid, Foid-bearing alkali feldspar syenite, Foid-bearing diorite, Foid-bearing gabbro, Foid-bearing monzodiorite, Foid-bearing monzogabbro, Foid-bearing monzonite, Foid-bearing syenite, Foidolite, Gabbronorite, Olivine clinopyroxenite, Olivine gabbronorite, Olivine hornblende pyroxenite, Olivine hornblendite, Olivine orthopyroxenite, Olivine pyroxene hornblendite, Olivine pyroxenite, Olivine websterite, Plagioclase-bearing hornblende pyroxenite, Plagioclase-bearing hornblendite, Plagioclase-bearing pyroxene hornblendite, Plagioclase-bearing pyroxenite, Pyroxene hornblende gabbronorite, Pyroxene hornblende peridotite, Pyroxene hornblendite, Quartz alkali feldspar syenite, Quartz monzodiorite, Quartz monzogabbro, Quartz syenite, Quartz-rich granitoid, Quartzolite, Syenitoid]

STRECKEISEN, A., 1974. How should charnockitic rocks be named?. *Centenaire de la Société Géologique de Belgique Géologie des domaines cristallins, Liège.* p.349–360. [Alkali feldspar charnockite]

STRECKEISEN, A., 1976. To each plutonic rock its proper name. *Earth Science Reviews. International Magazine for Geo-Scientists. Amsterdam.* Vol.12, p.1–33. [Cancrinite diorite, Cancrinite gabbro, Cancrinite monzodiorite, Cancrinite monzogabbro, Cancrinite monzosyenite, Cancrinite plagisyenite, Nephelinolite, Ultramafite, Ultramafitolite]

STRECKEISEN, A., 1978. IUGS Subcommission on the Systematics of Igneous Rocks. Classification and Nomenclature of Volcanic Rocks, Lamprophyres, Carbonatites and Melilite Rocks. Recommendations and Suggestions. *Neues Jahrbuch für Mineralogie. Stuttgart. Abhandlungen.* Vol.143, p.1–14. [Alkali feldspar foidite, Alkali feldspar rhyolite, Alkali feldspar trachyte, Basanitic foidite, Calcite carbonatite, Foid-bearing alkali feldspar trachyte, Foid-bearing latite, Foid-bearing trachyte, Foiditoid, Olivine melilitite, Olivine melilitolite, Olivine pyroxene melilitolite, Olivine uncompahgrite, Phonobasanite, Phonofoidite, Phonolitic basanite, Phonolitic foidite, Phonolitic tephrite, Pyroxene melilitolite, Pyroxene olivine melilitolite, Quartz alkali feldspar trachyte, Tephritic foidite, Tephritic phonolite, Ultramafitite]

STRENG, A., 1864. Bemerkungen über den Serpentinfels und den Gabbro von Neurode in Schlesien. *Neues Jahrbuch für Mineralogie, Geologie und Paläontologie. Stuttgart. Jahrgang.* Bd. 35, p.257–278. [Enstatitite]

STRENG, A. & KLOOS, J.H., 1877. Ueber die Krystallinischen Gesteine von Minnesota in Nord-Amerika. *Neues Jahrbuch für Mineralogie, Geologie und Paläontologie. Stuttgart. Jahrgang.* p.113–138. [Hornblende gabbro]

SUN, S-S., NESBIT, R.W. & SHARASKIN, A.YA, 1979. Geochemical characteristics of mid-ocean ridge basalts. *Earth and Planetary Science Letters.* Vol.44, p.119-138. [Mid-ocean ridge basalt]

SUTHERLAND, D.S., 1965. Nomenclature of the potassic–feldspathic rocks associated with carbonatites. *Bulletin of the Geological Society of America. New York.* Vol.76, p.1409-1412. [Orthoclasite]

SZÁDECZKY, J., 1899. Ein neues Ganggestein aus Assuan. *Földtani Közlöny. A Magyar Földtani Tarsulat Folyoirata. Budapest.* Vol.29, p.210–216. [Josefite]

SZENTPÉTERY, S.v., 1935. Petrologische Verhältnisse des Fehérkö-Berges und die detaillierte Physiographie seiner Eruptivgesteine. *Acta Litterarum ac Scientiarum Regiae Universitatis Hungaricae Francisco-Josephinae. Sectio Chemica, Mineralogica et Physica. Szeged.* Vol.4, p.18–123. [Plagiophyrite]

TARASENKO, V.E., 1895. O khimicheskom sostave porod semeistva gabbro iz Zhitomirskogo uezda Volynskoi gubernii [transliterated from Russian]. *Zapiski Kievskago Obshchestva Estestvoispytatelei. Kiev.* Vol.14, p.15. [Gabbrosyenite]

TARGIONI-TOZZETTI, G., 1768. Relazioni d'Alcuni Viaggi Fatti in Diversi Parti della Toscana. *Florence.* Vol.2, 2nd. Edition, p.1–540. [Gabbro]

TAYLOR, H.P. (JR.), FRECHEN, J. & DEGENS, E.T., 1967. Oxygen and Carbon Isotope studies of Carbonatites from the Laacher See District, West Germany and the Alnö District, Sweden. *Geochimica et Cosmochimica Acta. London.* Vol.31, p.407–430. [Boderite, Rodderite]

TAYLOR, S.R., 1969. Trace element chemistry

of andesites and associated calc-alkaline rocks. In: A.R. McBirney (Ed), Proceedings of the Andesite Conference. *Department of Geology and Mineral Industries. State of Oregon. Portland.* Bulletin 65, p.43–63. [High-K, Low-K]

TEALL, J.J.H., 1887. On the Origin of Certain Banded Gneisses. *Geological Magazine. London.* Vol.4, Decade 3, p.484–493. [Pyroclastic]

TEALL, J.J.H., 1888. British Petrography: with special reference to the igneous rocks. *Dulau, London.* 469pp. [Orthofelsite]

TERMIER, H., TERMIER, G. & JOURAVSKY, G., 1948. Une roche volcanique á gros grain de la famille des Ijolites: la Talzastite. *Notes et Mémoires. Service des Mines et de la Carte Géologique du Maroc.* Vol.71, p.81–120. [Talzastite]

THEOPHRASTUS, C., 320BC. De Lapidus (On Stones). *[various printed versions and translations are available].* [Basanite, Obsidian, Pumice]

THOMAS, H.H., 1911. The Skomer Volcanic Series (Pembrokeshire). *Quarterly Journal of the Geological Society of London.* Vol.67, p.175–214. [Marloesite, Skomerite]

THOMAS, H.H. & BAILEY, E.B., 1915. Notes on the Rock-Species Leidleite & Inninmorite. In: Anderson, E.M. & Radley, E.G. The Pitchstones of Mull and their Genesis. *Quarterly Journal of the Geological Society of London.* Vol.71, p.205–217. [Inninmorite, Leidleite]

THOMAS, H.H. & BAILEY, E.B., 1924. Tertiary and post-Tertiary geology of Mull, Loch Aline, and Oban. *Memoirs. Geological Survey of Scotland. Edinburgh.* p.1–445. [Craignurite]

THORARINSSON, S., 1944. Tefrokronologiska studier på Island. *Akademisk Avhandling, Stockholms Högskola.* p.1–217. [Tephra]

TILLEY, C.E., 1936. Enderbite, a new member of the Charnockite Series. *Geological Magazine. London.* Vol.73, p.312–316. [Enderbite]

TILLEY, C.E. & MUIR, I.D., 1963. Intermediate members of the oceanic basalt-trachyte association. *Geologiska Föreningens i Stockholm Förhandlingar. Stockholm.* Vol.85, p.436–444. [Benmoreite, Tristanite]

TOBI, A.C., 1971. The nomenclature of the charnockitic rock suite. *Neues Jahrbuch für Mineralogie. Stuttgart. Monatshefte.* p.193–205. [m-Charnockite, m-Enderbite, etc.]

TOBI, A.C., 1972. The nomenclature of the Charnockitic Rock Suite: Reply to a discussion. *IUGS Subcommission on Nomenclature in Igneous Petrology,10th. Circular, Contribution No. 25, May 1972.* [Charnoenberbite]

TOMKEIEFF, S.I., WALTON, E.K., RANDALL, B.A.O., BATTEY, M.H. & TOMKEIEFF, O., 1983. Dictionary of Petrology. *Wiley, New York.* 680pp. [Amphiboldite, Anortholite, Bizeul, Busonite, Ceramicite, Egerinolith, Eutectite, Florianite, Gamaicu, Ghiandone, Granitelle, Ilmen-granite, Ilvaite, Isophyre, Kammstein, Leckstone, Lhercoulite, Rhenoaplite, Riacolite, Smalto, Spinellite, Tabona, Terzontli, Zutterite]

TÖRNEBOHM, A.E., 1877a. Om Sveriges vigtigere diabas-och gabbro-arter. *Handlingar Kongliga Svenska Vetenskaps-Akademiens, Stockholm.* Vol.14, No.13, p.1–55. [Hyperitite]

TÖRNEBOHM, A.E., 1877b. Ueber die Wichtigeren diabas und gabbro gesteine Schwedens. *Neues Jahrbuch für Mineralogie, Geologie und Paläontologie. Stuttgart.* Jahrgang. p.379–393. [Gabbrodiorite]

TÖRNEBOHM, A.E., 1881. Beskrifning till blad no. 7 af Geologisk öfversigtskarta öfver mellersta Sveriges Bergslag. *Beckman, Stockholm.* Vol.7, p.1-36. [Gabbrogranite]

TÖRNEBOHM, A.E., 1883. Om den s.k. Fonoliten från Elfdalen, dess klyftort och förekomstsätt. *Geologiska Föreningens i Stockholm Förhandlingar. Stockholm.* Vol.6, p.383–405. [Cancrinite syenite]

TÖRNEBOHM, A.E., 1906. Katapleiit-syenit – en nyupptäckt varietet af nefelinsyenit i Sverige. *Sveriges Geologiska Undersökning. Afhandlingar och Uppsatser. Stockholm.* Ser C, No.199, p.1–54. [Lakarpite]

TRAVIS, C.B. (ED.), 1915. The igneous and pyroclastic rocks of the Berwyn Hills (North Wales). Cope Memorial Volume. *Proceedings of the Liverpool Geological Society. Liverpool.* p.1–115. [Hirnantite]

TRIMMER, J., 1841. Practical Geology and Mineralogy. *Parker, London.* 519pp. [Haüyne basalt]

TRÖGER, W.E., 1928. Alkaligesteine aus der Serra do Salitre im westlichen Minas Geraes, Brasilien. *Centralblatt für Mineralogie, Geologie und Paläontologie. Stuttgart. Abt.A.* p.202–204. [Bebedourite, Salitrite]

TRÖGER, W.E., 1931. Zur "Typenvermischung" bei Lamprophyren. *Fortschritte der Mineralogie, Kristallographie und Petrographie. Jena.* Vol.16, p.139–140. [Camptospessartite]

TRÖGER, W.E., 1935. Spezielle Petrographie der Eruptivgesteine. *Verlag der Deutschen Mineralogischen Gesellschaft, Berlin.* p.360. [Andesilabradorite, Anorthitissite, Aplosyenite, Dolomite carbonatite, Engadinite, Essexite-diorite, Essexite-foyaite, Evisite, Feldspattavite, Granatite, Haüyne basanite, Mimesite, Nosean basanite, Peléeite, Picotitite, Rongstockite, Yosemitite]

TRÖGER, W.E., 1938. Spezielle Petrographie der Eruptivgesteine. Eruptivgesteinsnamen. *Fortschritte der Mineralogie, Kristallographie und Petrographie. Jena.* Vol.23, p.41–90. [Kaulaite, Vredefortite]

TSCHERMAK, G., 1866. Felsarten von ungewöhnlicher zusammensetzung in den Umgebungen von Teschen und Neutitschein. *Sitzungsberichte der Kaiserlichen Akademie der Wissenschaften. Mathematisch-Naturwissenschaftliche Classe. Wien.* Vol.53, Abt.1, p.260–286. [Picrite]

TSUBOI, S., 1918. Notes on Miharite. *Journal of the Geological Society of Toyko.* Vol.25, No.302, p.47–58. [Miharaite]

TURNER, H.W., 1896. Further contributions to the Geology of the Siera Nevada. *Report of the United States Geological Survey. Washington.* Pt.1, p.521–740. [Albitite]

TURNER, H.W., 1899. The granitic rocks of the Sierra Nevada. *Journal of Geology. Chicago.* Vol.7, p.141–162. [Granolite]

TURNER, H.W., 1900. The Nomenclature of Feldspathic Granolites. *Journal of Geology. Chicago.* Vol.8, p.105–111. [Andesinite, Anorthitite, Labradite, Oligosite, Orthosite]

TURNER, H.W., 1901. Perknite (Lime – Magnesia Rocks). *Journal of Geology. Chicago.*

Vol.9, p.507–511. [Perknite]

TYRRELL, G.W., 1911. A Petrological Sketch of the Carrick Hills, Ayrshire. *Transactions of the Geological Society of Glasgow.* Vol.15, p.64–83. [Plagiophyre]

TYRRELL, G.W., 1912. The Late Palaeozoic Alkaline Igneous Rocks of the West of Scotland. *Geological Magazine. London.* Vol.9, Decade 5, p.120–131. [Kylite, Lugarite]

TYRRELL, G.W., 1917. Some Tertiary Dykes of the Clyde Area. *Geological Magazine. London.* Vol.4, Decade 6, p.305–315. [Cumbraite]

TYRRELL, G.W., 1926. The Principles of Petrology. *Methuen, London.* 349pp. [Quartz dolerite]

TYRRELL, G.W., 1931. Volcanoes. *Butterworth, London.* 252pp. [Agglutinate]

TYRRELL, G.W., 1937. Flood basalts and fissure eruptions. *Bulletin Volcanologique. Organe de l'Association de Volcanologie de l'Union Géodésique et Géophysique Internationale. Bruxelles. (Napoli).* Vol.1, 2nd. Ser., p.89–111. [Flood basalt]

UHLEMANN, A., 1909. Die Pikrite des sächsischen Vogtlandes. *Tschermaks Mineralogische und Petrographische Mitteilungen. Wien.* Vol.28, 2nd.Ser., p.415–472. [Schönfelsite]

USSING, N.V., 1912. Geology of the country around Julianehaab, Greenland. *Meddelelser om Grønland, Kjøbehavn.* Vol.38, p.1–426. [Agpaite, Kakortokite, Naujaite, Sodalitite]

VAKAR, V., 1931. Geologicheskie issledovaniya v basseine r. Berezovki Kolymskogo okruga [transliterated from Russian]. *Izvestiya Glavnogo Geologo-Razvedochnogo Upravleniya. Moskva.* Vol.65, p.1015–1032. [Antirapakivi]

VÄYRYNEN, H., 1938. Notes on the geology of Karelia and the Onega region in the summer of 1937. *Bulletin de la Commission Géologique de la Finlande. Helsingfors.* Vol.21, No.123, p.65–80. [Karjalite]

VERBEEK, R.D.M., 1905. Geologische Beschrijving van Ambon. *Jaarboek van het Mijnwezen in Nederlandsch Oost-Indië. Amsterdam.* Vol.34, p.1–308. [Ambonite]

VILJOEN, M.J. & VILJOEN, R.P., 1969. The geology and geochemistry of the Lower

Ultramafic Unit of the Onverwacht Group and a proposed new class of igneous rocks. *Special Publication of the Geological Society of South Africa*. Vol.2, p.55–85. [Basaltic komatiite, Komatiite, Peridotitic komatiite]

VIOLA, C., 1892. Nota preliminare sulla regione dei gabbri e delle serpentine nell'alta valle del Sinni in Basilicata. *Bollettino del Reale Comitato Geologico d'Italia. Firenze.* Vol.23, p.105–125. [Plagioclasite]

VIOLA, C., 1901. L'augitite anfibolica di Giumarra presso Rammacca (Sicilia). *Bollettino del Reale Comitato Geologico d'Italia. Firenze.* Vol.32, p.289–312. [Giumarrite]

VIOLA, C. & STEFANO, G. DI, 1893. La Punta delle Pietre Nere presso il lago di Lesina in provincia di Foggia. *Bollettino del Reale Comitato Geologico d'Italia. Firenze.* Vol.24, p.129–143. [Garganite]

VLODAVETS, V.I., 1930. Nefelin-apatitovye mestorozhdeniya v Khibinskikh tundrakh [transliterated from Russian]. *Trudy Instituta po Izucheniyu Severa. Moskva.* Vol.46, p.14–60. [Apaneite, Neapite]

VLODAVETS, V.I., 1952. O svoeobraznom kaltsitovom diabaze (kabitavite) Mongolii [transliterated from Russian]. *Doklady Akademii Nauk SSSR. Leningrad.* Vol.87, p.657–659. [Cabytauite]

VOGELSANG, H., 1872. Ueber die Systematik der Gesteinslehre und die Eintheilung der gemengten Silikatgesteine. *Zeitschrift der Deutschen Geologischen Gesellschaft. Berlin.* Vol.24, p.507–544. [Basalt-trachyte, Basite-porphyry, Felsophyre, Felsophyrite, Granophyre, Granophyrite, Leucite-basite, Nepheline basite, Vitrophyre, Vitrophyrite]

VOGELSANG, H.P.J., 1875. Die Krystalliten. *Cohen, Bonn.* 175pp. [Felsovitrophyre, Granofelsophyre]

VOGT, J.H.L., 1918. Jernmalm og jernverk. *Norges Geologiske Undersögelse. Kristiania.* No.85, p.1–181. [Rødberg]

VOGT, TH., 1915. Petrographisch-chemische Studien an einigen Assimilations-Gesteinen der nordnorwegischen Gebirgskette. *Skrifter udgit av Videnskabsselskabet i Kristiania. I. Math.-Nat. Klasse.* No.8, p.1–34. [Hortite]

WADE, A. & PRIDER, R.T., 1940. The Leucite-bearing Rocks of the west Kimberley area, Western Australia. *Quarterly Journal of the Geological Society of London.* Vol.96, p.39–98. [Cedricite, Fitzroyite, Mamilite, Wolgidite]

WADSWORTH, M.E., 1884. Lithological studies: a description and classification of the rocks of the Cordilleras. *Memoirs of the Museum of Comparitive Zoology, at Harvard College, Cambridge, Mass.* Vol.11, Pt.1, p.1–208. [Buchnerite, Cumberlandite, Saxonite, Shastalite]

WADSWORTH, M.E., 1887. Preliminary description of the peridotytes, gabbros, diabases and andesytes of Minnesota. *Bulletin of the Geological Survey of Minnesota.* Vol.2, p.1–159. [Zirkelyte]

*WADSWORTH, M.E., 1893. Sketch of the geology of the iron, gold and copper districts of Michigan. *Report of the State Board of Geological Survey of Michigan for the years 1891 and 1892. Lansing.* [Ferrolite, Lassenite, Metabolite]

WAGER, L.R. & DEER, W.A., 1939. Geological investigations in East Greenland. III. The petrology of the Skaergaard Intrusion, Kangerdlugssuaq, East Greenland. *Meddelelser om Grønland, Kjøbehavn.* Vol.105, No.4, p.1–352. [Ferrogabbro]

WAGER, L.R. & VINCENT, E.A., 1962. Ferrodiorite from the Isle of Skye. *Mineralogical Magazine and Journal of the Mineralogical Society. London.* Vol.33, p.26–36. [Ferrodiorite]

WAGER, L.R., BROWN, G.M. & WADSWORTH, W.J., 1960. Types of Igneous Cumulates. *Journal of Petrology. Oxford.* Vol.1, p.73–85. [Cumulate]

WAHL, W., 1907. Beiträge zur Geologie der Präkambrischen Bildungen im Gouvernement Olonez. II.3. Die Gesteine der Westküste des Onega-sees. *Fennia: Bulletin de la Société de Géographie de Finlande.* Vol.24, No.3, p.1–94. [Valamite]

WAHL, W., 1925. Die Gesteine des Wiborger Rapakiwigebietes. *Fennia: Bulletin de la Société de Géographie de Finlande.* Vol.45, No.20, p.1–127. [Pyterlite, Tirilite, Wiborgite]

WALKER, G.P.L. & CROASDALE, R., 1971. Characteristics of Some Basaltic Pyroclastics. *Bulletin Volcanologique. Organe de l'Association de Volcanologie de l'Union*

Géodésique et Géophysique Internationale. Bruxelles. (Napoli). Vol.35, 2nd.Ser., p.303–317. [Achnelith]

WALKER, T.L., 1931. Alexoite, a pyrrhotite-peridotite from Ontario. *University of Toronto Studies. Geological Series. Toronto.* No.30, p.5–8. [Alexoite]

WALTERSHAUSEN, W.S. VON, 1846. Ueber die submarinen vulkanischen Ausbrüche im Tertiär-Formation des Val di Noto. *Göttinger Studien.* p.371–431. [Palagonite tuff]

WALTERSHAUSEN, W.S. VON, 1853. Über die vulkanischen Gesteine in Sicilien und Island und ihre submarine Umbildung. *Dieterichschen Buchhandlung, Göttingen.* 532pp. [Sideromelane]

WARREN, C.H., 1912. Art. XXV – The Ilmenite Rocks near St. Urbain, Quebec; a new occurrence of Rutile and Sapphirine. *American Journal of Science. New Haven.* Vol.33, 4th.Ser., p.263–277. [Urbainite]

WASHINGTON, H.S., 1894. Art. XV – On the Basalts of Kula. *American Journal of Science. New Haven.* Vol.47, 3rd.Ser., p.114–123. [Kulaite]

WASHINGTON, H.S., 1896a. Italian Petrological Sketches. I. The Bolsena Region. *Journal of Geology. Chicago.* Vol.4, p.541–566. [Vulsinite]

WASHINGTON, H.S., 1896b. Italian Petrological Sketches. II. The Viterbo Region. *Journal of Geology. Chicago.* Vol.4, p.826–849. [Ciminite]

WASHINGTON, H.S., 1897a. Italian Petrological Sketches. III. The Bracciano, Cerveteri and Tolfa regions. *Journal of Geology. Chicago.* Vol.5, p.34–49. [Toscanite]

WASHINGTON, H.S., 1897b. Italian Petrological Sketches. V. Summary and conclusion. *Journal of Geology. Chicago.* Vol.5, p.349–377. [Santorinite]

WASHINGTON, H.S., 1900. The composition of kulaite. *Journal of Geology. Chicago.* Vol.8, p.610–620. [Kulaite]

WASHINGTON, H.S., 1901. The Foyaite-Ijolite Series of Magnet Cove: A chemical study in differentiation. *Journal of Geology. Chicago.* Vol.9, p.607–622. [Arkite, Covite]

WASHINGTON, H.S., 1906. The Roman Comagmatic Region. *Publications. Carnegie Institution of Washington.* Vol.57, p.1–199.

[Tavolatite, Trachivicoite, Vicoite, Viterbite]

WASHINGTON, H.S., 1913. The Volcanoes and Rocks of Pantelleria. *Journal of Geology. Chicago.* Vol.21, p.683–713. [Gibelite, Khagiarite, Ponzite]

WASHINGTON, H.S., 1914a. Art. VII – An Occurrence of Pyroxenite and Hornblendite in Bahia, Brazil. *American Journal of Science. New Haven.* Vol.38, 4th.Ser., p.79–90. [Bahiaite]

WASHINGTON, H.S., 1914b. The analcite basalts of Sardinia. *Journal of Geology. Chicago.* Vol.22, p.742–753. [Ghizite]

WASHINGTON, H.S., 1920. Art. IV – Italite, a new Leucite Rock. *American Journal of Science. New Haven.* Vol.50, 4th.Ser., p.33–47. [Albanite, Italite, Vesbite]

WASHINGTON, H.S., 1923. Petrology of the Hawaiian Islands. I. Kohala and Mauna Kea, Hawaii. *American Journal of Science. New Haven.* Vol.5, 5th.Ser., p.465–502. [Oligoclasite]

WASHINGTON, H.S., 1927. The Italite Locality of Villa Senni. *American Journal of Science. New Haven.* Vol.14, 5th.Ser., p.173–198. [Biotitite]

WASHINGTON, H.S. & KEYES, M.G., 1928. Petrology of the Hawaiian Islands: VI. Maui. *American Journal of Science. New Haven.* Vol.15, 5th.Ser., p.199–220. [Oligoclase andesite]

WASHINGTON, H.S. & LARSEN, E.S., 1913. Magnetite basalt from North Park, Colorado. *Journal of the Washington Academy of Sciences.* Vol.3, p.449–452. [Arapahite]

WATSON, T.L., 1907. Mineral Resources of Virginia. *Bell, Lynchberg, Virginia.* 618pp. [Nelsonite]

WATSON, T.L., 1912. Art. XLIV – Kragerite, a Rutile-bearing Rock from Krageroe, Norway. *American Journal of Science. New Haven.* Vol.34, 4th.Ser., p.509–514. [Kragerite]

WATSON, T.L. & TABER, S., 1910. The Virginia Rutile Deposits. In: C.W. Hayes & W. Lindgren, Contributions to Economic Geology (Short papers and preliminary reports) 1909. *Bulletin of the United States Geological Survey, Washington.* No.430, p.200–213. [Gabbro-nelsonite]

WATSON, T.L. & TABER, S., 1913. Geology of

the Titanium and Apatite Deposits of Virginia. *Bulletin. Virginia Geological Survey, University of Virginia.* Vol.3A, p.1–308. [Amherstite]

WATTS, W.W., 1925. The Geology of South Shropshire. *Proceedings of the Geologists' Association, London.* Vol.36, p.321–363. [Ercalite]

WATTS, W.W. ET AL., 1921. Report of the Committee on British Petrographic Nomenclature. *Mineralogical Magazine and Journal of the Mineralogical Society. London.* Vol.19, p.137-147. [Olivine leucitite]

WEED, W.H. & PIRSSON, L.V., 1895a. Highwood Mountains of Montana. *Bulletin of the Geological Society of America. New York.* Vol.6, p.389–422. [Shonkinite]

WEED, W.H. & PIRSSON, L.V., 1895b. Art. LII – Igneous Rocks of Yogo Peak, Montana. *American Journal of Science. New Haven.* Vol.50, 3rd.Ser., p.467–479. [Yogoite]

WEED, W.H. & PIRSSON, L.V., 1896. Art. XLVI – Missourite, a new Leucite Rock from the Highwood Mountains of Montana. *American Journal of Science. New Haven.* Vol.2, 4th.Ser., p.315–323. [Missourite]

WEINSCHENK, E., 1891. Beiträge zur Petrographie Japans. *Neues Jahrbuch für Mineralogie, Geologie und Paläontologie. Stuttgart. Beilagebände.* Vol.7, p.133–151. [Sanukite]

WEINSCHENK, E., 1895. Beiträge zur Petrographie der östlichen Centralalpen speciell des Gross-Venedigerstockes. *Abhandlungen der Mathematisch-Physikalischen Classe der Königlich Bayerischen Akademie der Wissenschaften. München.* Vol.18, p.651–713. [Stubachite]

WEINSCHENK, E., 1899. Zur Kenntnis der Graphitlagerstätten. Chemisch-geologische Studien. *Abhandlungen der Mathematisch-Physikalischen Classe der Königlich Bayerischen Akademie der Wissenschaften. München.* Vol.19, p.509–564. [Bojite]

WELLS, M.K., 1948. In: J.M. Cotelo Neiva, Serpentines and Serpentinisation – Discussion. *International Geological Congress. Report of the Eighteenth Session. Great Britain, 1948. Part II. Proceedings of Section A. Problems of Geochemistry.* p.88–95. [Serpentinite]

WENTWORTH, 1935. The terminology of coarse sediments. *Bulletin of the National Research Council. Washington.* No.98, p.225-246. [Block]

WENTWORTH, C.K. & WILLIAMS, H., 1932. The Classification and Terminology of the Pyroclastic Rocks. *Bulletin of the National Research Council. Washington.* Vol.89, p.19–53. [Bomb, Reticulite]

WERNER, A.G., 1787. Kurze Klassifikationen und Beschreibung der verschiedenen Gebirgsarten. *Dresden.* 28pp. [Klingstein, Mandelstein, Porphyry]

WESTERMANN, J.H., 1932. The geology of Aruba. *Geographische en Geologische Mededeelingen. Utrecht.* Vol.7, p.1–129. [Hooibergite]

WHITE, A.J.R., 1979. Sources of Granite Magma. *Abstracts of papers to be presented at the Annual Meetings of the Geological Society of America and Associated Societies, San Diego, California, November 5-8, 1979.* Vol.11, p.539. [M-type granite]

*WILLEMS, H.W.V., 1934. *Ingénieur in Nederlandsch-Indië. Bandoeng.* Vol.1. [Astridite]

WILLIAMS, G.H., 1886. Art. 4 – The peridotites of the "Cortlandt Series" on the Hudson River near Peekskill, N.Y. *American Journal of Science. New Haven.* Vol.31, 3rd.Ser., p.26–41. [Cortlandtite]

WILLIAMS, G.H., 1890. The non-feldspathic intrusive rocks of Maryland and the course of their alteration. *American Geologist. Minneapolis.* Vol.6, p.35–49. [Bronzitite, Websterite]

WILLIAMS, H., 1941. Calderas and their origin. *University of California Publications in Geological Sciences. Berkeley.* Vol.25, No.6, p.239–346. [Ash-stone, Aso lava]

WILLIAMS, J.F., 1891. The igneous rocks of Arkansas. *Annual Report of the Geological Survey of Arkansas. Little Rock.* Vol.2, p.1–391. [Fourchite, Pulaskite]

WILLMANN, K., 1919. Die Redwitzite, eine neue Gruppe von granitischen Lamprophyren. *Zeitschrift der Deutschen Geologischen Gesellschaft. Berlin.* Vol.71, p.1–33. [Redwitzite]

WINCHELL, A.N., 1912. Geology of the National mining district, Nevada. *Mining and Science Press, San Francisco.* Vol.105, p.655–659. [Auganite]

WINCHELL, A.N., 1913. Rock classification on three co-ordinates. *Journal of Geology. Chicago.* Vol.21, p.211. [Peralkaline, Rhyodacite]

WINKLER, G.G., 1859. Allgovit (Trapp) in den Allgäuer Alpen Bayerns. *Neues Jahrbuch für Mineralogie, Geognosie, Geologie und Petrefaktenkunde. Stuttgart.* Vol.30, p.641–670. [Allgovite]

WOLF, H., 1866. Trachytsammlungen aus Ungarn. *Verhandlungen der Geologischen Reichsanstalt (Staatanstalt-Landesanstalt). Wien.* Vol.16, p.33–34. [Microtinite]

WOLFF, F. VON, 1899. Beiträge zur Geologie und Petrographie Chile's. *Zeitschrift der Deutschen Geologischen Gesellschaft. Berlin.* Vol.51, p.471–555. [Andennorite]

WOLFF, F. VON, 1914. Der Vulkanismus. *Enke, Stuttgart.* Vol.1, 711pp. [Duckstein]

WOOLLEY, A.R., 1982. A discussion of carbonatite evolution and nomenclature, and the generation of sodic and potassic fenites. *Mineralogical Magazine and Journal of the Mineralogical Society. London.* Vol.46, p.13–17. [Magnesiocarbonatite]

WOOLLEY, A.R. & KEMPE D.R.C., in press. Carbonatites: nomenclature, average chemical compositions and element distribution. In: K. Bell (Ed.), Carbonatite Genesis and Evolution. *Unwin Hyman, London.* [Calciocarbonatite]

WORTH, R.N., 1884. On Trowlesworthite, and certain granitoid rocks near Plymouth. *Transactions of the Royal Geological Society of Cornwall. Penzance.* Vol.10, p.177. [Trowlesworthite]

WYLLIE, B.K.N. & SCOTT, A., 1913. The Plutonic Rocks of Garabal Hill. *Geological Magazine. London.* Vol.10, Decade 5, p.499–508. [Davainite]

WYLLIE, P.J., 1967. Ultramaifc and Related Rocks. *Wiley, New York.* 464pp. [Clinopyroxenite, Hornblende peridotite, Hornblende pyroxenite, Orthopyroxenite, Pyroxene peridotite]

YACHEVSKII, L., 1909. Mestorozhdeniya khrizotila na khrebte Bis-Tag v Minusinskom okruge Eniseiskoi gubernii [transliterated from Russian]. *Geologicheskiya Izsledovaniya v "Zolotonosnykh Oblastyakh" Sibiri.*

S.-Peterburg. Vol.8, p.31–50. [Bistagite]

YASHINA, R.M., 1957. Shchelochnye porody yugo-vostochnoi Tuvy [transliterated from Russian]. *Izvestiya Akademii Nauk SSSR, Seriya Geologicheskaya.* No.5, p.17–36. [Tuvinite]

YODER, H.S. & TILLEY, C.E., 1962. Origin of Basalt Magmas: An Experimental Study of Natural and Synthetic Rock Systems. *Journal of Petrology. Oxford.* Vol.3, p.342–532. [Tholeiite]

YOUNG, G.A., 1906. The Geology and Petrography of Mount Yamaska. *Annual Report. Geological Survey of Canada. Montreal and Ottawa.* Pt.H, p.1–43. [Yamaskite]

ZAVARITSKII, A.N., 1955. Izverzhennye gornye porody [transliterated from Russian]. *Izdatel' stvo Akademii Nauk SSSR, Moskva.* 479pp. [Bereshite, Plagiogranite]

ZAVARITSKII, A.N. & KRIZHANOVSKII, V.I., 1937. Gosudarstvennyi Ilmenskii mineralogicheskii zapovednik [transliterated from Russian]. *XXVII International Geological Congress, Moskva. Uralskaya ekskursiya, yuzhnyi marshrut.* p.5–17. [Sandyite]

ZHIDKOV, A. YA., 1962. Slozhnaya Synnyrskaya intruziya sienitov Severo-Baikalskoi shchelochnoi provintsii [transliterated from Russian]. *Geologiya i geofyzika.* Vol.9, p.29–40. [Synnyrite]

ZIRKEL, F., 1866a. Lehrbuch der Petrographie. *Marcus, Bonn.* Vol.1, 607pp. [Ditroite]

ZIRKEL, F., 1866b. Lehrbuch der Petrographie. *Marcus, Bonn.* Vol.2, 635pp. [Corsite, Quartz diorite, Teschinite]

ZIRKEL, F., 1870. Untersuchungen über die mikroskopische Zusammensetzung und Structur der Basaltgesteine. *Moncur, Bonn.* 208pp. [Feldspar-basalt, Leucite basalt]

ZIRKEL, F., 1894a. Lehrbuch der Petrographie. *Engelmann, Leipzig.* Vol.2, 2nd. Edition, 941pp. [Felsogranophyre, Plethorite]

ZIRKEL, F., 1894b. Lehrbuch der Petrographie. *Engelmann, Leipzig.* Vol.3, 2nd. Edition, 833pp. [Therolite]

ZLATKIND, Z.G., 1945. Olivinovye turyaity (kovdority) – novye glubinnye melilitovye porody Kolskogo poluostrova [transliterated from Russian]. *Sovetskaya Geologiya. Moskva.* Vol.7, p.88–95. [Kovdorite]

Appendix A. List of Circulars

This is a list of all the Circulars distributed by the Subcommission. Copies of these documents are now housed in the Department of Mineralogy at the British Museum (Natural History), London so that they are freely available for future use.

CIRCULAR NO.1 – MARCH 1970
Communications by the Chairman.

CIRCULAR NO.2 – MARCH, 1970
Contribution No.1. A. Streckeisen: On the Principles of Classification of Igneous Rocks. With Questionnaire. 8pp.
Contribution No.2. On the Classification and Nomenclature of Granitic Rocks (Granitoids). 6pp.
Contribution No.3. S. Subba Rao: Need for Establishment of Central Petrological Museums. 2pp.

CIRCULAR NO.3 – AUGUST, 1970
Communications by the Chairman: Future work of the Subcommission – Establishing of Working Groups. With Questionnaire. 5pp.
Report of the Terminological Commission of the Petrographic Committee of the USSR, 1969: Russian text. Die derzeitige Lage der Terminologie der Eruptivgesteine. German translation (by G. Pantó), 17pp. Figures and Tables (in German), 14pp.
Contribution No.4. Statement concerning the Proposal of S. Subba Rao. 8pp.
Contribution No.5. Remarks on the Principles of Classification of Igneous Rocks. With Questionnaire. (P.C. Bateman, Chairman). 9pp.
Contribution No.6. Remarks on the Classification and Nomenclature of Granitoids. (P.C. Bateman, W.B. Dallwitz, K.R. Mehnert, W.P. de Roever, M.E. Teruggi, B. Zanettin, A. Streckeisen). 10pp.

Contribution No.7. Preliminary Remarks on the Classification and Nomenclature of Plutonic Ultramafic Rocks. (F. Rost, E.D. Jackson). 4pp.

CIRCULAR NO.4 – JANUARY, 1971
Communication by the Chairman. Future work of the Subcommission. Working Groups Meetings. 5pp.
T.F.W. Barth: IUGS Policy. 1pp.
Zd. Vejnar – N. Neužilová: Application of Streckeisen's classification to plutonic rocks of the Bohemian Massif. 1pp.
Contribution No.8. H. de la Roche: Objection au groupement de l'albite avec le feldspath potassique pour la classification des roches granitiques. 7pp.
Annex: Reply to Questionnaire of Contribution No.1.
Contribution No.9. A. Streckeisen: Classification and Nomenclature of Quartz-Feldspathic Rocks. With Questionnaire. 22pp.
Annex: Reply by A. Dudek to Circulars No.2 and 3.
H. de la Roche: Reply to proposal of S. Subba Rao.

CIRCULAR NO.5 – MARCH, 1971
Communications by the Chairman (Charnockitic Rocks). 9pp.
Contribution No.10. A.C. Tobi: The Nomenclature of the Charnockitic Rock Suite. 13pp.
Contribution No.10a. A.C. Tobi: Appendix. Remarks on the alternative proposal of July 1968 and to Contribution No.9.

INTERMEDIARY CIRCULAR
Information on the present state of the discussion. (Chairman). 4pp.

CIRCULAR NO.6 – SEPTEMBER, 1971
Contribution No.11. A. Streckeisen: On the Systematics of Quartzo-Feldspathic Rocks. 24pp. Includes remarks by A. Dudek, W.P. de

Roever, B. Zanettin, H.G.F. Winkler, P.A. Sabine, P.C. Bateman, F. Chayes. Results of Questionnaire of 15.5.1971 (54 replies)

Contribution No.12. A.K. Wells and M.K. Wells: Igneous Rock Classification and Nomenclature. 7pp.
Reply by A. Streckeisen. 3pp.

Contribution No.13. Zentrales Geologisches Institut der Deutschen Demokratischen Republik, Berlin: Magmatische Gesteine. Petrographische Gesteins-bezeichnung. Zusammenfassung, 7pp. Original text 20pp. including 3 tables.

Contribution No.14. H. Sørensen: Introduction to a discussion of the nomenclature and systematics of alkaline rocks. 18pp.

Contribution No.15. Comments on charnockite nomenclature (P.C. Cooray, H.G.F. Winkler, A. Dudek, W.P. de Roever, F. Chayes).

CIRCULAR NO.7 – DECEMBER, 1971

Contribution No.16. A. Streckeisen: Classification and Nomenclature of Coarse-Grained Basic and Ultramafic Rocks. 22pp.

Contribution No.17. A. Streckeisen: On the Systematics of Granitoids. 4pp.
A. Davidson: On the Nomenclature and Systematics of Granitoid Rocks. Reply to Questionnaire No.7. Appendix to Contribution No.11, 3pp.
H. Ivanov: On the Systematics of Quartzo-Feldspathic Plutonic Rocks. 4pp.

CIRCULAR NO.8 – FEBRUARY, 1972

Contribution No.18. Remarks on the Classification and Nomenclature of Alkaline Rocks. (Comments to Contribution No.14 by the Chairman, D.S. Barker, A.D. Edgar, P.A. Sabine, L. Kopecky, O.A. Vorobieva, A. Streckeisen, the late Harry von Eckermann). 20pp.

Contribution No.19. M.K. Bose: Discussion on Nomenclature and Classification of Alkaline Rocks. A Rejoinder to Contribution No.14. 6pp.
M.K. Bose: Comments to A. Streckeisen's Final Report of 1967. 3pp.
M.J. Le Bas: Carbonatites. 9pp.

CIRCULAR NO.9 – MARCH, 1972

Contribution No.20. On the Systematics of Basic and Ultramafic Rocks. (Working Group Basic and Ultramafic Rocks, Communication No.2.).
A. Dudek: Reply to Contribution No.16. 2pp; E.D. Jackson: Comments to Contribution No.16. 7pp.
A.J. Naldrett: Remarks on Contribution No.16. 2pp.
G. Pantó: Zur Frage des Gebrauchs oder Missbrauchs des Gesteinsnamens "Wehrlit". 2pp.
K. Smulikowski: Comments to I. Ultramafic Rocks; II. Gabbroid to Ultramafic Rocks. 3pp.
B. Zanettin: Reply to Contribution No.16. 3pp.
Comments by S.E. Ellis and H.P. Zeck.

Contribution No.21. On the Classification and Nomenclature of Granitoids. (Working Group on Granitoids.).
Inquiry of P.C. Bateman on the Nomenclature of Granitoids. 6pp.
K. Smulikowski: On the Classification and Nomenclature of Quartzo-Feldspathic Rocks. 3pp.
Comments on Contribution No.16 by P.A. Sabine, B. Zanettin, G. Pantó, A. Dudek, R. Ivanov. 2pp.

CIRCULAR NO.10 – MAY, 1972

Rules for the Mode of Procedure in the Meetings. Adopted by the Subcommission by way of correspondence. The Preliminary Meeting was held in Berne on April 12-14, 1972. The Circular presents the various contributions that were received before the Meeting. The Minutes of the corresponding Working Group are annexed to any section.

Contribution No.22. Working Group on Granitoids.
G.D. Afanass'yev, On the Classification of Granitoids on the Triangular Diagram. 2pp.
S.V. Efremova, Remark on the Nomenclature of Granitoids.
W. Pälchen, Stellungnahme zu einigen Fragen der Magmatit-Nomenklatur. 4pp.
R. Ivanov, Remarks on the Classification of Igneous Rocks. 1pp.

B.K. L'vov, Personal proposal on the Classification of Granitoids. 2pp.

A.C. Tobi: A Simple System for the Determination of Igneous Rocks. 2pp.

H. de la Roche: Diagram on the Classification of Granitoids. 1pp.

K. Smulikowski: Remarks on the Nomenclature of Granitoids.

Contribution No.23. Working Group on Basic and Ultramafic Rocks. Communication No.3.

F. Rost und G. Lensch: Zur Klassifikation und Nomenklatur ultramafischer plutonischer Magmatite. 22pp.

N.P. Mikhailov, N. Sobolev, S.F. Sobolev: Classification of Ultrabasic and Basic Rocks. 2pp.

S.F. Sobolev: The Classification of Garnet Peridotites. 3pp.

Contribution No.23a. Minutes of the Working Group on Basic and Ultramafic Rocks. (F. Rost). 19pp.

Contribution No.24. Working Group on Alkaline and Feldspathoidal Rocks.

M.J. Le Bas: Reply to the "Comments by the Chairman". 1pp.

L. Kopecky: Summarised Remarks and Suggestions to Contributions Nos. 18 and 19. 5pp.

T. Miller (USA): Comments on Professor Sørensen's Classification of Alkaline Rocks. 2pp.

Minutes of the Working Group on Alkaline and Feldspathoidal Rocks. Communication No.4 (H. Sørensen). 4pp.

Contribution No.25. Working Group on Charnockitic Rocks. Communication No.3.

H.G.F. Winkler and S.K. Sen: Granulites and Charnockites, a Proposal for the Unified Nomenclature. 7pp.

T. Torske: The Nomenclature of the Charnockitic Rock Suite. 3pp.

A.C. Tobi: The Nomenclature of the Charnockitic Rock Suite: Reply to a Discussion.

V.M. Shemyakin und K.A. Shurkin: Ueber Nomenklatur und Klassifikationsprinzipien der Gesteinsgruppe der "Charnockite" (translated by G. Pantó). 5pp.

Report of the Working Group on Charnockitic Rocks (A.C. Tobi). Communication No.4, 6pp.

Contribution No.26. Minutes of the Preliminary Meeting. April 11-14, 1972, held in Berne. 13pp.

CIRCULAR NO.11 – OCTOBER, 1972

Minutes of the Ordinary Meeting held in Montreal on August 23 and 25, 1972. 14pp. Annex: Summary of Recommendations on which the Subcommission has agreed to. 7pp.

CIRCULAR NO.12 – DECEMBER, 1972

Contribution No.27. Working Group on Lamprophyres. Communication No.1.

K.L. Currie: Thoughts on the Nomenclature of Lamprophyres. 2pp.

W. Wimmenauer: Zur .Begriffsbestimmung der Lamprophyre. 8pp.

A. Streckeisen: Nomenclature of Hypabyssal Rocks. 2pp.

Contribution No.28. Working Group on Charnockitic Rocks. Communication No.5.

Chairman: Introduction to the Questionnaire on the Classification and Nomenclature of Charnockitic Rocks. 2pp. E.H. Dahlberg: Comment. 1pp.

K.A. Shurkin: Proposals for the Charnockitic Rocks to be Discussed at Montreal. 2pp.

A. Watznauer: Comments. 2pp.

H.P. Zeck: The Classification and Nomenclature of the Charnockitic Rock Suite. 2pp.

A.C. Tobi: Comments on Various Contributions. 2pp.

H.G.F. Winkler and S.K. Sen: Comments. June 5, 1972. 3pp.

H.G.F. Winkler: Granulite – Charnockite Nomenclature Again. July 14, 1972. 8pp.

D. de Waard: Classification and Nomenclature of the Charnockitic Rock Suite. 12pp.

Questionnaire on Charnockitic Rocks. 3pp.

Contribution No.29. Working Group on Effusive and Pyroclastic Rocks. Communication No.1. Terminological Commission of the Petrographic Committeee of the USSR: Die derzeitige Lage der Terminologie und Nomenklatur der Eruptivgesteine (translated from Russian by G. Pantó).

H. Pichler and A. Rittmann: On the Classification and Nomenclature of Volcanic Rocks. 22pp.

A. Streckeisen: Nomenclature of Volcanic Rocks. (1967). 9pp.
Questionnaire for a First Exchange of Views.

CIRCULAR NO.13 – JUNE, 1973
Working Group on Effusive and Pyroclastic Rocks. Communication No.2.

Contribution No.30. Results of the Inquiry of Contribution No.29. Comments by G.C. Amstutz, P.C. Bateman, F. Chayes, R.V. Fisher, D. Jung, E. Lehmann, A.R. McBirney, E.A.K. Middlemost, M. Mittempergher, D.L. Peck, H. Pichler, G. Tischendorf, V. Trommsdorff, A.C. Tobi, T.G. Vallance, M. Vergara, J.F.G. Wilkinson, B. Zanettin. 22pp.

Contribution No.31. M.A. Petrova and T.I. Frolova: Petrochemical Classification of the Basic, Ultrabasic and Intermediate Volcanic Rocks. 4pp. (Moscow).
M.A. Petrova: Petrochemical Classification of the Acid Volcanic Rocks. 4pp. (Moscow)

Contribution No.32. B.A. Markovsky, V.L. Massaitis, L.A. Polunina, V.K. Rotman, N.A. Rumiantseva (Leningrad): Principles of Classification and Nomenclature of Effusive Basic Rocks. 5pp.
V.S. Gladkikh (Leningrad): Basic Principles of Classification of Alkali Volcanic Rocks. 6pp.

Contribution No.33. On the Terms Dolerite, Diabase, and Melaphyre. Citations and Remarks from H. Rosenbusch, A. Johannsen, A. Streckeisen, W.E. Tröger, H. Williams et al., J.F.G. Wilkinson, Hatch-Wells-Wells, M. Vuagnat, P. Bearth, H. von Eckermann, E.K. Ustiev. 6pp.

CIRCULAR NO.14 – AUGUST, 1973
Working Group on Charnockitic Rocks. Communication No.6.

Contribution No.34. Results on the Inquiry of Communication No.5, Contribution No.28. 8pp.
Comments by P.C. Bateman, J.-C. Duchesne, A.F. Laurin and K.N.M. Sharma, O.H. Leonardos Jr., J. Martignole, J. Michot, H.G.F. Winkler. 9pp.
Questionnaire. Classification and Nomenclature of Igneous Rocks, with special regard to Volcanic Rocks. This documentation was prepared for the Volcanological Congress, held in Bucharest in September, 1973.

CIRCULAR NO.15 – MARCH, 1974
Working Group on Charnockitic Rocks. Communication No.7.

Contribution No.35. Results of the Inquiry of Communication No.6, Contribution No.34. 6pp.
Glossary of Terms. 2pp.
Comments by P.C. Cooray, E.H. Dahlberg, J.-C. Duchesne, A. Dudek, J. Ferguson, A.F. Laurin and K.N.M. Sharma, O.H. Leonardos Jr., J. Martignole, K.H. Mehnert, J. Michot, H. de la Roche, S.K. Sen, K. Smulokowski, A.C. Tobi, T. Torske, D. de Waard. 11pp.

CIRCULAR NO.16 – APRIL, 1974
Contribution No.36. Plutonic Rocks: Classification and Nomenclature. Provisional Text. The text was later published under the title "To each plutonic rock its proper name". 1976.

Contribution No.37. Working Group on Charnockitic Rocks. Communication No.8. Points on which definitive precisions are needed.
Modified Proposal.
Comments by P.C. Cooray, A. Dudek, A.F. Laurin and K.N.M. Sharma, O.H. Leonardos Jr., J. Michot, S.K. Sen, A.C. Tobi, D. de Waard. 8pp.

CIRCULAR NO. 17 & 18 – NOVEMBER, 1974
Contribution No.38. Working Group on Charnockitic Rocks. Communication No.9. Chairman: Results of the Discussion.
Comments by P.C. Cooray, J.-C. Duchesne, O.H. Leonardos Jr., J. Martignole, G. Mathé, E.W.F. de Roever, S.K. Sen, A.C. Tobi, D. de Waard. 8pp.
A. Streckeisen: How Should Charnockitic Rocks be Named? (Soc. Géol. de Belgique: Streckeisen, 1974).

Contribution No.39. Working Group on Effusive and Pyroclastic Rocks. Communication No.3.
A. Streckeisen: Comments and Suggestions. 6pp.

Contributions and Comments by Y.K. Bentor, F. Chayes, A. Dudek, T.I. Frolova and M.A. Petrova, H. Kräutner, H. Kunz and P. Müller, E. Lehmann, A.R. McBirney, E.A.K. Middlemost, H. de la Roche, K. Smulikowski, R.N. Sukheswala, M. Seclaman, M. Vergara, B. Zanettin. 31pp.

I.D. Muir: Glossary of Volcanic Rock Names. 9pp.

Contribution No.40. Working Group on Dolerite, Diabase, Melaphyre, Spilite. Communication No.2.

A. Streckeisen: Comments and Suggestions. Contributions and Comments by P. Bearth, V. Dietrich, A. Dudek, J. Ferguson, R.K. Harrison, K. Kunz and P. Müller, E. Lehmann, A.R. McBirney, E.A.K. Middlemost, D. Mucke – L. Pfeiffer – G. Röllig, H. Pichler, K. Smulikowski, A.C. Tobi, T.G. Vallance, M. Vergara, M. Vuagnat. 10pp.

Contribution No.41. Working Group on Lamprophyres. Communication No.2. Characterisation and classification of lamprophyres.

Tentative classification.

Glossary. 8pp.

Contributions and comments by H. Carstens, K.L. Currie, S.V. Efremova, W. Kramer, W. Pälchen and G. Röllig, H. de la Roche, A. Streckeisen, A.C. Tobi, J.F.G. Wilkinson, B. Zanettin. 13pp.

CIRCULAR NO.19 – MARCH, 1975

Working Group on Effusive and Pyroclastic Rocks. Communication No.4.

Contribution No.42. A. Streckeisen: On the classification and nomenclature of volcanic rocks. Remarks and suggestions. 7pp.

Replies by Central Geol., Institute of GDR, A.C. Tobi, D.L. Peck, H. Sørensen, A. Dudek. Selected terms of alkaline volcanic rocks. 4pp.

Contribution No.43. M.E. Teruggi, M.M. Mazzoni, L.A. Spalletti, R.R. Andreis: Propuesta de Sistematizacion de las Rocas Piroclasticas. 35pp.

M.E. Teruggi: A Bi-Lingual Glossary of Pyroclastic Terms. 9pp.

Contribution No.44. L.A.J. Williams: Notes on the Classification and Nomenclature of Kenya Volcanic Rocks. 18pp.

Contribution No.45. Working Group Dolerite – Diabase – Melaphyre – Spilite. Communication No.3.

Comments by L. Pfeiffer – G. Röllig – J. Mädler – H.D. Huebscher – G. Tischendorf, F. Chayes, H. Sørensen, A. Watznauer, A. Dudek, A.C. Tobi. 3pp.

Contribution No.46. Working Group on Charnockitic Rocks. Communication No.10. Comments by A. Watznauer and E.H. Dahlberg. 2pp.

Contribution No.47. Working Group on Lamprophyres. Communication No.10. Comments by W. Kramer, A. Watzhauer, A.C. Tobi, K.L. Currie. 3pp.

CIRCULAR NO.20 – OCTOBER, 1975

Working Group on Effusive and Pyroclastic Rocks. Communication No.5.

Contribution No.48. Remarks on the Classification and Nomenclature of Volcanic Rocks. W.R.A. Baragar, F. Chayes, B.N. Church, D. Giusca, S. Karamata, E.A.K. Middlemost, S.R. Nockolds, Petrographic Committee of the USSR, K. Smulikowski, A. Streckeisen, K.H. Wedepohl. 17pp.

Contribution No.49. Chemical and Modal Data of Selected Rocks. (At its Meeting at Grenoble, 2-4 September, 1975, the Subcommission decided to circulate a set of chemical and modal data (216 specimens, plutonic and volcanic) in order to test the relationships between modal and mineral contents and the various methods that are suggested for the classification of volcanic rocks.) 22pp.

Contribution No.50. Comité pétrographique de l'USSR, Commission terminologique, représentée par N.P. Mikhailov, Chairman: Proposition sue la classification et la nomenclature des roches volcaniques. 4pp, 2 diagrammes.

Contribution No.51. H. de la Roche et J. Leterrier: Transposition du tétraèdre minéralogique de Yoder et Tilley dans un diagramme chimique de classification des roches basaltiques. 3pp, 7 diagrammes.

Contribution No.52. R. Ivanov: On the Classification of Volcanic Rocks. 3pp, 4 diagrams.

Contribution No.53. A. Streckeisen: Classification of Common Igneous Rocks by Chemical Composition. A Provisional Attempt. 19pp.

Contribution No.54. Minutes of the Preliminary Meeting, Grenoble, September 2-4, 1975.

Contribution No.55. Petrographic Committee of the USSR (L.S. Borodin and E.D. Andreeva): On the Terminology of Alkaline Rocks. 4pp.

CIRCULAR NO.21 – OCTOBER, 1975

Working Group on Effusive and Pyroclastic Rocks.

Contribution No.56. Guidelines for a Pyroclastic Rock Classification. 1pp.

Proposal for the Classification of Pyroclastic Rocks. 2pp.

Questionnaire. 5pp.

Nationalkomitee für geologische Wissenschaften der DDR (GDR), H.D. Huebscher. U. Seemann: Stellungnahme zur Nomenklatur der Pyroklastite. 5pp.

CIRCULAR NO.22 – FEBRUARY, 1976

Working Group on Effusive and Pyroclastic Rocks. Communication No.6.

Contribution No.57. Chemical and Modal Data of Selected Rocks: Results.

Remarks and Comments. Z1-3.

Errata to the Circular No.20. Z4.

List of Selected Rocks. References. C1-9.

Modes, Rittmann norms, Cation Norms. A1-12.

Parameters for plotting into the diagrams, Silica-alkali (Mikhailov), QAPF, M (modes, Rittmann and Cation Norms), X-Y (de la Roche), An-Or (Streckeisen). B1-12.

QAPF diagrams of modes. B13-14.

QAPF diagrams of Rittmann norms. B15-16.

Silica-alkali diagrams according to Mikhailov. C1-3.

X-Y diagrams according to de la Roche: $X = 4Si - 11(Na+K) - 2(Fe+Ti)$; $Y = 6Ca + 2Mg + Al$. D0-5.

Parameters calculated from the Cation Norm. E1-5.

Or'-Ab'-An' diagrams of Cation Norms according to Streckeisen. E6-9.

Comparison of Modes, Rittmann norms and Cation Norms by the figures Q, Foid, Fsp, and M (c.i.). P1-3.

Annex: Feldspar diagram for plutonic rocks, by A. Rittmann (1975).

A. Streckeisen: Classification of the Common Igneous Rocks by Means of their Chemical Composition. N. Jb. Miner. Mh. 1976, 1-15.

CIRCULAR NO.23 – MARCH, 1976

Working Group on Effusive and Pyroclastic Rocks.

Contribution No.58. Results of the first questionnaire (Contribution No.56) on Pyroclastic Rocks. Second questionnaire. 18pp.

List of colleagues who replied to second questionnaire. Summarised results of the first questionnaire. Comments on "Guidelines" of Contribution No.56.

Revised definitions. Table I: Granulometric definition of pyroclastic rocks. Table II: Suggestions regarding the classification of poorly sorted consolidated pyroclastic rocks. Table III: Suggestions regarding the nomenclature of epiclastic volcanic sediments.

CIRCULAR NO.24 – MAY, 1976

Working Group on Effusive and Pyroclastic Rocks. Communication No.7.

Contribution No.59. A. Streckeisen: Effusive Rocks – Classification and Nomenclature. Questionnaire. 11pp.

Comments by K. Smulikowski (5pp), G. von Gruenewaldt, C.P. Thornton (2pp), M.J. Le Bas, C.D. Branch.

Glossary of Volcanic Rock Names, I.D. Muir, 1973.

H. Pichler: Veraltete Eruptivgesteins-Namen. Aus: Italienische Vulkangebiete I, p. 247 f., 1970.

H. de la Roche: Sur la contribution des données chimiques à une systematique générale des roches ingées, avec en appendice, "A diagram for a chemical classification of igneous rocks referred to their mineral contents". Sciences de la Terre, vol. 21, 1976, 17-35.

CIRCULAR NO.25 – OCTOBER, 1976

Contribution No.60. Minutes of the Ordinary Meeting held in Sydney on August 17, 19, 23 and 24, 1976.

The Subcommission discussed classification and nomenclature of volcanic rocks and the topic dolerite – diabase – melaphyre – spilite, without taking definitive decisions.

A meeting is planned for autumn, 1977.

Contribution No.61. Results of the Inquiry on volcanic rock names (Questionnaire of Contribution No.59). 25 replies were received.

Comments were presented by D. Giusca, D. Jung, M.J. Le Bas, A.R. McBirney, E.A.K. Middlemost, K. Smulikowski, H. Sørensen, P. Thurston, G. Tischendorf, B. Zanettin. Questionnaire.

Contribution No.62. Petrographic Committee of the USSR: Suggestions made by the Effusive Subcommission on classification and nomenclature of volcanic rocks. 5pp.

Contribution No.63. Missing.

CIRCULAR NO.26 – DECEMBER, 1976

Working Group on Effusive and Pyroclastic Rocks. 10pp.

Contribution No.64. Summarised results of the second questionnaire. 38 replies were received. 1. Definition of fundamental terms. Revised definitions. 2. Granulometric classification of pyroclastic rocks. 3. Classification and terminology of deposits in which pyroclasts are mixed with epiclasts.

Classification of mixed rocks.

Suggestion of the ad hoc working group.

Third questionnaire.

CIRCULAR NO.27

Contribution No.65. Distinction of Different Volcanic Rock Types of Fields 10 and 9. 15pp.

CIRCULAR NO.28 – JUNE, 1977

Working Group on Effusive and Pyroclastic Rocks. 19pp.

Contribution No.66. Results of the Inquiry of Contribution No.61. 31 replies. Distinction andesite/basalt. Subdivision of basalt. Subdivision of andesite. Alkali rhyolite/trachyte. Paleo-volcanic terms. Dolerite, diabase, melaphyre, spilite, keratophyre. Glass-bearing rocks.

Contributions and comments by F. Chayes

(5pp), B.N. Church, T.J. Frolova and E.B. Iacovleva, H. Henschel (3pp), A.R. McBirney, D. Mucke, I.D. Muir, W. Pälchen, P.A. Sabine, A. Streckeisen, Ch.P. Thornton (2pp), A.C. Tobi, E. Wenk.

Contribution No.67. Working Group on Lamprophyres.

Proposal by the Chairman.

W. Wimmenauer: Remarks on the Discussion on the Nomenclature of Lamprophyres. 7pp.

I.K. Pjatenko: About Nomenclature and Classification of Alkaline Lamprophyres. 3pp.

N.M.S. Rock: The Lamprophyre Problem. 11pp.

K.L. Currie: Comments.

W. Wimmenauer: Remarks on the contribution by N.M.S. Rock.

W. Kramer: Zum Questionnaire: Lamporphyr-Nomenklatur.

Contribution No.68. H. Sørensen: Carbonatite Nomenclature. 2pp.

Contribution No.69. H. Sørensen: The classification and nomenclature of melilite bearing rocks. 3pp.

List of 52 melilite-bearing rocks.

CIRCULAR NO.29 – OCTOBER, 1977 – AUGUST, 1978

Contribution No.70. Minutes of the Working Meeting held at Prague on 26-29 September, 1977.

Contribution No.71. Working Group on Pyroclastic Rocks. Final Questionnaire on Pyroclastic and Mixed-Pyroclastic Rocks. 5pp.

Contribution No.72. Classification and Nomenclature of Volcanic rocks. Second version.

Annexed are texts of Lamprophyres, Carbonatites and Melilite-bearing Igneous Rocks.

Subsequently, remarks and comments were received by A. Dudek, M.J. Le Bas, R.W. Le Maitre, P.A. Sabine, R. Schmid, K. Smulikowski, H. Sørensen, A. Streckeisen, V. Széky-Fux, W. Wimmenauer, B. Zanettin and the National Committee of the GDR (DDR).

Contribution No.73. Summary of Remarks and Comments. Questionnaire. 6pp.

Comments by K. Smulikowski, W. Pälchen and G. Röllig, A.C. Tobi (2pp. on keratophyre).

Contribution No.74. Classification and Nomenclature of Volcanic Rocks, Lamprophyres, Carbonatites and Melilitic Rocks. Recommendations and Suggestions. Final draft.

CIRCULAR NO.30 – FEBRUARY, 1979

Contribution No.75. Terminological Commission of the Petrographic Committee of the USSR: Systematics of Magmatic Rocks. Text translated by A. Dudek (9pp.). Tables and diagrams in Russian, partly translated. (The contribution was received on April 8, 1978, but delayed for translation).

Minutes of the Ordinary Meeting held at Padova on April 30-May 5, 1979.

CIRCULAR NO.31 – NOVEMBER, 1979

Contribution No.76. Nomenclature and Classification of Descriptive Nomenclature and Classification of Pyroclastic Deposits and Fragments. First draft. 6pp.

Contribution No.77. F. Chayes: Distribution of basalt, basanite, andesite and dacite in a normative equivalent of the Q-A-P-F double triangle. 14pp text. 8pp tables and diagrams.

Contribution No.78. A. Streckeisen: The QAPF double triangle: Modal or normative? 10pp text, 10pp tables and diagrams.

Contribution No.79. E.F. Malejev, M.A. Petrova, V.T. Frolov, L.V. Chvorova, M.N. Schtschebakova: Die heutigen Vorstellungen über die Klassifikation der vulkanoklastichen und gemischt-vulkonoklastisch-sedimentären Gesteine. 6pp text, 4pp tables.

CIRCULAR NO.32 – MAY, 1980

Contribution No.80. Comments on the Contributions by F. Chayes (Contribution No.77) and A. Streckeisen (Contribution No.78): B. Zanettin, P.A. Sabine, M.J. Le Bas, K. Smulikowski, A. Dudek, H. de la Roche, J.W. Cole, R.W. Le Maitre, C.P. Snyman, R. Schmid, V. Széky-Fux.

Reply by F. Chayes.

Reply by A. Streckeisen. 13pp.)

Contribution No.81. Some replies to the

Suggestions Submitted by R.W. Le Maitre: S. V. Efremova (on behalf of the Termin. Comm. of USSR); K. Smulikowski; A. Dudek.

Contribution No.82. F. Chayes: A further plea for modification of the basalt-basanite field boundary in the QAP and Q'A'P' diagram. 6pp.

Contribution No.83. A. Streckeisen: A plea for retaining the present basalt-basanite boundary. 10pp.

Contribution No.84. R. Schmid. Descriptive Nomenclature and Classification of Pyroclastic Deposits and Fragments. Revised draft. 5pp.

Minutes of the Ordinary Meeting held in Paris on July 8-10, 1980. It was decided to test the following methods for the chemical classification of effusive rocks:- SiO_2 vs $Na_2O + K_2O$ (Harker diagram); R_1-R_2 (de la Roche); Q'A'P'F' (Chayes); and Q'/F' – ANOR (Streckeisen and Le Maitre).

CIRCULAR NO.33 – OCTOBER, 1980

Contribution No.85. A. Streckeisen: On the Use of the Normative Q'A'P'F' Diagram. 5pp.

CIRCULAR NO.34 – JULY, 1981

Contribution No.86. F. Chayes: The names of the lavas of Etna.

Contribution No.87. G. Bellieni, E.M. Piccirillo and B. Zanettin: Classification and Nomenclature of Basalts. First draft. 9pp text, 12pp tables and diagrams.

Contribution No.88. R.W. Le Maitre: (a) CLAIR and PETROS Plots. (b) Classification Comparisons.

Contribution No.89. P.A. Sabine and R.S. Lawson: Note on the plotting of analysed igneous rocks. 7pp and 10 tables.

Contribution No.90. A. Streckeisen: Provisional Remarks on Chemical Classifications. 4pp.

Contribution No.91. M.J. Le Bas: Plots of basic and ultramafic rocks from:- East African Rift Valley Ultramafic Rocks; West Kenyan and East Ugandan Tertiary Volcanic and Plutonic Rocks.16 diagrams.

Contribution No.92. E. Ancochea and J.L. Brändle: Basaltic Rocks from Campos de Calatrava (Central Spain); (a) Basalts; (b)

Basanites; (c) Olivine nephelinites; (d) Olivine melilitites. 3pp text and 32pp tables and diagrams.

Contribution No.93. Terminological Commission of USSR. Classification and nomenclature of magmatic rocks. Textbook in Russian.

Contribution No.94. A. Streckeisen: Remarks on the Lavas of Mt Etna. 9pp.

Contribution No.95. S.R. Nockolds, 1978, 78-89. Latites and latite-andesites. 120-123. Basalts and dolerites.

Contribution No.96. A. Streckeisen: Test of various methods. 10pp.

Contribution No.97a. H. de la Roche: Naming and/or classifying the igneous rocks?

Contribution No.97b. M. Tégyey: Note on the plotting of various volcanic rocks. 30 diagrams; (a) rhyolites-dacites; (b) andesites; (c) laves basiques; (d) trachytes-latites; (e) phonolites.

Comments: S. Peltz: Chemical classification of volcanic rocks from the Calimani-Gurghiu-Harghita Mts (East Carpathians).

CIRCULAR NO.35 – NOVEMBER, 1981.

Minutes of the Working Meeting on September, 21-24, in Cambridge, England.

CIRCULAR NO.36 – JANUARY, 1983

Contribution No.98. A. Streckeisen: On the Use of the Normative QAPF Diagram for the Chemical Classification of Igneous Rocks. Text 7pp, tables and diagrams 11pp. (Plutonic rocks of the Lachlan Mobile Belt, N.S.W., are tested by various methods.)

Contribution No.99. F. Chayes: Do rock Names Have Usefully Exclusive Chemical Referents? 5pp.

Contribution No.100. R.W. Le Maitre, G. Bellieni, M.J. Le Bas, R. Schmid, A. Streckeisen, B. Zanettin, E.M. Piccirillo and E. Justin-Vistenin: A Proposal for a Definitive Chemical Classification of Volcanic Rocks Based on the Total Alkali-Silica Diagram (TAS diagram). 24pp text, 12pp diagrams.

CIRCULAR NO.37 – MAY, 1983

Contribution No.101. P.A. Sabine, R.S. Lawson and R.K. Harrison: The Nomenclature and Classification of British Volcanic and Hypabyssal Rocks: The Total Alkali Oxides : SiO_2 Diagram. Text 11pp, tables and diagrams 25pp.

Comments regarding Contributions No.98, 99 and 100:-
1. A. Dudek.
2. S.V. Efremova.
3. M.J. Le Bas.
4. E.A.K. Middlemost.
5. J.F.G. Wilkinson

CIRCULAR NO.38 – NOVEMBER, 1983

Further comments to the TAS diagram (Circular No.36, Contribution No.100):-
6. F. Chayes: On the desirability of deleting existing rock names from the TAS partition of SIlica-Alkali space.
7. A. Dudek.
8. H. Sørensen.
9. A. Cundari.
10. S.V. Efremova.
11. A. Streckeisen: Suggestion for Fields of Foidal Rocks in the TAS Diagram. 5pp.
12. F. Chayes: Use of the names hawaiite, mugearite, benmoreite, and tristanite in the proposed TAS classification.
13. F. Chayes: Relative frequency of models in published descriptions of chemically analysed igneous rocks.
14. F. Chayes.
15. V. Széky-Fux.
16. R.W. Le Maitre.

Contribution No.102. G. Bellieni, E. Justin-Visentin, R.W. Le Maitre, E.M. Piccirillo and B. Zanettin: Proposal for a Division of the Basaltic (B) Field for the TAS Diagram. Text 6pp, 6 diagrams.

Minutes of the Working Meeting at Granada, 4-8 September, 1983

CIRCULAR NO.39 – MARCH, 1984

Comments to Circular No.38:-
1. A. Streckeisen: On the distinction between rhyolite (trachyte) sensu stricto and alkali rhyolite (trachyte). Foidal rocks of fields U1-U3. 6pp.
2. S. Peltz.
3. A. Dudek.

4. F. Chayes (8pp).

5. A. Streckeisen.

6. M.J. Le Bas.

7. G. Bellieni, E. Justin-Visentin, R.W. Le Maitre, E.M. Piccirillo, B. Zanettin: Proposal for a division of the basaltic (B) fields of the TAS. Reply to Comments on Contribution No.102 "Proposal for a division of the basaltic (B) field of the TAS diagram".

New text of the Total Alkali Silica Diagram (TAS).

Voting form.

CIRCULAR NO.40 – JUNE, 1984

Results of voting on TAS. Observations and proposals on TAS. Information on the Moscow Congress. Comments: J.L. Brändle: Remarks on TAS.

Contribution No.103. P.A. Sabine and D.S. Sutherland with contributions by R.B. Elliott, R.K. Harrison, J. Preston, C.J. Stillman: Petrography of British Igneous Rocks. In, D.S. Sutherland: Igneous Rocks of the British Isles, 1982, Wiley,pp. 479-544.

CIRCULAR NO.41 – SEPTEMBER, 1984

Minutes of the Ordinary Meeting in Moscow, 8-10 July, 1984.

Comments: A. Streckeisen: Comments to Circular No.40 (7.7.1984)

CIRCULAR NO.42A – MARCH, 1985

B. Zanettin: Proposed new chemical classification of volcanic rocks. Episodes, vol. 7, No.4, December, 1984, 2pp.

Division of trachyandesite field in TAS – result of vote.

Basalt nomenclature – result of vote.

Contribution No.104. M.J. Le Bas: Basalt Nomenclature.

Contribution No.105. M.J. Le Bas: Juxtaposition of dacite and trachyte in TAS. 7pp.

Contribution No.106. A.R. Woolley and M.J. Le Bas: Use of the qualifiers sodic and potassic. 8pp.

Contribution No.107. M.J. Le Bas: Basanite-tephrite field in the TAS diagram.

Contribution No.108. S.V. Efremova: Classification of igneous rocks. Comments by H. de la Roche.

Contribution No.109. O.A. Bogatikov, N.P. Mikhailov, V.I. Gon'shakova (Editors): Classification and nomenclature of magmatic rocks. Nedra, Moscow, 1981. In Russian.

Contribution No.110. S.V. Efremova: On the present state of the classification of magmatic rocks. Nauka, Moscow, 1983.

Contribution No.111. Terminological Commission of the USSR: Extract from "Magmatic Rocks". Nauka, Moscow, 1983. In French. 8pp.

Contribution No.112. S.V. Efremova: Commentary on the principles of the classification in the USSR. (February, 1985) 6pp.

Contribution No.113. A. Streckeisen: A very minor suggestion.

CIRCULAR NO.42B

Contribution No.114. M. Stefanova: On the unreliability of silica and total alkali percentages as main classification parameters in petrology. 5pp.

Contribution No.115. M. Stefanova: On the principles of igneous rock classification. 5pp.

Contribution No.116. E.A.K. Middlemost: Comments on Circular No.42. 3pp.

Contribution No.117. D.S. Barker: Comments on Contribution No.104-113. 4pp.

Contribution No.118. S.V. Efremova: Comments on Circular No.42.

Contribution No.119. A. Streckeisen: Remarks on Contribution No.104-107.

Contribution No.120. K.G. Cox: Comments on trachyandesites and picrite basalts.

Contribution No.121. J.L. Brändle: Rocas volcanicas de Campos de Calatrava. 7pp.

Contribution No.122. V. Széky-Fux: Remarks to Circular No.42. 2pp.

CIRCULAR NO.43 – AUGUST, 1985

Minutes of the London Meeting, May 22-24, 1985:-

1. Division of the trachyandesite field in TAS. Result of vote.

2. Basalt nomenclature and result of vote.

3. Juxtaposition of dacite and trachyte in the TAS diagram.

4. The use of the qualifiers sodic and potassic.

5. The basanite-tephrite field.

6. Classification of ultrabasic and ultra-alka-

line rocks.

7. Classification of volcanic cumulate rocks.

8. Glossary.

9. Other topics. (a) The terms alkali and peralkaline; (b) version from cumulate diagram from Mikhailov.

10. (c) Proposal by the Terminological Commission of the USSR for a nomenclatorial hierarchy; (d) Use of the latest version of TAS in forthcoming books.

Contribution No.123. R.W. Le Maitre: Remarks on (a) trachydacite-dacite field; (b) low-K, medium-K, high-K; (c) hawaiite and trachybasalt.

Contribution No.124. M.J. Le Bas: Comments on Contribution No.123: (a) trachydacite-dacite; (b) sodic-potassic; (c) trachybasalt. Questionnaire on (a) trachydacite; (b) high-K, medium-K, low-K.

CIRCULAR NO.44 – DECEMBER, 1985

Analysis of answers to Questionnaire circulated in August, 1985.

Contribution No.125. Remarks on Questionnaire by A. Streckeisen, J. Keller, H. de la Roche.

Contribution No.126. Y. Yasef: Suggestion for a chemical classification of the acid volcanic rocks: vertical limits at SiO_2 52, 57, 63, 68, 73%; a horizontal limit at $Na_2O + K_2O = 8\%$ separates a subalkaline series (O) from a midalkaline series (S). 4pp.

Contribution No.127. M. Stefanova: Principles of a combined chemical-mineralogical classification of igneous rocks according to the parameters k_{Si}-k_K. 5pp.

Contribution No.128. E.A.K. Middlemost: Remarks on limits in the TAS diagram: (a) alkali basalt; (b) trachydacite; (c) terms potassic, sodic, ultrasodic; (d) foidites. 4pp.

CIRCULAR NO.45 – JULY, 1986

Contribution No.129a. M.J. Le Bas: Ultra-alkaline and ultrabasic volcanic rocks. 2pp., 6 diagrams. (July, 1986).

Contribution No.129b. M.J. Le Bas: The same: Figs 2, 3, 4 are revised (September, 1986).

CIRCULAR NO.46 – SEPTEMBER, 1986

Contribution No.130. A. Streckeisen: An

Alkali Granite and Related Rocks. On the term 'alkali'. 6pp.

Contribution No.131. E.A.K. Middlemost: Various suggestions to the TAS diagram.

Contribution No.132. D.S. Barker: Ultra-alkaline Rocks. 5pp.

Contribution No.133. A. Dudek: Ultra-alkaline Rocks. 7pp.

Contribution No.134. J.L. Brändle: Ultra-alkaline Rocks. 3pp.

CIRCULAR NO.47A – MARCH, 1987

Minutes of the Meeting held at the University of Freiburg i.B., F.R.G., October 23-28, 1986:-

1. Answers to the Questionnaire circulated with Circular 43, August, 1985.

2. The terms low-K, medium-K, and high-K.

3. TAS – final considerations.

4. Ultramafic and ultra-alkaline rocks.

5. Volcanic cumulate rocks.

6. The terms 'alkali', 'alkaline', and 'peralkaline'.

7. Relationship of volcanic classification to plutonic classification.

8. Other rock groups for inclusion in the Glossary.

9. Glossary.

10. Hierarchy of nomenclature.

Contribution No.135. M.J. Le Bas and A.R. Woolley: Carbonatite nomenclature.

Questionnaire: (a) ultra-alkaline rocks; (b) komatiite; (c) lamproite.

CIRCULAR NO.47B – SEPTEMBER, 1987

Replies to the Questionnaire regarding the classification of (a) ultra-alkaline rocks; (b) komatiites; (c) lamproites.

Contribution No.136. A. Streckeisen: Lamproites. Part of a recently published paper by Bergman on the classification of lamproites. A report on the visit of M.J. Le Bas and A.R. Woolley in the USSR, April 19 – May 2, 1987.

CIRCULAR NO.48 – APRIL, 1988

Contribution No.137. E.A.K. Middlemost: Iron oxidation ratios, norms, and the classification of volcanic rocks.

Contribution No.138. E.A.K. Middlemost:

Lamproites, ultrabasic rocks in TAS, and the sub-root name 'potassic trachybasalt'. 2pp.

Contribution No.139. N.M.S. Rock: Summary of arguments for classifying kimberlites and lamproites as lamprophyres. 2pp.

Contribution No.140. S.V. Efremova: Method proposed by V.E. Gendler for the petrochemical standardisation of igneous rocks. 5pp.

Contribution No.141. Some general remarks on classification of igneous rocks as adopted by the Terminological Commission of the USSR. 3pp.

Contribution No.142. A.R. Woolley: The nomenclature of carbonatites. 6pp.

Draft sections of the Glossary.

CIRCULAR NO.49 – JUNE, 1988

Minutes of the Subcommission Meeting held in Copenhagen, 13-17 May, 1988:-

1. Ultrabasic Rocks.
2. Lamprophyre, lamproite and kimberlite.
3. Carbonatites.
4. Nephelinites.
5. Other rock groups.
6. Glossary.

Contribution No.143. G. Bellieni, E. Justin-Visentin and B. Zanettin: On the necessity to use the root names 'peralkaline trachytes and rhyolites' instead of 'alkali trachytes and rhyolites' in the TAS diagram.

Contribution No.144. A. Streckeisen: Volcanic rocks: Principles of classification. 2pp.

Contribution No.145. B. Zanettin, G. Bellieni and E. Justin-Visentin: Distinction of pantellerite and comendite.

K. Smulikowski: Some comments on Classification and Nomenclature of volcanic rocks. Archivum Mineralogiczne, t.37, 1981.

Appendix B. Lists of Participants

This appendix contains the names of all those people who have helped the Subcommission in various ways and consists of two lists.

The **first list** of participants is arranged alphabetically by the country which the participant represented or the country in which they were residing when they corresponded with the Subcommission. The name of each participant is followed in brackets by their affiliation with the Subcommission using the following abbreviations:-

M – a Member of the Subcommission acting as a representative of the country.

W – a member of one of the Working groups who participated by attending meetings and/or by correspondence.

P – a member of the Pyroclastic working group only.

C – a Correspondent with the Subcommission who contributed with written submissions in response to Subcommission circulars.

G – a Guest at one or more of the meetings.

The **second list** is simply arranged alphabetically by name, followed in brackets by their country.

1. PARTICIPANTS LISTED ALPHABETICALLY BY COUNTRY

ARGENTINA (4)
Andreies, R.R. (C)
Mazzoni, M.M. (C)
Spalletti, L.A. (C)
Teruggi, M.E. (M)

AUSTRALIA (21)
Branch, C.D. (W)
Cundari, A. (W)
Dallwitz, W.B. (M)
Ewart, A. (C)
Flinter, B. (G)
Green, D.H. (W)
Harris, P.G. (C)
Hensel, H.D. (C)
Joyce, E.B. (G)
Le Maitre, R.W. (M)
Mackenzie, D.E. (C)
Matthison, C.J. (C)
Middlemost, E.A.K. (W)
Neall, V. (G)
Rock, N.M.S. (W)
Taylor, S.R. (P)

Vallance, T.G. (W)
Wass, S.Y. (C)
White, A.J.R. (C)
Wilkinson, J.F.G. (M)
Wilson, A.P. (W)

AUSTRIA (4)
Daurer, A. (C)
Exner, C. (C)
Richter, W. (C)
Wieseneder, H. (C)

BELGIUM (3)
Duchesne, J.-C. (W)
Michot, J. (W)
Michot, P. (W)

BOLIVIA (1)
Ploskonka, E. (P)

BRAZIL (2)
Coutinho, J.M.V. (W)
Leonardos, Jr., O.H. (W)

BULGARIA (5)
Boyadjiev, St.G. (G)
Georgiev, G.K. (C)
Ivanov, R. (W)
Stefanova, M. (W)
Yasef, Y. (W)

CANADA (22)
Armstrong, R.L. (C)
Ayres, L.D. (M)
Baragar, W.R.A. (P)
Church, B.N. (W)
Currie, K.L. (W)
Davidson, A. (W)
Edgar, A.D. (W)
Frisch, Th. (C)
Gittins, J. (W)
Goodwin, A.M. (M)
Lambert, M.B. (P)
Laurent, R. (C)
Laurin, A.F. (W)
Martignole, J. (W)
McCutcheon, S.R. (C)

Moore, J.M. (C)
Naldrett, A.J. (W)
Ruitenberg, A.A. (C)
Sharma, K.N.M. (W)
Thurston, P.C. (P)
Woodsworth, G.J. (C)
Woussen, G. (C)

CHILE (1)
Vergara, M. (W)

CHINA (3)
Heng Sheng Wang (M)
Sung Shu Ho (C)
Wang, B. (M)

COLOMBIA (1)
Schaufelberger, P. (C)

CZECHOSLOVAKIA (8)
Dudek, A. (M)
Fediuk, F. (C)
Fiala, F. (W)
Kopecky, L. (W)
Neužilová, N. (C)
Palivcova, M. (C)
Suk, M. (C)
Vejnar, Zd. (C)

DENMARK (6)
Berthelsen, A. (C)
Jakobsson, S. (G)
Noe Nygaard, A. (M)
Sørensen, H. (M)
Wilson, J.R. (C)
Zeck, H.P. (W)

EIRE (1)
Mohr, P. (C)

FINLAND (3)
Härme, M. (C)
Laitakati, I. (C)
Vartiainen, H. (C)

FRANCE (17)
Alsac, C. (P)
Bordet, P. (G)
Brousse, R. (C)
Chenevoy, M. (C)

Colin, F. (C)
de la Roche, H. (M)
Forestier, F.H. (W)
Giraud, P. (W)
Juteau, Th. (W)
Lameyre, J. (M)
Leterrier, J. (G)
Raguin, E. (C)
Roques, M. (G)
Soler, E. (P)
Tégyey-Dumait, M. (P)
Velde, D. (W)
Vincent, P.M. (W)

FRG (34)
Amstutz, G.C. (W)
Arndt, N. (G)
Bambauer, H.U. (C)
Chudoba, K. ()
Correns, C.W. (W)
Förster, H. (C)
Frechen, J. (C)
Hentschel, H. (W)
Jung, D. (W)
Keller, J. (M)
Knetsch, G. (C)
Kunz, K. (W)
Lehmann, E. (W)
Lensch, G. (W)
Lorenz, V. (W)
Martin, H. (W)
Matthes, S. (C)
Mehnert, K.R. (M)
Mihm, A. (C)
Morteani, G. (C)
Müller, P. (W)
Murawski, H. (C)
Okrusch, M. (C)
Paulitsch, P.D. (C)
Pichler, H. (W)
Rost, F. (W)
Schmincke, H.U. (C)
Stengelin, R. (W)
Troll, G. (C)
Vieten, K. (P)
Wedepohl, K.H. (W)
Wimmenauer, W. (M)
Winkler, H.G.F. (W)
Wurm, A. (C)

GDR (11)
Huebscher, H.D. (C)
Kramer, W. (W)
Mädler, J. (C)
Mathé, G. (C)
Mucke, D. (W)
Pälchen, W. (W)
Pfeiffer, L. (W)
Röllig, G. (W)
Seim, R. (C)
Tischendorf, G. (M)
Watznauer, A. (W)

GUYANA (1)
Singh, S. (W)

HUNGARY (3)
Kubovics, I. (G)
Pantó, G. (M)
Széky-Fux, V. (M)

ICELAND (1)
Thorarinsson, S. (P)

INDIA (8)
Bose, M.K. (W)
Chatterjee, S.C. (M)
Sen, S.K. (W)
Subba Rao, S. (C)
Subbarao, K.V. (P)
Subramaniam, A.P. (W)
Subramaniam, K.V. (G)
Sukheswala, R.N. (M)

ISRAEL (1)
Bentor, Y.K. (W)

ITALY (23)
Balconi, M. (C)
Bellieni, G. (M)
Bertolani, M. (C)
Boriani, A. (G)
Callegari, E. (C)
Cristofolini, R. (C)
D'Amico, Cl. (P)
Dal Piaz, G.V. (C)
De Vecchi, Gp. (G)
Deriu, M. (C)
Galli, M. (C)
Innocenti, F. (C)

Justin-Visentin, E. (W)
Malaroda, R. (C)
Mittempergher, M. (W)
Morbidelli, L. (C)
Piccirillo, E.M. (W)
Rittmann, A. (W)
Sassi, F. (C)
Simboli, G. (W)
Venturelli, A. (C)
Villari, L. (P)
Zanettin, B. (M)

JAPAN (3)
Aoki, K.J. (M)
Aramaki, S. (M)
Katsui, Y. (M)

KENYA (1)
Williams, L.A.J. (W)

MALAYSIA (1)
Hutchinson, C.S. (C)

MARTINIQUE (1)
Westercamp, D. (P)

NETHERLANDS (6)
de Roever, W.P. (M)
Den Tex, E. (C)
Oen, I.S. (C)
Tobi, A.C. (M)
Uytenbogaardt, W. (C)
van der Plas, L. (C)

NEW ZEALAND (3)
Cole, J.W. (M)
Coombs, D.S. (C)
Neal, V.E. (W)

NORWAY (8)
Barth, T.F.W. (W)
Bryhni, I. (C)
Carstens, H. (W)
Dons, J.A. (C)
Griffin, W.L. (C)
Oftedahl, C. (C)
Strand, T. (C)
Torske, T. (W)

POLAND (2)
Nowakowski, A. (C)
Smulikowski, K. (M)

PORTUGAL (3)
Aires Barros, L. (P)
Mourazmiranda, A. (C)
Schermerhorn, L.J.G. (C)

ROMANIA (9)
Berbeleac, I. (W)
Bercia, J. (G)
Dimitrescu, R. (C)
Giusca, D. (P)
Kräutner, H. (W)
Peltz, S. (P)
Radulescu, D. (M)
Savu, H. (C)
Seclaman, M. (W)

SAUDI ARABIA (1)
Wolf, K.H. (P)

SOUTH AFRICA (6)
Ferguson, J. (M)
Frick, G. (C)
Moore, A.C. (C)
Snyman, C.P. (M)
van Biljon, S. (C)
von Gruenewaldt, G. (M)

SPAIN (7)
Ancochea, E. (W)
Brändle, J.H. (M)
Corretgé, L. (C)
Fuster, J.M. (C)
Muñoz, M. (W)
Pascual, E. (W)
Puga, E. (C)

SURINAM (2)
Dahlberg, E.H. (W)
de Roever, E.W.F. (W)

SWEDEN (5)
Gorbatschev, R. (C)
Koark, H.J. (C)
Lundqvist, T. (P)
Persson, L. (P)
von Eckerman, H. (C)

SWITZERLAND (18)
Ayrton, S. (C)
Bearth, P. (W)
Burckhardt, C.E. (P)
Dietrich, V. (C)
Hänny, R. (C)
Hügi, T. (C)
Kübler, B. (C)
Maggetti, M. (W)
Nickel, E. (C)
Niggli, E. (W)
Persoz, F. (C)
Schmid, R. (M)
Seemann, U. (P)
Streckeisen, A. (M)
Trommsdorff, V. (M)
Vuagnat, M. (W)
Wenk, E. (W)
Woodtli, R. (C)

TAIWAN (2)
Juan, V.C. (C)
Veichow, C.J. (C)

TURKEY (5)
Aslaner, M. (C)
Ataman, G. (C)
Irdar, E. (C)
Kraëff, A. (C)
Kumbazar, I. (C)

UGANDA (1)
Macdonald, R. (W)

UK (29)
Baker, P.E. (C)
Berridge, N.G. (C)
Collins, G.H. (C)
Cox, K.G. (W)
Dearnley, R. (C)
Elliot, R.B. (C)
Ellis, S.E. (W)
Fitch, F.J. (C)
Francis, E.H. (P)
Harrison, R.K. (W)
Howells, M. (P)
Howie, R.A. (W)
King, B.C. (W)
Lawson, R.I. (W)
Le Bas, M.J. (M)

MacKenzie, W.S. (C)
Muir, I.D. (W)
Nockolds, S.R. (C)
Preston, J (C)
Rea, W.J. (P)
Sabine, P.A. (M)
Stillman, C.J. (C)
Upton, B.J.G. (W)
Wadsworth, W.J. (W)
Weaver, S.D. (P)
Wells, A.K. (W)
Wells, M.K. (W)
Woolley, A.R. (M)
Wright, A.E. (W)

USA (75)
Barker, D.S. (W)
Bateman, P.C. (M)
Best, M.G. (C)
Boone, E.M. (C)
Brooks, E.R. (C)
Chayes, F. (M)
Cheney, J.T. (C)
Chesterman, C.W. (C)
Christiansen, R.L. (C)
Coats, R.R. (P)
Coveney, R.M. (C)
de Waard, D. (W)
Dietrich, R.V. (P)
Duffield, W. (G)
Earton, R.M. (P)
Effimoff, I. (C)
Ekblaw, S.E. (C)
Emerson, D.O. (C)
Ernst, G. (C)
Evans, B.W. (C)
Fisher, R.V. (M)
Forbes, R.B. (C)
Gastil, G. (C)
Guidotti, C.V. (C)
Harrison, J.E. (C)
Hawkin, J.W. (C)
Hays, J.F. (C)
Heiken, G. (P)
Herz, N. (C)
Hobbs, S.W. (C)
Hollister, L.S. (C)
Honnorez, J. (P)
Hyndman, D.W. (C)

Jackson, E.D. (W)
Krauskopf, K. (C)
Lofgren, G. (M)
Lyons, P.C. (C)
Macdonald, G.A. (W)
Mattson, P.H. (C)
McBirney, A.R. (M)
McCoy, F.W. (P)
McHone, J.G. (C)
Miller, T. (W)
Misch, P. (C)
Miyashiro, A. (C)
Mohr, P.A. (P)
Moore, G.E. (C)
Murray, A.D. (C)
Mutschler, F.E. (C)
O'Connor, J.T. (C)
Peck, D.L. (M)
Peterson, D.W. (W)
Putman, G.W. (C)
Rankin, D.W. (C)
Reitan, P.H. (C)
Scott, R.B. (C)
Sheridan, M.F. (P)
Simkin, T. (C)
Snavely, P. (P)
Snyder, G.L. (W)
Taubeneck, W.H. (C)
Thornton, C.P. (M)
Tilling, R.I. (P)
Toulmin III, P. (C)
Tweto, O. (C)
Vidale, R.J. (C)
Vitaliano, C.J. (C)
Walker, G.P.L. (P)
Waters, A.C. (C)
Whitney, J.A. (C)
Whitten, E.H.T. (C)
Wilcox, R.E. (C)
Williams, H. (C)
Wones, D.R. (C)
Wyllie, P.J. (C)

USSR (42)
Afanass'yev, G.D. (W)
Andreeva, E.D. (W)
Bogatikov, O.A. (C)
Borodaevskaya, M.R. (W)
Borodin, L.S. (W)

Chvorova, L.V. (C)
Daminova, A.M. (W)
Dmitriev, Yu.I. (P)
Efremova, S.V. (M)
Frolova, T.I. (W)
Gladkikh, V.S. (W)
Gon'shakova, V.I. (C)
Gorshkov, G.S. (W)
Gurulev, S.A. (G)
Iakovleva, E.B. (W)
Kononova, V.A. (G)
Kravchenko, S. (G)
L'vov, B.K. (W)
Lazko, E. (W)
Luchitsky, J. (G)
Malejev, E.F. (P)
Markovsky, B.A. (W)
Massaitis, V.L. (W)
Mikhailov, N.P. (M)
Orlova, M.P. (W)
Pertsev, N.N. (W)
Petrova, M.A. (W)
Pjatenko, I.K. (W)
Polunina, L.A. (W)
Rotman, V.K. (W)
Ruminantseva, N.A. (W)
Schtschebakova, M.N. (C)
Sharpenok, L. (G)
Shemyakin, V.M. (W)
Shirinian, H.G. (G)
Shurkin, K.A. (W)
Sobolev, N. (W)
Sobolev, S.F. (W)
Sobolev, V.S. (G)
Sveshnikova, S.V. (W)
Vorobieva, O.A. (M)
Yashina, R. (G)

VENEZUELA (1)
Urbani, F. (C)

YUGOSLAVIA (3)
Karamata, S. (W)
Majer, V. (C)
Obradovic, J. (P)

ZAMBIA (1)
Cooray, P.G. (W)

2. PARTICIPANTS LISTED ALPHABETICALLY (WITH COUNTRY)

Afanass'yev, G.D. (USSR)
Aires Barros, L. (Portugal)
Alsac, C. (France)
Amstutz, G.C. (FRG)
Ancochea, E. (Spain)
Andreeva, E.D. (USSR)
Andreies, R.R. (Argentina)
Aoki, K.J. (Japan)
Aramaki, S. (Japan)
Armstrong, R.L. (Canada)
Arndt, N. (FRG)
Aslaner, M. (Turkey)
Ataman, G. (Turkey)
Ayres, L.D. (Canada)
Ayrton, S. (Switzerland)
Baker, P.E. (UK)
Balconi, M. (Italy)
Bambauer, H.U. (FRG)
Baragar, W.R.A. (Canada)
Barker, D.S. (USA)
Barth, T.F.W. (Norway)
Bateman, P.C. (USA)
Bearth, P. (Switzerland)
Bellieni, G. (Italy)
Bentor, Y.K. (Israel)
Berbeleac, I. (Romania)
Bercia, J. (Romania)
Berridge, N.G. (UK)
Berthelsen, A. (Denmark)
Bertolani, M. (Italy)
Best, M.G. (USA)
Bogatikov, O.A. (USSR)
Boone, E.M. (USA)
Bordet, P. (France)
Boriani, A. (Italy)
Borodaevskaya, M.R. (USSR)
Borodin, L.S. (USSR)
Bose, M.K. (India)
Boyadjiev, St.G. (Bulgaria)
Branch, C.D. (Australia)
Brändle, J.H. (Spain)
Brooks, E.R. (USA)
Brousse, R. (France)
Bryhni, I. (Norway)
Burckhardt, C.E. (Switzer-
   land)
Callegari, E. (Italy)

Carstens, H. (Norway)
Chatterjee, S.C. (India)
Chayes, F. (USA)
Chenevoy, M. (France)
Cheney, J.T. (USA)
Chesterman, C.W. (USA)
Christiansen, R.L. (USA)
Chudoba, K. (FRG)
Church, B.N. (Canada)
Chvorova, L.V. (USSR)
Coats, R.R. (USA)
Cole, J.W. (New Zealand)
Colin, F. (France)
Collins, G.H. (UK)
Coombs, D.S. (New Zealand)
Cooray, P.G. (Zambia)
Correns, C.W. (FRG)
Corretgé, L. (Spain)
Coutinho, J.M.V. (Brazil)
Coveney, R.M. (USA)
Cox, K.G. (UK)
Cristofolini, R. (Italy)
Cundari, A. (Australia)
Currie, K.L. (Canada)
D'Amico, Cl. (Italy)
Dahlberg, E.H. (Surinam)
Dal Piaz, G.V. (Italy)
Dallwitz, W.B. (Australia)
Daminova, A.M. (USSR)
Daurer, A. (Austria)
Davidson, A. (Canada)
de la Roche, H. (France)
de Roever, E.W.F. (Surinam)
de Roever, W.P. (Netherlands)
De Vecchi, Gp. (Italy)
de Waard, D. (USA)
Dearnley, R. (UK)
Den Tex, E. (Netherlands)
Deriu, M. (Italy)
Dietrich, R.V. (USA)
Dietrich, V. (Switzerland)
Dimitrescu, R. (Romania)
Dmitriev, Yu.I. (USSR)
Dons, J.A. (Norway)
Duchesne, J.-C. (Belgium)
Dudek, A. (Czechoslovakia)
Duffield, W. (USA)

Earton, R.M. (USA)
Edgar, A.D. (Canada)
Effimoff, I. (USA)
Efremova, S.V. (USSR)
Ekblaw, S.E. (USA)
Elliot, R.B. (UK)
Ellis, S.E. (UK)
Emerson, D.O. (USA)
Ernst, G. (USA)
Evans, B.W. (USA)
Ewart, A. (Australia)
Exner, C. (Austria)
Fediuk, F. (Czechoslovakia)
Ferguson, J. (South Africa)
Fiala, F. (Czechoslovakia)
Fisher, R.V. (USA)
Fitch, F.J. (UK)
Flinter, B. (Australia)
Forbes, R.B. (USA)
Forestier, F.H. (France)
Förster, H. (FRG)
Francis, E.H. (UK)
Frechen, J. (FRG)
Frick, G. (South Africa)
Frisch, Th. (Canada)
Frolova, T.I. (USSR)
Fuster, J.M. (Spain)
Galli, M. (Italy)
Gastil, G. (USA)
Georgiev, G.K. (Bulgaria)
Giraud, P. (France)
Gittins, J. (Canada)
Giusca, D. (Romania)
Gladkikh, V.S. (USSR)
Gon'shakova, V.I. (USSR)
Goodwin, A.M. (Canada)
Gorbatschev, R. (Sweden)
Gorshkov, G.S. (USSR)
Green, D.H. (Australia)
Griffin, W.L. (Norway)
Guidotti, C.V. (USA)
Gurulev, S.A. (USSR)
Hänny, R. (Switzerland)
Härme, M. (Finland)
Harris, P.G. (Australia)
Harrison, J.E. (USA)
Harrison, R.K. (UK)

Hawkin, J.W. (USA)
Hays, J.F. (USA)
Heiken, G. (USA)
Heng Sheng Wang, (China)
Hensel, H.D. (Australia)
Hentschel, H. (FRG)
Herz, N. (USA)
Hobbs, S.W. (USA)
Hollister, L.S. (USA)
Honnorez, J. (USA)
Howells, M. (UK)
Howie, R.A. (UK)
Huebscher, H.D. (GDR)
Hügi, T. (Switzerland)
Hutchinson, C.S. (Malaysia)
Hyndman, D.W. (USA)
Iakovleva, E.B. (USSR)
Innocenti, F. (Italy)
Irdar, E. (Turkey)
Ivanov, R. (Bulgaria)
Jackson, E.D. (USA)
Jakobsson, S. (Denmark)
Joyce, E.B. (Australia)
Juan, V.C. (Taiwan)
Jung, D. (FRG)
Justin-Visentin, E. (Italy)
Juteau, Th. (France)
Karamata, S. (Yugoslavia)
Katsui, Y. (Japan)
Keller, J. (FRG)
King, B.C. (UK)
Knetsch, G. (FRG)
Koark, H.J. (Sweden)
Kononova, V.A. (USSR)
Kopecky, L.(Czechoslovakia)
Kraëff, A. (Turkey)
Kramer, W. (GDR)
Krauskopf, K. (USA)
Kräutner, H. (Romania)
Kravchenko, S. (USSR)
Kübler, B. (Switzerland)
Kubovics, I. (Hungary)
Kumbazar, I. (Turkey)
Kunz, K. (FRG)
L'vov, B.K. (USSR)
Laitakati, I. (Finland)
Lambert, M.B. (Canada)
Lameyre, J. (France)
Laurent, R. (Canada)
Laurin, A.F. (Canada)

Lawson, R.I. (UK)
Lazko, E. (USSR)
Le Bas, M.J. (UK)
Le Maitre, R.W. (Australia)
Lehmann, E. (FRG)
Lensch, G. (FRG)
Leonardos, Jr., O.H. (Brazil)
Leterrier, J. (France)
Lofgren, G. (USA)
Lorenz, V. (FRG)
Luchitsky, J. (USSR)
Lundqvist, T. (Sweden)
Lyons, P.C. (USA)
Macdonald, G.A. (USA)
Macdonald, R. (Uganda)
Mackenzie, D.E. (Australia)
MacKenzie, W.S. (UK)
Mädler, J. (GDR)
Maggetti, M. (Switzerland)
Majer, V. (Yugoslavia)
Malaroda, R. (Italy)
Malejev, E.F. (USSR)
Markovsky, B.A. (USSR)
Martignole, J. (Canada)
Martin, H. (FRG)
Massaitis, V.L. (USSR)
Mathé, G. (GDR)
Matthes, S. (FRG)
Matthison, C.J. (Australia)
Mattson, P.H. (USA)
Mazzoni, M.M. (Argentina)
McBirney, A.R. (USA)
McCoy, F.W. (USA)
McCutcheon, S.R. (Canada)
McHone, J.G. (USA)
Mehnert, K.R. (FRG)
Michot, J. (Belgium)
Michot, P. (Belgium)
Middlemost, E.A.K. (Australia)
Mihm, A. (FRG)
Mikhailov, N.P. (USSR)
Miller, T. (USA)
Misch, P. (USA)
Mittempergher, M. (Italy)
Miyashiro, A. (USA)
Mohr, P. (Eire)
Mohr, P.A. (USA)
Moore, A.C. (South Africa)
Moore, G.E. (USA)

Moore, J.M. (Canada)
Morbidelli, L. (Italy)
Morteani, G. (FRG)
Mourazmiranda, A.(Portugal)
Mucke, D. (GDR)
Muir, I.D. (UK)
Müller, P. (FRG)
Muñoz, M. (Spain)
Murawski, H. (FRG)
Murray, A.D. (USA)
Mutschler, F.E. (USA)
Naldrett, A.J. (Canada)
Neal, V.E. (New Zealand)
Neall, V. (Australia)
Neuzilová, N. (Czechoslovakia)
Nickel, E. (Switzerland)
Niggli, E. (Switzerland)
Nockolds, S.R. (UK)
Noe Nygaard, A. (Denmark)
Nowakowski, A. (Poland)
O'Connor, J.T. (USA)
Obradovic, J. (Yugoslavia)
Oen, I.S. (Netherlands)
Oftedahl, C. (Norway)
Okrusch, M. (FRG)
Orlova, M.P. (USSR)
Pälchen, W. (GDR)
Palivcova, M. (Czechoslovakia)
Pantó, G. (Hungary)
Pascual, E. (Spain)
Paulitsch, P.D. (FRG)
Peck, D.L. (USA)
Peltz, S. (Romania)
Persoz, F. (Switzerland)
Persson, L. (Sweden)
Pertsev, N.N. (USSR)
Peterson, D.W. (USA)
Petrova, M.A. (USSR)
Pfeiffer, L. (GDR)
Piccirillo, E.M. (Italy)
Pichler, H. (FRG)
Pjatenko, I.K. (USSR)
Ploskonka, E. (Bolivia)
Polunina, L.A. (USSR)
Preston, J (UK)
Puga, E. (Spain)
Putman, G.W. (USA)
Radulescu, D. (Romania)

Raguin, E. (France)
Rankin, D.W. (USA)
Rea, W.J. (UK)
Reitan, P.H. (USA)
Richter, W. (Austria)
Rittmann, A. (Italy)
Rock, N.M.S. (Australia)
Röllig, G. (GDR)
Roques, M. (France)
Rost, F. (FRG)
Rotman, V.K. (USSR)
Ruitenberg, A.A. (Canada)
Ruminantseva, N.A. (USSR)
Sabine, P.A. (UK)
Sassi, F. (Italy)
Savu, H. (Romania)
Schaufelberger, P. (Colombia)
Schermerhorn, L.J.G. (Portugal)
Schmid, R. (Switzerland)
Schmincke, H.U. (FRG)
Schtschebakova,   M.N. (USSR)
Scott, R.B. (USA)
Seclaman, M. (Romania)
Seemann, U. (Switzerland)
Seim, R. (GDR)
Sen, S.K. (India)
Sharma, K.N.M. (Canada)
Sharpenok, L. (USSR)
Shemyakin, V.M. (USSR)
Sheridan, M.F. (USA)
Shirinian, H.G. (USSR)
Shurkin, K.A. (USSR)
Simboli, G. (Italy)
Simkin, T. (USA)
Singh, S. (Guyana)
Smulikowski, K. (Poland)
Snavely, P. (USA)
Snyder, G.L. (USA)
Snyman, C.P. (South Africa)
Sobolev, N. (USSR)
Sobolev, S.F. (USSR)
Sobolev, V.S. (USSR)
Soler, E. (France)
Sørensen, H. (Denmark)

Spalletti, L.A. (Argentina)
Stefanova, M. (Bulgaria)
Stengelin, R. (FRG)
Stillman, C.J. (UK)
Strand, T. (Norway)
Streckeisen, A. (Switzerland)
Subba Rao, S. (India)
Subbarao, K.V. (India)
Subramaniam, A.P. (India)
Subramaniam, K.V. (India)
Suk, M. (Czechoslovakia)
Sukheswala, R.N. (India)
Sung Shu Ho, (China)
Sveshnikova, S.V. (USSR)
Széky-Fux, V. (Hungary)
Taubeneck, W.H. (USA)
Taylor, S.R. (Australia)
Tégyey-Dumait, M. (France)
Teruggi, M.E. (Argentina)
Thorarinsson, S. (Iceland)
Thornton, C.P. (USA)
Thurston, P.C. (Canada)
Tilling, R.I. (USA)
Tischendorf, G. (GDR)
Tobi, A.C. (Netherlands)
Torske, T. (Norway)
Toulmin III, P. (USA)
Troll, G. (FRG)
Trommsdorff, V. (Switzerland)
Tweto, O. (USA)
Upton, B.J.G. (UK)
Urbani, F. (Venezuela)
Uytenbogaardt, W. (Netherlands)
Vallance, T.G. (Australia)
van Biljon, S. (South Africa)
van der Plas, L. (Netherlands)
Vartiainen, H. (Finland)
Veichow, C.J. (Taiwan)
Vejnar, Zd. (Czechoslovakia)
Velde, D. (France)
Venturelli, A. (Italy)
Vergara, M. (Chile)
Vidale, R.J. (USA)
Vieten, K. (FRG)

Villari, L. (Italy)
Vincent, P.M. (France)
Vitaliano, C.J. (USA)
von Eckerman, H. (Sweden)
von Gruenewaldt, G. (South Africa)
Vorobieva, O.A. (USSR)
Vuagnat, M. (Switzerland)
Wadsworth, W.J. (UK)
Walker, G.P.L. (USA)
Wang, B. (China)
Wass, S.Y. (Australia)
Waters, A.C. (USA)
Watznauer, A. (GDR)
Weaver, S.D. (UK)
Wedepohl, K.H. (FRG)
Wells, A.K. (UK)
Wells, M.K. (UK)
Wenk, E. (Switzerland)
Westercamp, D. (Martinique)
White, A.J.R. (Australia)
Whitney, J.A. (USA)
Whitten, E.H.T. (USA)
Wieseneder, H. (Austria)
Wilcox, R.E. (USA)
Wilkinson, J.F.G. (Australia)
Williams, H. (USA)
Williams, L.A.J. (Kenya)
Wilson, A.P. (Australia)
Wilson, J.R. (Denmark)
Wimmenauer, W. (FRG)
Winkler, H.G.F. (FRG)
Wolf, K.H. (Saudi Arabia)
Wones, D.R. (USA)
Woodsworth, G.J. (Canada)
Woodtli, R. (Switzerland)
Woolley, A.R. (UK)
Woussen, G. (Canada)
Wright, A.E. (UK)
Wurm, A. (FRG)
Wyllie, P.J. (USA)
Yasef, Y. (Bulgaria)
Yashina, R. (USSR)
Zanettin, B. (Italy)
Zeck, H.P. (Denmark)

Appendix C. Approved IUGS Names

This is a list of all the rock names and terms that have been defined and recommended for use by the IUGS Subcommission i.e. all those terms in the glossary that have been printed in bold capital letters.

Acid
Agglomerate
Alaskite
Alkali
Alkali basalt
Alkali feldspar charnockite
Alkali feldspar granite
Alkali feldspar rhyolite
Alkali feldspar syenite
Alkali feldspar trachyte
Alnöite
Alvikite
Analcime basanite
Analcime diorite
Analcime gabbro
Analcime monzodiorite
Analcime monzogabbro
Analcime monzosyenite
Analcime phonolite
Analcime plagisyenite
Analcime syenite
Analcimite
Andesite
Andesitoid
Anorthosite
Ash tuff
Ash, Ash grain
Basalt
Basaltic andesite
Basaltic trachyandesite
Basaltoid
Basanite
Basanitic foidite
Basic
Beforsite
Benmoreite
Block
Bomb

Boninite
Calciocarbonatite
Calcite carbonatite
Camptonite
Cancrinite diorite
Cancrinite gabbro
Cancrinite monzodiorite
Cancrinite monzogabbro
Cancrinite monzosyenite
Cancrinite plagisyenite
Cancrinite syenite
Carbonatite
Charno-enberbite
Charnockite
Clinopyroxene norite
Clinopyroxenite
Colour index
Comendite
Crystal tuff
Dacite
Dacitoid
Diabase
Diorite
Dioritoid
Dolerite
Dolomite carbonatite
Dunite
Dust grain
Dust tuff
Enderbite
Enstatitite
Essexite
Fenite
Fergusite
Ferrocarbonatite
Foid diorite
Foid dioritoid
Foid gabbro

Foid gabbroid
Foid monzodiorite
Foid monzogabbro
Foid monzosyenite
Foid plagisyenite
Foid syenite
Foid syenitoid
Foid-bearing alkali feldspar syenite
Foid-bearing alkali feldspar trachyte
Foid-bearing diorite
Foid-bearing gabbro
Foid-bearing latite
Foid-bearing monzodiorite
Foid-bearing monzogabbro
Foid-bearing monzonite
Foid-bearing syenite
Foid-bearing trachyte
Foidite
Foidolite
Gabbro
Gabbroid
Gabbronorite
Granite
Granitoid
Granodiorite
Harzburgite
Haüyne basanite
Haüyne phonolite
Haüynite
Hawaiite
High-K
Hornblende gabbro
Hornblende peridotite
Hornblende pyroxenite
Hornblendite
Igneous rock

191

Ignimbrite
Ijolite
Intermediate
Italite
Jotun-norite
Jotunite
Kersantite
Kimberlite
Komatiite
Kugdite
Lamproite
Lamprophyre
Lapilli
Lapilli tuff
Larvikite
Latite
Leucite basanite
Leucite phonolite
Leucite tephrite
Leucitite
Leucocratic
Lherzolite
Limburgite
Liparite
Lithic tuff
Low-K
m-Charnockite, m-Enderbite, etc.
Magnesiocarbonatite
Malignite
Mangerite
Medium-K
Meimechite
Melanocratic
Melilitite
Melilitolite
Melteigite
Miaskite
Minette
Missourite
Monchiquite
Monzodiorite
Monzogabbro
Monzogranite
Monzonite
Monzonorite
Mugearite
Natrocarbonatite
Nepheline basanite
Nepheline diorite

Nepheline gabbro
Nepheline monzodiorite
Nepheline monzogabbro
Nepheline monzosyenite
Nepheline plagisyenite
Nepheline syenite
Nepheline tephrite
Nephelinite
Nephelinolite
Norite
Nosean basanite
Noseanite
Obsidian
Olivine basalt
Olivine clinopyroxenite
Olivine gabbro
Olivine gabbronorite
Olivine hornblende pyroxenite
Olivine hornblendite
Olivine melilitite
Olivine melilitolite
Olivine norite
Olivine orthopyroxenite
Olivine pyroxene hornblendite
Olivine pyroxene melilitolite
Olivine pyroxenite
Olivine uncompahgrite
Olivine websterite
Opdalite
Orthopyroxene gabbro
Orthopyroxenite
Oversaturated
Pantellerite
Pegmatite
Peralkaline
Peridotite
Phonobasanite
Phonofoidite
Phonolite
Phonolitic basanite
Phonolitic foidite
Phonolitic tephrite
Phonolitoid
Phonotephrite
Picrite
Picrobasalt
Plagioclase-bearing hornblende pyroxenite
Plagioclase-bearing hornblendite

Plagioclase-bearing pyroxene hornblendite
Plagioclase-bearing pyroxenite
Plagioclasite
Plagiogranite
Plutonic
Polzenite
Porphyry
Potassic trachybasalt
Pyroclastic
Pyroclastic deposit
Pyroclasts
Pyroxene hornblende gabbronorite
Pyroxene hornblende peridotite
Pyroxene hornblendite
Pyroxene melilitolite
Pyroxene olivine melilitolite
Pyroxene peridotite
Pyroxenite
Quartz alkali feldspar syenite
Quartz alkali feldspar trachyte
Quartz anorthosite
Quartz diorite
Quartz dolerite
Quartz gabbro
Quartz latite
Quartz monzodiorite
Quartz monzogabbro
Quartz monzonite
Quartz norite
Quartz syenite
Quartz trachyte
Quartz-rich granitoid
Quartzolite
Rhyolite
Rhyolitoid
Sannaite
Saturated
Shonkinite
Shoshonite
Silicocarbonatite
Sodalite diorite
Sodalite gabbro
Sodalite monzodiorite
Sodalite monzogabbro
Sodalite monzosyenite
Sodalite plagisyenite

Sodalite syenite
Sodalitite
Sövite
Spessartite
Spilite
Subalkali
Subalkali basalt
Syenite
Syenitoid
Syenogranite
Tephra
Tephriphonolite
Tephrite
Tephritic foidite

Tephritic phonolite
Tephritoid
Teschenite
Theralite
Tholeiitic basalt
Tonalite
Trachyandesite
Trachybasalt
Trachydacite
Trachyte
Trachytoid
Troctolite
Trondhjemite
Tuff

Tuffite
Ultrabasic
Ultramafic
Ultramafitite
Ultramafitolite
Uncompahgrite
Undersaturated
Urtite
Verite
Vitric tuff
Vogesite
Volcanic
Websterite
Wehrlite